D0871118

DIFFERENTIAL EQUATIONS for ENGINEERS and SCIENTISTS

C. G. LAMBE & C. J. TRANTER

DOVER PUBLICATIONS, INC.
Mineola, New York

Bibliographical Note

This Dover edition, first published in 2018, is an unabridged republication of the work originally published by the English Universities Press, Ltd., London, in 1961. The General Editor's Foreword by Sir Graham Sutton from the original edition has been omitted.

International Standard Book Number

ISBN-13: 978-0-486-82408-6
ISBN-10: 0-486-82408-X

Manufactured in the United States by LSC Communications
82408X01 2018
www.doverpublications.com

PREFACE

THE study of ordinary and partial differential equations is an essential part of the training of engineers and scientists and this book is primarily designed for students reading for a first degree in those subjects. The authors have endeavoured to provide a text which covers the elementary theory of differential equations with emphasis on the application of the methods of solution to practical problems. More advanced analytical and numerical techniques are also introduced in so far as they have a bearing on modern scientific theory. Nearly nine hundred worked examples and exercises, with answers, are included.

The first four chapters are mainly concerned with the solution of first order equations, linear equations with constant coefficients and simultaneous equations. In the next two chapters the method of solution by infinite series is followed by an account of the more important special functions of mathematical physics. Partial differential equations are treated in Chapter 7 and solutions of Laplace's and other equations are obtained in terms of the special functions, while in the following chapter the method of integral transforms is discussed and applied to the solution of ordinary and partial equations. The next two chapters deal with methods of obtaining approximate solutions by graphical and numerical processes and by the method of relaxation. The final chapter is an introduction to the important modern theory of non-linear differential equations.

The authors wish to express their special thanks to their colleague, Dr. W. N. Everitt, who read the manuscript with great care and made many valuable suggestions. They are also grateful to the Senate of the University of London and the Department of Engineering, Cambridge, for permission to use examination questions, and to the editor and publishers of the *Philosophical Magazine* for allowing them to reproduce Figures 72, 73 and 74.

C. G. LAMBE
C. J. TRANTER

Royal Military College of Science
Shrivenham
1961

CONTENTS

CHAPTER 1

PRELIMINARY IDEAS AND DIRECT METHODS

1.1 Introduction

Most problems in science and engineering have to be "idealized" before their solution can be attempted. This idealization is generally necessary to bring the problem into a form capable of solution by known mathematical techniques, but it is essential, of course, that the actual and idealized problems should bear a close resemblance to one another. The solution then normally starts from well-established physical laws which are of two main types:

(*a*) statements which can usually be stated as mathematical equations, and

(*b*) assertions, not necessarily of a mathematical character, which provide a set of rules for the selection of physically acceptable solutions.

Typical of the first type of physical law is the knowledge that the drag of the air on an aeroplane (over a certain range of speeds) is proportional to the square of its velocity. As a mathematical equation this law can be expressed in the form $D = kv^2$, where D is the drag, v the velocity, and k is a constant whose value can be determined experimentally. An example of a law of the type (*b*) is the principle that mechanical systems cannot exist in which energy is created or destroyed. Such principles are useful in choosing an appropriate solution to a physical problem, since the mathematician can often produce a whole set of solutions each of which is correct in the sense that no mathematical error has been perpetrated.

The essentials in specifying an idealized problem are therefore provided by a number of physical laws and some rules which exclude unsuitable solutions. The only other requirement is a principle for the incorporation of these essentials into a mathematical equation. This is provided by the so-called doctrine of *determinism*. Put simply, this doctrine asserts that the state of a system at any instant is uniquely linked to, or determined by, the succession of states which precedes it. It is generally only possible to apply this principle to neighbouring states of the system, that is, those which differ infinitesimally in respect to time and space. Hence differential coefficients are usually involved and the resulting equations are known as *differential equations*. These equations have to be solved (or integrated) to give expressions relating states of the system separated by finite intervals of time and space.

There are two main classes of differential equations, *ordinary* and *partial*. An ordinary differential equation is one in which there is only one independent variable. For example, if an aeroplane of mass m is flying horizontally with propeller thrust P and drag D, its acceleration (which is the rate of change of its velocity v at time t) is given by the application of Newton's law and the relation $D = kv^2$ by the equation

$$m\frac{dv}{dt} = P - kv^2. \tag{1}$$

The equation

$$\frac{\partial^2 V}{\partial x^2}+\frac{\partial^2 V}{\partial y^2}+\frac{\partial^2 V}{\partial z^2} = \frac{1}{k}\frac{\partial V}{\partial t} \tag{2}$$

is a partial differential equation in which the unknown function V depends on the time t and the coordinates (x, y, z) of a representative point. It arises, for instance, in problems in the conduction of heat in a solid body.

The differential equations arising in physical problems are statements of physical laws and indicate features which may well be common to several problems. Besides appearing in heat conduction problems, equation (2) occurs in problems in the diffusion of matter in still air and in the drag of a surface in contact with a moving fluid. It is necessary therefore to possess additional information to specify any particular problem.

This information is given by *initial* and *boundary* conditions. Suppose, for example, a rectangular block of metal is heated on one face. The subsequent temperature in the block is determined by equation (2) in conjunction with a knowledge of the initial temperature of the block and the conditions on its other faces. The differential equation expresses the way in which the heat flows in *any* solid and the initial and boundary conditions specify the particular case under consideration.

Some differential equations can be solved quickly and easily by comparatively elementary methods, but the solution of many equations involves advanced mathematical techniques, while some can only be solved by numerical methods. A systematic study of standard methods of solving differential equations is of great importance. Solutions are not obtained by guesswork or by trial and error but by a systematic procedure and the student will develop ability to recognize the type of equation involved and to choose an appropriate method of solution. This ability is most readily acquired by working through large numbers of examples.

In this opening chapter we give some definitions and consider such preliminary ideas as are necessary for a study of the subject. We also give some examples to show how a physical problem can be expressed as a differential equation with appropriate initial and/or boundary conditions. We further show how a differential equation can be formed by elimination of constants from a mathematical equation and consider the reverse process of finding a solution of a differential equation when this can be done by direct integration.

1.2 Definitions

The *order* of a differential equation is defined as the order of the highest differential coefficient present. The *degree* of a differential equation is the degree of the highest differential coefficient present when the equation has been made rational and integral as far as the differential coefficients and the dependent variable are concerned.

A *linear differential equation* is one which is linear in the dependent variable and all its derivatives. The general form of a linear differential equation of the nth order is

$$p_0(x)\frac{d^n y}{dx^n}+p_1(x)\frac{d^{n-1}y}{dx^{n-1}}+\ldots+p_{n-1}(x)\frac{dy}{dx}+p_n(x)y = q(x), \tag{1}$$

where the functions $p_r(x)$, $r = 0, 1, 2, \ldots, n$, and $q(x)$ are functions of x only. The coefficient of $d^n y/dx^n$ in this equation is often taken as unity without loss of generality. A linear equation is, of course, of the first degree, but the term *non-linear* is used to describe an equation in which y or any of its derivatives is of degree higher than the first. Considering the examples,

$$x\frac{d^2y}{dx^2}+\frac{dy}{dx}+xy = \sin x, \tag{2}$$

$$\frac{d^2y}{dx^2}-\left\{1+\left(\frac{dy}{dx}\right)^2\right\}^{\frac{3}{2}} = 0, \tag{3}$$

$$\frac{d^2y}{dx^2}+y^3 = 0, \tag{4}$$

(2) is ordinary linear of the second order and first degree, (3) is ordinary non-linear of the second order and second degree and (4) is ordinary non-linear of the second order and first degree.

A *solution* (or *integral*) of a differential equation is a relation between the variables, not involving differential coefficients, which satisfies the differential equation. Thus the well-known differential equation of simple harmonic motion

$$\frac{d^2x}{dt^2}+\omega^2 x = 0, \qquad (\omega \text{ constant}), \tag{5}$$

has solution $x = C \sin(\omega t + \phi)$ where C and ϕ are any constants. This is easily verified, since we have

$$x = C\sin(\omega t+\phi), \tag{6}$$

$$\frac{dx}{dt} = \omega C\cos(\omega t+\phi), \tag{7}$$

$$\frac{d^2x}{dt^2} = -\omega^2 C\sin(\omega t+\phi) = -\omega^2 y. \tag{8}$$

Equation (6) furnishes the *general solution* of the differential equation (5). In an actual problem we usually require a particular solution which satisfies certain initial conditions. Thus we may be given that $y = 2$ and $dy/dt = 3\omega$ when $t = 0$. If the constants C and ϕ are such that the general solution satisfies these conditions, we have from (6) and (7),

$$2 = C\sin\phi,$$
$$3\omega = \omega C\cos\phi.$$

Hence $C = \sqrt{13}$ and $\tan\phi = 2/3$. Thus the initial conditions determine the amplitude and phase (but not the frequency) of the motion.

Notation. It is sometimes convenient to use primes to denote derivatives. Thus

$$y' \equiv \frac{dy}{dx}, \quad y'' \equiv \frac{d^2y}{dx^2} \quad \text{and} \quad y^{(n)} \equiv \frac{d^ny}{dx^n}.$$

When the time t is the independent variable it is customary to denote differentiation with respect to t by dots. Thus

$$\dot{x} \equiv \frac{dx}{dt}, \quad \ddot{x} \equiv \frac{d^2x}{dt^2}, \text{ etc.}$$

1.3 Formation of differential equations

Some remarks on the translation of a physical problem into a mathematical one have been made in the introductory paragraphs of this chapter. Here we illustrate by examples some of the methods used in forming differential equations from the physical data.

Example 1. *A particle moves in a straight line, being attracted to a fixed point by a force which is proportional to its distance from the point. Form the differential equation of the motion.*

Let x be the distance of the particle from the fixed point at time t; then the force is μx directed towards the fixed point, μ being a constant. The velocity v at any instant is the rate of change of x with respect to t and the acceleration f is the rate of change of v with respect to t. That is

$$v = \frac{dx}{dt}, \quad f = \frac{dv}{dt} = \frac{d^2x}{dt^2}.$$

Equating the product of the mass m of the particle and its acceleration to the force, we have the differential equation

$$m\frac{d^2x}{dt^2} = -\mu x.$$

This differential equation has the same form as equation (5) of §1.2 and a solution is

$$x = C\sin(\omega t+\phi),$$

where $\omega^2 = \mu/m$.

Example 2. *In a chemical reaction $A \to B$, the velocity of reaction is proportional to the amount remaining of A. Form the differential equation of the reaction and verify that if a is the initial amount of A and k the constant of proportionality, the amount of A at time t is ae^{-kt}.*

If x is the amount of A remaining at time t the rate of change of x with respect to t is negative and the velocity of reaction is $-\dot{x}$. We have therefore

$$-\frac{dx}{dt} = kx.$$

It is easily verified that if $x = Ce^{-kt}$, where C is any constant,

$$\frac{dx}{dt} = -kCe^{-kt} = -kx.$$

Thus, since $x = a$ when $t = 0$, it follows that $C = a$ and the solution is $x = ae^{-kt}$.

Example 3. *A uniform beam of length l and weight wl is supported at its ends in a horizontal position. Form the differential equation for the deflexion y of the centre line of the beam and find a solution which satisfies the end conditions.*

The bending moment at a distance x from one end of the beam is $-\frac{1}{2}wlx$ due to the end reaction, and $\frac{1}{2}wx^2$ due to the distributed load. A general expression for the bending moment of a deflected beam is EIy'', where E is Young's modulus for the material and I is the second moment of area of the cross-section of the beam about its neutral axis. We have therefore

$$EI\frac{d^2y}{dx^2} = \tfrac{1}{2}wx^2-\tfrac{1}{2}wlx.$$

This equation may be integrated twice with respect to x, giving

$$EI\frac{dy}{dx} = \tfrac{1}{6}wx^3-\tfrac{1}{4}wlx^2+C,$$

$$EIy = \tfrac{1}{24}wx^4-\tfrac{1}{12}wlx^3+Cx+D,$$

where C and D are any constants. If now we consider the initial conditions that $y = 0$ when $x = 0$ and when $x = l$ we find that $D = 0$ and $C = wl^3/24$, and the solution becomes

$$24EIy = wx(l-x)(l^2+lx-x^2).$$

1.4 Elimination of constants

The solution of an ordinary differential equation is a relation between two variables and will involve one or more arbitrary constants. When a relation between the variables is given, constants may be eliminated by differentiating and forming a differential equation.

For example, the equation $y = x(a-x)$ is the equation of a parabola whose axis is parallel to the y-axis and whose vertex is at the point $(\frac{1}{2}a, \frac{1}{4}a^2)$. Differentiating we have

$$\frac{dy}{dx} = a-2x.$$

Eliminating the constant a between the two equations we have

$$\frac{dy}{dx} = x+\frac{y}{x}-2x,$$

that is,

$$x\frac{dy}{dx}-y+x^2 = 0. \tag{1}$$

The equation (1) is a differential equation whose solution is $y = x(A-x)$, where A is an arbitrary constant, and thus the differential equation represents a family of parabolas whose axes are parallel to the y-axis. It is easily seen that if an equation involves n constants the constants can, in general, be eliminated between the original equation and the equations which give the first n derivatives of y with respect to x thus forming a differential equation of the nth order. We would expect, therefore, the solution of a differential equation of the nth order to contain n arbitrary constants.

We may, therefore, think of the solution of an ordinary differential equation as being the equation of a curve which passes through certain fixed points determined by the initial conditions. Such a curve is called an *integral curve* of the differential equation. The constants determined by the initial conditions are eliminated in the differential equation which thus represents a whole family of curves of similar shape but passing through different points.

For example, the equation $y^2 = 3x+7$ is the equation of a parabola. It is also an integral curve of the differential equation

$$\frac{d^2}{dx^2}(y^2) = 0.$$

The general solution of this differential equation is $y^2 = Ax+B$, which can represent any one of a family of parabolas.

Example 4. *Form a differential equation to eliminate the constants a, b and r in the equation of the circle* $(x-a)^2+(y-b)^2 = r^2$.

By repeated differentiation we have

$$(x-a)+(y-b)\frac{dy}{dx} = 0,$$

$$1+(y-b)\frac{d^2y}{dx^2}+\left(\frac{dy}{dx}\right)^2 = 0,$$

$$(y-b)\frac{d^3y}{dx^3}+3\frac{dy}{dx}\frac{d^2y}{dx^2} = 0.$$

Eliminating b between the last two equations we have

$$\left\{1+\left(\frac{dy}{dx}\right)^2\right\}\frac{d^3y}{dx^3}-3\frac{dy}{dx}\left(\frac{d^2y}{dx^2}\right)^2 = 0. \tag{2}$$

This differential equation which represents all circles in the x, y plane may be obtained more simply by stating that the curvature of a circle is constant, that is

$$\frac{\dfrac{d^2y}{dx^2}}{\left\{1+\left(\dfrac{dy}{dx}\right)^2\right\}^{3/2}} = \text{constant}. \tag{3}$$

Differentiation of this equation leads to the differential equation (2). If the constant is taken as $1/r$, the equation (3) is the differential equation of all circles of radius r and could be obtained by eliminating a and b between the original equation and its first two derivatives.

1.5 Taylor series expansion of solutions

Suppose that the solution of a linear differential equation is such that the independent variable y can be expressed as a Taylor series in $(x-a)$ near $x = a$, that is

$$y = (y)_a+(x-a)\left(\frac{dy}{dx}\right)_a+\frac{(x-a)^2}{2!}\left(\frac{d^2y}{dx^2}\right)_a+\ldots, \tag{1}$$

where $(y)_a$, $(dy/dx)_a$, etc. denote the values of y, dy/dx, etc. when $x = a$.

Let the differential equation be of the nth order of the form

$$p_0(x)\frac{d^ny}{dx^n}+p_1(x)\frac{d^{n-1}y}{dx^{n-1}}+\ldots+p_n(x)y = q(x), \tag{2}$$

where the coefficients $p_i(x)$, $i = 0, 1, \ldots, n$, and $q(x)$ have finite differential coefficients of all orders for $x = a$ and $p_0(a) \neq 0$. Putting $x = a$ in (2) we have

$$p_0(a)\left(\frac{d^ny}{dx^n}\right)_a+p_1(a)\left(\frac{d^{n-1}y}{dx^{n-1}}\right)_a+\ldots+p_n(a)(y)_a = q(a). \tag{3}$$

Equation (3) gives the value of $(d^ny/dx^n)_a$ in terms of $(d^{n-1}y/dx^{n-1})_a$ and lower derivatives. Differentiating equation (2) and putting $x = a$ we have

$$p_0(a)\left(\frac{d^{n+1}y}{dx^{n+1}}\right)_a+(p_0'(a)+p_1(a))\left(\frac{d^ny}{dx^n}\right)_a+\ldots+p_n'(a)(y)_a = q'(a). \tag{4}$$

Equations (4) and (3) determine the value of $(d^{n+1}y/dx^{n+1})_a$ in terms of $(d^{n-1}y/dx^{n-1})_a$ and lower derivatives. By repeated differentiation all further coefficients of the Taylor series can be obtained in terms of $(y)_a$ and the first $n-1$ derivatives of y at $x = a$. Thus a solution may be obtained in which these n quantities are chosen arbitrarily. In other words, *the general solution of the* nth *order linear differential equation contains n arbitrary constants.*

This theorem is of great importance. In particular, it enables us to recognize that we have obtained the most general solution by seeing that it contains the requisite number of arbitrary constants.

In advanced treatises on differential equations it is proved that the most general solution of a differential equation of the nth order, whether linear or not, contains n arbitrary constants. Theorems, known as *existence theorems*, are proved giving the precise analytical conditions under which the equation has a solution.

The solution of an ordinary differential equation of order n which contains n arbitrary constants is called the *complete primitive* or the *general solution* of the equation.

For most differential equations any solution that can be found to satisfy the equation can be derived from the general solution by giving suitable values to the arbitrary constants. In certain cases, however, it is possible to find a solution which cannot be derived in this way and such a solution is called a *singular solution*. Some singular solutions of differential equations are considered in §2.11.

Example 5. *Find a Taylor series expansion valid near the origin for the solution of the differential equation* $\dfrac{dy}{dx} = xy$.

Let the value of y when $x = 0$ be a. The value of $(y')_0$ is evidently zero. We can differentiate both sides of the differential equation n times, and we have

$$\frac{d^{n+1}y}{dx^{n+1}} = \frac{d^n}{dx^n}(xy)$$

$$= x\frac{d^ny}{dx^n}+n\frac{d^{n-1}y}{dx^{n-1}}, \text{ using Leibnitz' theorem.}$$

Therefore

$$\left(\frac{d^{n+1}y}{dx^{n+1}}\right)_0 = n\left(\frac{d^{n-1}y}{dx^{n-1}}\right)_0, \text{ for } n \geqq 1$$

and, in particular,

$$\left(\frac{d^2y}{dx^2}\right)_0 = (y)_0 = a,$$

$$\left(\frac{d^4y}{dx^4}\right)_0 = 3\left(\frac{d^2y}{dx^2}\right)_0 = 3a,$$

$$\left(\frac{d^6y}{dx^6}\right)_0 = 5\left(\frac{d^4y}{dx^4}\right)_0 = 15a,$$

and so on.

Since

$$\left(\frac{dy}{dx}\right)_0 = (xy)_0 = 0,$$

it is evident that all the odd derivatives are zero and hence we have

$$y = a\left\{1+\frac{x^2}{2!}+\frac{3x^4}{4!}+\frac{15x^6}{6!}+\ldots\right\}$$

$$= a\left\{1+(x^2/2)+\frac{(x^2/2)^2}{2!}+\frac{(x^2/2)^3}{3!}+\ldots\right\}$$

$$= ae^{x^2/2}.$$

Exercises 1(*a*)

1. Write down the order and degree of the following differential equations:

(*a*) $\dfrac{dy}{dx} = \dfrac{1}{y^{1/2}(1+x^{1/2})}$,

(*b*) $y\dfrac{d^2y}{dx^2}+\left(\dfrac{dy}{dx}\right)^2 = 0$,

(*c*) $\dfrac{d^2}{dx^2}\left\{\left(\dfrac{d^2y}{dx^2}\right)^{-3/2}\right\} = 0$,

(*d*) $\dfrac{\partial^2 y}{\partial x^2}-\dfrac{1}{h^2}\dfrac{\partial y}{\partial t} = 0$.

2. Find the ordinary differential equations of which the following are the general solutions, A and B being arbitrary constants:

(*a*) $y = A\cosh nx+B\sinh nx$,
(*b*) $y = e^{-kt}(A\cos nt+B\sin nt)$,
(*c*) $y = (A+Bx)e^{mx}$,
(*d*) $y^2 = Ax+B$,
(*e*) $y^2 = Ax+A^2$.

3. Prove that $y = \dfrac{Ax+B}{Cx+D}$ is the general solution of the differential equation

$$2\frac{dy}{dx}\frac{d^3y}{dx^3}-3\left(\frac{d^2y}{dx^2}\right)^2 = 0.$$

4. Find the differential equation of all circles that pass through the origin.

5. In successive chemical reactions $A \to B \to C\ldots$, the velocity of each reaction is proportional to the amount remaining of each concentration. Prove that if x is the concentration of A and y the concentration of C at time t, y being initially zero,

$$\frac{dx}{dt} = -k_1x; \qquad \frac{dy}{dt} = k_2(b+a-x-y),$$

where a and b are the initial concentrations of A and B respectively.

6. A particle moves on a smooth horizontal plane in a medium which causes a retardation kv^2, where v is the velocity. Show that if x is the displacement at time t

$$\frac{d^2x}{dt^2}+k\left(\frac{dx}{dt}\right)^2 = 0,$$

and verify that the general solution of this equation is

$$kx = \log_e (A+Bt).$$

Show that if the initial velocity is V and the initial displacement is zero, $A = 1$ and $B = kV$.

7. Find a Maclaurin expansion in ascending powers of x for the solution of the differential equation

$$\frac{dy}{dx} - ny = 0.$$

8. Show that if $y = A \sin^{-1} x + B$,

$$(1-x^2)\frac{d^2y}{dx^2} - x\frac{dy}{dx} = 0.$$

By finding a Maclaurin expansion for the solution in ascending powers of x, prove that $\sin^{-1} x = x + \frac{1^2}{3!}x^3 + \frac{1^2 3^2}{5!}x^5 + \ldots$.

9. Prove Euler's theorem on homogeneous functions, that is, if $z = f(x, y)$, where $f(x, y)$ is a homogeneous function of x and y of degree n, then z satisfies the differential equation

$$x\frac{\partial z}{\partial x} + y\frac{\partial z}{\partial y} = nz.$$

10. If $\surd(1-x^2)y = \sin^{-1} x$, prove that $(1-x^2)\frac{dy}{dx} - xy = 1$. By substituting $y = a_0 + a_1 x + a_2 x^2 + \ldots$ in the above differential equation, or by any other method, obtain the first three non-vanishing terms in the expansion of y in ascending powers of x. (L.U.)

1.6 The differential equation $y^{(n)} = f(x)$

(a) First order equations

The differential equation

$$\frac{dy}{dx} = f(x),$$

in which y is explicitly absent, has clearly the solution

$$y = \int f(x)dx + A,$$

where A is an arbitrary constant. The solution may be written as a definite integral

$$y = \int_a^x f(t)\,dt.$$

Here a is an arbitrary constant and is the value of x corresponding to the value $y = 0$.

The differential equation

$$\frac{dy}{dx} = f(y),$$

in which x is explicitly absent, is solved in the same way since

$$\frac{dx}{dy} = \frac{1}{f(y)},$$

and the solution is

$$x = \int \frac{dy}{f(y)} + A$$

$$= \int_a^y \frac{dt}{f(t)}.$$

Example 6. *A particle is projected vertically upwards with velocity V and the air resistance causes a retardation kv, where v is the velocity at time t. Find the height reached and the time to reach this height.*

The retardation is $g+kv$, where v is the upward velocity, and $v = \frac{dx}{dt}$ where x is the height above the ground. The acceleration is dv/dt and we have

$$\frac{dv}{dt} = -g-kv,$$

$$\frac{dt}{dv} = -\frac{1}{g+kv},$$

$$t = -\frac{1}{k}\log_e(g+kv)+A.$$

The solution must satisfy the initial condition that $v = V$ when $t = 0$, therefore

$$0 = -\frac{1}{k}\log_e(g+kV)+A,$$

giving

$$-kt = \log_e\frac{g+kv}{g+kV}. \tag{1}$$

The velocity becomes equal to zero and the particle has reached its highest point when

$$t = \frac{1}{k}\log_e\left(1+\frac{kV}{g}\right).$$

From (1)

$$v = \frac{dx}{dt} = -\frac{g}{k}+\left(\frac{g}{k}+V\right)e^{-kt}.$$

Therefore

$$x = -\frac{gt}{k}-\left(\frac{g}{k^2}+\frac{V}{k}\right)e^{-kt}+B,$$

and the initial condition that $x = 0$ when $t = 0$ gives the value of B and the solution

$$x = -\frac{gt}{k}+\frac{1}{k^2}(g+kV)(1-e^{-kt}).$$

At the highest point $kt = \log_e(1+kV/g)$ and $e^{-kt} = g/(g+kV)$, therefore

$$x = -\frac{g}{k^2}\log_e\left(1+\frac{kV}{g}\right)+\frac{V}{k}.$$

(b) Non-linear equations

Non-linear differential equations of the first order sometimes arise in which one of the variables is explicitly absent. A method of solving such equations may be seen from the following example.

Example 7. *Solve the differential equation* $\left(\frac{dy}{dx}\right)^2 - y^2 = 1$.

We have

$$\frac{dy}{dx} = \pm\sqrt{(y^2+1)},$$

which can be written

$$\frac{dx}{dy} = \pm\frac{1}{\sqrt{(y^2+1)}}.$$

This gives, on integration,

$$x = \pm \tan^{-1} y + A,$$

that is

$$y = \pm \tan (x - A),$$

and the general solution is $y^2 = \tan^2 (x-A)$.

(c) Equations of higher order

The differential equation

$$\frac{d^n y}{dx^n} = f(x),$$

is solved in a similar manner by repeated integration. We have

$$\frac{d^{n-1} y}{dx^{n-1}} = \int f(x)\, dx + A,$$

$$\frac{d^{n-2} y}{dx^{n-2}} = \int dx \int f(x)\, dx + Ax + B,$$

and so on, an arbitrary constant being introduced at each integration. Differential equations of this type occur in the theory of deflexions of a beam where a general expression for the bending moment is $EI\, d^2y/dx^2$, and for the loading $\frac{d^2}{dx^2}\left(EI\frac{d^2y}{dx^2}\right)$. Here y is the deflexion, assumed small, at distance x along the beam, I is the second moment of area of the cross-section about its neutral axis and E is Young's modulus for the material. If the bending moment or the loading at any point can be expressed as a function of x the resulting differential equation can be solved by direct integration, the constants of integration being determined by the slope, bending moment, etc. at the ends of the beam.

Example 8. *A light beam of uniform cross-section and length l is supported at its ends in a horizontal position. The ends are clamped horizontally and a load is placed on the beam in such a way that the intensity of loading increases uniformly from zero at one end to w at the other. Find an expression for the deflexion of the beam at any point of its length.*

The loading at distance x from one end is wx/l and we have the differential equation

$$EI\frac{d^4y}{dx^4} = \frac{wx}{l}.$$

Integrating we have

$$EI\frac{d^3y}{dx^3} = \frac{wx^2}{2l}+A,$$

$$EI\frac{d^2y}{dx^2} = \frac{wx^3}{6l}+Ax+B,$$

$$EI\frac{dy}{dx} = \frac{wx^4}{24l}+\tfrac{1}{2}Ax^2+Bx+C,$$

$$EIy = \frac{wx^5}{120l}+\tfrac{1}{6}Ax^3+\tfrac{1}{2}Bx^2+Cx+D.$$

The initial conditions to be satisfied by the solution are

(i) $y = 0$ when $x = 0$,

(ii) $y = 0$ when $x = l$,

(iii) $\dfrac{dy}{dx} = 0$ when $x = 0$,

(iv) $\dfrac{dy}{dx} = 0$ when $x = l$,

the last two conditions expressing the fact that the beam is horizontal at the ends. The conditions (i) and (iii) make $C = D = 0$, and (ii) and (iv) give

$$\tfrac{1}{24}wl^3+\tfrac{1}{2}Al^2+Bl = 0,$$

$$\tfrac{1}{120}wl^4+\tfrac{1}{6}Al^3+\tfrac{1}{2}Bl^2 = 0.$$

Hence, $A = -3wl/20$, $B = wl^2/30$, and we have the solution

$$120lEIy = wx^2(x-l)(x^2+lx-2l^2).$$

Exercises 1(*b*)

1. Solve the differential equation $\dfrac{dy}{dx} = \dfrac{x}{\sqrt{(1-x)}}$, with the initial condition that $y = 0$ when $x = 0$.

2. Solve the differential equation $\dfrac{dy}{dx} = 1+\dfrac{1}{y^2}$, with the initial condition that $y = 0$ when $x = 1$.

3. In a chemical reaction $A \to B+C$, the reaction rate at time t is proportional to the concentration x of A and $x = a$ when $t = 0$. Prove that $\dfrac{dx}{dt} = -kx$ and hence that $x = ae^{-kt}$.

4. In a second order chemical reaction $A+B\rightarrow C+D$, the concentrations of A and B are initially a and b respectively and x is the concentration change at time t. Prove that

$$\frac{dx}{dt}=k(x-a)(x-b).$$

Show that if $b\neq a$, $x=ab\{e^{(b-a)kt}-1\}/\{be^{(b-a)kt}-a\}$.

5. A particle is projected vertically upwards with velocity V and the retardation is $k(v^2+v_0^2)$, where v is the velocity at time t and $kv_0^2=g$. Show that

$$\frac{dv}{dt}=-k(v_0^2+v^2),$$

and hence that

$$v=v_0\frac{V-v_0\tan v_0kt}{v_0+V\tan v_0kt}.$$

6. Solve the equation $\left(\frac{dy}{dx}\right)^2=1-y^2$, with the condition that $y=0$ when $x=0$.

7. Solve the equation $\left(\frac{dy}{dx}\right)^2=1-x^2$, with the condition that $y=1$ when $x=0$.

8. Solve the equation $\left(\frac{dy}{dx}\right)^2=\frac{(y^2+1)^3}{y^2}$.

9. Solve the equation $\left(\frac{dy}{dx}\right)^2=\frac{1}{x^2(x^2+1)}$, with the condition that $y=1$ when $x=1$.

10. Solve the equation $\left(\frac{dy}{dx}-1\right)^2=\frac{1}{1-x^2}$, with the condition that $y=\frac{\pi}{2}$ when $x=0$.

11. Solve the equation $\left(\frac{dy}{dx}\right)^3=1-y$, with the condition that $y=1$ when $x=0$.

12. Solve the equation $\left(\frac{dy}{dx}\right)^2-2x\frac{dy}{dx}=1$, with the condition that $y=0$ when $x=0$.

13. The bending moment in a uniform cantilever of length l which projects horizontally from a wall is $\frac{1}{2}w(l-x)^2$ at distance x from the wall. Show that the deflexion y of the centre line is given by the equation

$$24EIy=wx^2(x^2-4xl+6l^2).$$

14. A uniform cantilever of length l and weight wl projects horizontally from a wall and its free end is propped level with the wall end. Show

that the deflexion y of the centre line at distance x from the wall is given by the equation

$$48EIy = wx^2(x-l)(2x-3l).$$

15. A particle of mass 3 lb. is acted upon by a force which diminishes uniformly from $\frac{2}{5}$ lb. wt. to $\frac{1}{10}$ lb. wt. in $\frac{1}{2}$ min. If it starts from rest find its greatest velocity in this half minute and the distance traversed. Find, also, the velocity of the particle when the force on it becomes zero. (L.U.)

1.7 Exact equations

Let the solution of a first order differential equation be $\phi(x, y) = A$, a constant. Differentiating this equation with respect to x we have

$$\frac{\partial\phi}{\partial x}+\frac{\partial\phi}{\partial y}\frac{dy}{dx} = 0,$$

that is

$$\frac{dy}{dx}+\frac{P(x, y)}{Q(x, y)} = 0, \tag{1}$$

where P and Q are functions of x and y and

$$P(x, y) = \frac{\partial\phi}{\partial x}, \qquad Q(x, y) = \frac{\partial\phi}{\partial y}. \tag{2}$$

Conversely, if a function $f(x, y)$ is expressed as a quotient of two functions P and Q, so that $f(x, y) = P(x, y)/Q(x, y)$, a solution of the general equation of the first order and first degree

$$\frac{dy}{dx} = f(x, y) \tag{3}$$

is $\phi(x, y) = A$, provided that $\phi(x, y)$ satisfies the relations (2). Given P and Q, the necessary and sufficient condition for this is that

$$\frac{\partial P}{\partial y} = \frac{\partial Q}{\partial x}. \tag{4}$$

The necessity of the condition is clear, since each of the quantities will be equal to $\frac{\partial^2\phi}{\partial x\partial y}$. The sufficiency of the condition is seen from the method of solving the equation (3) when the condition is satisfied.

Let

$$P(x, y) = \frac{\partial}{\partial x}\{\psi(x, y)\},$$

that is, a function $\psi(x, y)$ can be obtained by integrating P with respect to x, treating y as a constant. Then from (4)

$$\frac{\partial Q}{\partial x} = \frac{\partial^2}{\partial y \partial x}\{\psi(x, y)\},$$

and integrating with respect to x

$$Q = \frac{\partial}{\partial y}\{\psi(x, y)\} + f(y),$$

where $f(y)$ is a function of y alone. Then the solution is $\phi(x, y) = A$, where

$$\phi(x, y) = \psi(x, y) + \int f(y)\, dy,$$

since the function ϕ thus defined is such that $\dfrac{\partial \phi}{\partial x} = P$ and $\dfrac{\partial \phi}{\partial y} = Q$.

The differential equation (1) is usually written in the form

$$P\, dx + Q\, dy = 0, \tag{5}$$

and is called an *exact equation* if P and Q satisfy the relation (4). The solution is $\phi(x, y) = A$, where

$$\phi(x, y) = \int P(x, y)\, dx + g(y),$$

and also

$$\phi(x, y) = \int Q(x, y)\, dy + h(x),$$

$g(y)$ being a function of y only and $h(x)$ a function of x only. The precise form of the solution is found by seeing that the two values of $\phi(x, y)$ are identical.

An explicit formula can be found for $\phi(x, y)$ as follows.

Let

$$\phi(x, y) = \int P(x, y)\, dx + g(y).$$

Then

$$\frac{\partial \phi}{\partial y} = \frac{\partial}{\partial y}\int P(x, y)\, dx + \frac{d}{dy}\{g(y)\} = Q(x, y),$$

and

$$g(y) = \int\left\{Q(x, y) - \frac{\partial}{\partial y}\int P(x, y)\, dx\right\} dy.$$

Thus

$$\phi(x, y) = \int P(x, y)\,dx + \int Q(x, y)\,dy - L(x, y), \tag{6}$$

where

$$L(x, y) = \int\left\{\frac{\partial}{\partial y}\int P(x, y)\,dx\right\}dy. \tag{7}$$

Similarly,

$$L(x, y) = \int\left\{\frac{\partial}{\partial x}\int Q(x, y)\,dy\right\}dx. \tag{8}$$

Example 9. *Solve the differential equation* $\dfrac{dy}{dx}+\dfrac{2x+y}{2y+x} = 0.$

Writing the equation in the form

$$(2x+y)\,dx+(2y+x)\,dy = 0,$$

we have $\dfrac{\partial P}{\partial y} = \dfrac{\partial}{\partial y}(2x+y) = 1$, $\dfrac{\partial Q}{\partial x} = \dfrac{\partial}{\partial x}(2y+x) = 1$, and hence the equation is exact. Then if $\phi(x, y) = A$ is the solution,

$$\frac{\partial \phi}{\partial x} = 2x+y, \quad \phi = x^2+xy+g(y),$$

$$\frac{\partial \phi}{\partial y} = 2y+x, \quad \phi = y^2+xy+h(x).$$

Since the two values of ϕ must be identical we have $\phi = x^2+y^2+xy$ and the solution is

$$x^2+y^2+xy = A.$$

Alternatively, using the formulae (6) and (7) we have

$$\int P\,dx = x^2+xy,$$

$$\int Q\,dy = y^2+xy.$$

Then

$$\frac{\partial}{\partial y}\int P\,dx = x,$$

$$L(x, y) = xy,$$

and

$$\phi(x, y) = x^2+y^2+xy.$$

A differential equation $Pdx+Qdy = 0$ which is not exact can be made exact by multiplying by a function $F(x, y)$, called an *integrating factor*. Assuming that there is a solution $\phi(x, y) = A$, one integrating factor is evidently $\dfrac{\partial \phi}{\partial x}\Big/P = \dfrac{\partial \phi}{\partial y}\Big/Q$. Multiplying by an integrating factor is equivalent to writing the function $f(x, y)$ in (3) as RP/RQ, so that any first order equation can be written as an exact equation and solved if the correct quotient is found.

If the equation $RP\,dx+RQ\,dy = 0$ is exact we have

$$\frac{\partial}{\partial y}(RP) = \frac{\partial}{\partial x}(RQ),$$

that is

$$P\frac{\partial R}{\partial y}-Q\frac{\partial R}{\partial x} = R\left(\frac{\partial Q}{\partial x}-\frac{\partial P}{\partial y}\right).$$

This partial differential equation for R is not readily soluble, but a value of R which satisfies this equation may sometimes be obtained by inspection. An infinite number of integrating factors exist, since if R is an integrating factor so also are $\phi\,.\,R$, $\phi^2\,.\,R$, etc.

Example 10. *Solve the differential equation* $2xy\,dx+(y^2-x^2)\,dy = 0$.

It is easily seen that this equation is not exact. If $R(x, y)$ is an integrating factor, we have

$$\frac{\partial}{\partial y}(R.2xy) = \frac{\partial}{\partial x}\{R(y^2-x^2)\},$$

that is

$$2xy\frac{\partial R}{\partial y}+(x^2-y^2)\frac{\partial R}{\partial x} = -4xR.$$

If we try a value of R independent of x, we find

$$2xy\frac{dR}{dy}+4xR = 0,$$

and a solution found by inspection is $R = 1/y^2$. Then the exact equation is

$$\frac{2x}{y}\,dx+\left(1-\frac{x^2}{y^2}\right)dy=0,$$

and if $\phi(x, y) = A$ is the solution

$$\phi = \frac{x^2}{y}+g(y) = y+\frac{x^2}{y}+h(x).$$

Hence, $h(x) = 0$, $g(y) = y$, and the solution is $x^2+y^2 = Ay$.

Exercises 1(*c*)

Show that the following equations are exact and find their solutions:

1. $y\,dx+x\,dy = 0$.
2. $x^{-1}\,dx-\tan y\,dy = 0$.
3. $(3x^2+y^2)\,dx+2xy\,dy = 0$.
4. $\cosh x \cosh y\,dx+\sinh x \sinh y\,dy = 0$.
5. $\log_e y\,dx+xy^{-1}\,dy = 0$.
6. $(2x^3+xy^2)\,dx+(2y^3+x^2y)\,dy = 0$.
7. $x(x^2+y^2)\,dx+y(x^2+y^2)\,dy = 0$.
8. $(3x^2-2y^2)\,dx-(4xy-2y)\,dy = 0$.

9. $(ax+by+c)\,dx+(bx+dy+e)\,dy = 0.$

10. Show that *Jacobi's equation*

$$(a_1+b_1x+c_1y)(x\,dy-y\,dx)-(a_2+b_2x+c_2y)\,dy+(a_3+b_3x+c_3y)\,dx = 0$$

is an exact equation if $2a_1 = b_2+c_3$ and $b_1 = c_1 = 0$, and find its solution in this case.

11. Show that the equation $(2x-y)\,dx+(2y+x)\,dy = 0$ can be made exact by the integrating factor $1/(x^2+y^2)$ and find its solution.

12. Using the integrating factor $\cos x \cos y$, solve the equation

$$\tan y\,dx+\tan x\,dy = 0.$$

13. Using the integrating factor $1/(x+y)^2$, solve the equation

$$(x^2+2xy-y^2)\,dx+(y^2+2xy-x^2)\,dy = 0.$$

14. Find an integrating factor which is a function of x only and solve the equation $y(1-x)\,dx-x\,dy = 0.$

15. Show that possible integrating factors of the equation $y\,dx-x\,dy = 0$ are x^{-2}, y^{-2}, $x^{-2}+y^{-2}$, $(x-y)^{-2}$, and find the solution.

16. Show that $(x-y)^{-2}$ is an integrating factor of the equation

$$(x^2-2xy-y^2)\,dx+(x^2+2xy-y^2)\,dy = 0$$

and find the solution.

17. Find an integrating factor which is a function of x only and solve the equation $(2y-x^2)\,dx+x\,dy = 0.$

18. Find the value of n so that the differential equation

$$(x^2+y^2)^n\,(xy^2\,dx-x^2y\,dy) = 0$$

may be exact, and find the solution for this value of n. (L.U.)

CHAPTER 2

FIRST ORDER DIFFERENTIAL EQUATIONS

2.1 Introductory

In this chapter we show how to obtain solutions of the following standard types of differential equation of the first order:

(i) equations with variables separable;
(ii) linear first order equations;
(iii) homogeneous equations;
(iv) some non-linear equations.

The study of non-linear equations leads to a brief consideration of singular solutions and orthogonal trajectories, and we conclude the chapter by showing the equivalence of the general second order linear equation to one which is of the first order and non-linear.

First order equations of types (i) and (ii) above occur very frequently in scientific problems. Numerous examples and exercises have been given so that the reader can become familiar with both the setting up and the solving of these types of equation from given physical data.

2.2 Equations with variables separable

The general equation of the first order and first degree

$$\frac{dy}{dx} = f(x, y), \tag{1}$$

is said to have *separable variables* if $f(x, y)$ can be expressed as a quotient (or product) of a function of x only and a function of y only, that is $f(x, y) = g(x)/h(y)$. Thus the functions

$$x^2y^2, \quad x^2y^2+x^2, \quad x^2 \log_e(1+y^3), \quad e^{2x+3y}, \quad y \sin x/\sin y,$$

can all be expressed as quotients of functions of x and y in this way. In the particular case where one of the variables is explicitly absent from the differential equation, one of the functions in the quotient will be unity.

If the variables are separable we may write the equation (1) in the form

$$\frac{dy}{dx} = \frac{g(x)}{h(y)},$$

or

$$g(x)\,dx - h(y)\,dy = 0.$$

This equation is exact, since $\frac{\partial}{\partial y}\{g(x)\} = \frac{\partial}{\partial x}\{h(y)\} = 0$, and can therefore be integrated directly. The solution is

$$\int g(x)\,dx - \int h(y)\,dy = A, \quad (A \text{ constant}). \tag{2}$$

It is clear that when the variables have been separated the solution of the differential equation is merely an exercise in integration. Indeed, it is usual to say that the equation has been solved when the solution has been written in the form (2), even though the integration cannot be carried out explicitly.

Example 1. *Solve the equation* $\frac{dy}{dx} = x^2y + x^2$.

Here we have on separating the variables

$$x^2\,dx - \frac{1}{y+1}\,dy = 0,$$

and on integrating

$$\frac{x^3}{3} - \log_e(1+y) = A, \quad (A \text{ constant}).$$

The solution can also be written as

$$\frac{x^3}{3} = \log_e B(y+1), \quad \text{where } B \text{ is a constant,}$$

or

$$y = Ce^{x^3/3} - 1, \quad \text{where } C \text{ is a constant.}$$

Here, $A = \log_e B$ and $C = 1/B$, but as the constant is arbitrary it may be taken as A, B or C.

Sometimes when the variables are not separable a differential equation may be reduced to one in which they are separable by a change of variable. Thus in the equation

$$\frac{dy}{dx} = f(ax+by),$$

let $z = ax+by$. Then, substituting for y, we have

$$\frac{dz}{dx} - a = bf(z),$$

giving

$$\frac{dz}{a+bf(z)} - dx = 0,$$

and hence the equation can be solved.

Example 2. *Solve the equation* $\frac{dy}{dx} = (x+y)^2$.

Let $z = x+y$, then

$$\frac{dz}{dx} = 1+\frac{dy}{dx}$$

and we have

$$\frac{dz}{dx} = 1+z^2,$$

and separating the variables

$$\frac{dz}{1+z^2} - dx = 0.$$

This gives

$$\tan^{-1} z - x = A,$$

leading to

$$z = \tan(x+A),$$
$$y = \tan(x+A) - x.$$

Exercises 2(*a*)

Solve the differential equations:

1. $x\frac{dy}{dx}+y^2 = 1.$
2. $(x+a)\frac{dy}{dx}-y = b.$
3. $x\frac{dy}{dx}-y = xy.$
4. $xy\frac{dy}{dx}+(x^2+1)(y^2-1) = 0.$
5. $y^2(x^2-1)\frac{dy}{dx}-x^2(y^2-1) = 0.$
6. $xy(1+x^2)\frac{dy}{dx}-y^2 = 1.$
7. $(1-x^2)^{1/2}\frac{dy}{dx}+(1-y^2)^{1/2} = 0.$
8. $\sin x \cos y\frac{dy}{dx}+\cos x \sin y = 0.$
9. $\cos^2 y \cos^2 2x\frac{dy}{dx}+\sin^2 y \sin^2 2x = 0.$
10. $\log_e y\frac{dy}{dx}+xy\log_e(1+x) = 0.$
11. $\frac{dy}{dx}+2x\cosh x \cosh y = 0.$
12. $\frac{dy}{dx}+x+2y = 0.$

13. $\dfrac{dy}{dx}+(x+4y)^2 = 0.$

14. $x+y\dfrac{dy}{dx} = x^2+y^2$, by substituting $z = x^2+y^2$.

15. $(1+x^2)^{1/2}\dfrac{dy}{dx} = x\,e^y$, given that $y = 0$ for $x = 0$. (L.U.)

2.3 Applications of equations with separable variables

The solutions of a large variety of problems in science and engineering are found by setting up and solving differential equations with separable variables. Some of these problems are considered in the following examples and exercises.

Example 3. *Neglecting the resistance of the atmosphere, find the velocity with which a particle must be projected from the earth's surface in order to escape from the earth's gravitational field.*

The acceleration due to the earth's attraction is inversely proportional to the square of the distance x from the centre of the earth when $x > R$, the earth's radius. When $x = R$ the acceleration is g, hence for $x > R$ it is gR^2/x^2. If v is the velocity vertically upwards when the distance is x, the acceleration upwards is $v\,dv/dx$ and we have the differential equation

$$v\frac{dv}{dx} = -\frac{gR^2}{x^2}.$$

Hence

$$\int v\,dv = -gR^2\int\frac{dx}{x^2},$$

giving

$$\tfrac{1}{2}v^2 = \frac{gR^2}{x}+A, \quad (A \text{ constant}).$$

If the initial velocity is V, that is, $v = V$ when $x = R$, we have

$$\tfrac{1}{2}V^2 = gR+A,$$

and hence

$$v^2 = \frac{2gR^2}{x}+(V^2-2gR).$$

The particle will escape from the earth's attraction if v does not become zero for very large values of x, that is if $V^2 \geqq 2gR$. Taking $R = 4000\times 5280$ ft. and $g = 32$ ft./sec^2. this gives $V \geqq 36{,}770$ ft./sec. $= 6{\cdot}96$ miles per sec.

Example 4. *Find the pressure at height h in the atmosphere, assuming that the pressure p and the volume per unit mass v are connected by the adiabatic relation $pv^\gamma =$ constant.*

By considering the equilibrium of a vertical column of air of unit cross-section we find that the decrease of pressure due to a small increase of height δh is the weight of air between heights h and $h+\delta h$, that is, $\rho\,\delta h$ where ρ is the density. Therefore, since $\rho = 1/v$,

$$-\delta p = \frac{1}{v}\,\delta h.$$

In the limit we have

$$\frac{dh}{dp} = -v.$$

If p_0 and v_0 be the pressure and volume at $h = 0$, $pv^\gamma = p_0v_0^\gamma$ and we have

$$\frac{dh}{dp} = -v_0p_0^{1/\gamma}p^{-1/\gamma}.$$

The solution of this differential equation is easily found to be

$$h = -\frac{v_0p_0^{1/\gamma}}{1-1/\gamma}p^{1-1/\gamma}+\text{constant},$$

and since $p = p_0$ when $h = 0$ the constant is $\gamma v_0p_0/(\gamma-1)$. Therefore

$$h = \frac{\gamma v_0p_0}{\gamma-1}\{1-(p/p_0)^{1-1/\gamma}\},$$

that is

$$p = p_0\left\{1-\frac{(\gamma-1)h}{\gamma v_0p_0}\right\}^{\gamma/(\gamma-1)}.$$

Example 5. *A circuit consisting of a resistance of R ohms and an inductance L henries is connected to a battery of constant voltage E. Find the current, i ampères, at time t after the circuit is closed.*

An inductance and a resistance each causes a drop in voltage. The drop due to the resistance is Ri according to Ohm's law, and that due to the inductance is $L(di/dt)$. Therefore, the voltage supplied by the battery is equal to the voltage drop due to the inductance and resistance, that is

$$L\frac{di}{dt}+Ri = E.$$

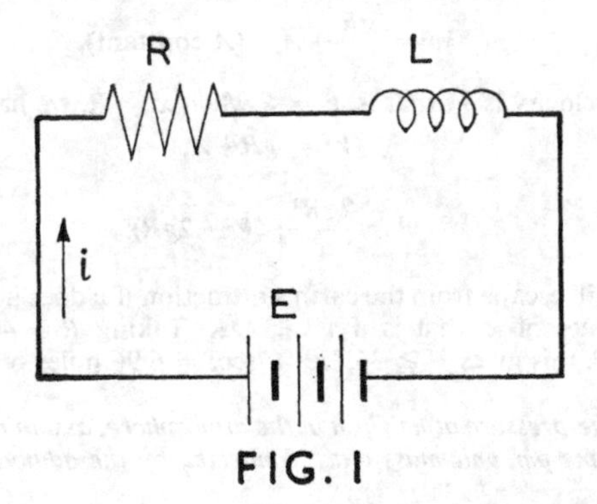

FIG. I

The circuit is shown in Fig. 1. Separating the variables we have

$$\frac{di}{E-Ri} = \frac{dt}{L},$$

giving

$$-\frac{1}{R}\log_e(E-Ri) = \frac{t}{L}+A, \quad (A\text{ constant}).$$

Since $i = 0$ when $t = 0$, we have

$$A = -\frac{1}{R}\log_e E,$$

and hence

$$\log_e\left(\frac{E-Ri}{E}\right) = -\frac{Rt}{L},$$

so that

$$1-\frac{Ri}{E} = e^{-Rt/L},$$

and hence

$$i = \frac{E}{R}(1-e^{-Rt/L}).$$

Thus the current increases eventually to the value E/R.

Exercises 2(*b*)

1. The acceleration of a particle moving in a straight line is k^2x away from a point O when x is its distance from O. By expressing the acceleration as $v\,dv/dx$, form the differential equation of the motion. Show that the solution consistent with the velocity being zero when $x = a$ is $v = k(x^2-a^2)^{1/2}$ and that if $x = a$ when $t = 0$, $x = a\cosh kt$.
2. A particle is let fall from a great height h above the earth. Neglecting air resistance, prove that it reaches the earth with velocity
$$\{2ghR/(R+h)\}^{1/2},$$
where R is the earth's radius.
3. The decrease of the intensity I of a beam of light passing through a medium in a distance δx along the beam is $\mu I\delta x$, where μ is the absorption coefficient. If $\mu = ae^{-bx}$, form the differential equation for the intensity at distance x and show that $I = I_0\exp\{a(e^{-bx}-1)/b\}$, where I_0 is the intensity where $x = 0$.
4. A tank of uniform cross-sectional area A containing a volume AH of liquid is being emptied through an orifice of effective cross-sectional area a in its base. Show that when the height of liquid in the tank is h the velocity of the emerging liquid is $\sqrt{(2gh)}$ and hence that
$$A\,dh/dt = -a\sqrt{(2gh)}.$$
Deduce that the tank will empty itself in time $2AH/\{a\sqrt{(2gH)}\}$.
5. Show that the pressure p at height h in the atmosphere satisfies the differential equation $dh/dp = -v$, where v is the volume per unit mass. Show also that in an isothermal layer in which $pv = RT$, R and T being constants, $p = p_0e^{-h/(RT)}$, where p_0 is the pressure for $h = 0$.
6. The ideal cable for a barrage balloon would taper so that the stress across every section had the same value. If w is the weight per unit volume of the material of the cable, f the constant stress and r the radius at distance x below the balloon, show that $dr/dx = -wr/(2f)$. Find a formula for r in terms of w, f and the tension T at $x = 0$, and show that the total weight of the cable, of length L, is $T(1-e^{-wL/f})$. (L.U.)

7. If a rocket of mass m is moving vertically upwards with velocity v and mass is being continuously projected backwards with relative velocity u, show that $m\,dv/dt + u\,dm/dt = -mg$, air resistance being neglected. If the rocket has mass M when full and after time t its mass is $M-ct$ and it burns for time T, show that the velocity attained is $u \log_e \{M/(M-cT)\} - gT$.

8. The rate of decay of a substance is kx, where x is the amount of the substance remaining. Show that the half-life of the substance is $(1/k) \log_e 2$.

9. A uniform circular tube of inner radius a and outer radius b contains liquid kept at a constant temperature T_1 and the outer surface of the tube is maintained at a temperature $T_0 (< T_1)$. Prove that, if K is the thermal conductivity of the material of the tube and T its temperature at distance r from its axis, the outward heat flow Q is $-K2\pi r(dT/dr)$. Hence show that in the steady state

   (i) $Q \log_e (b/a) = 2\pi K(T_1 - T_0)$,
   (ii) $(T_1 - T)/(T_1 - T_0) = \log_e (r/a)/\log_e (b/a)$.

10. By Newton's law of cooling the surface temperature T at time t of a sphere in isothermal surroundings at temperature T_0 is given by the equation $dT/dt = -k(T-T_0)$, where k is a constant. Show that if $T = T_1$ at time $t = 0$, $T - T_0 = (T_1 - T_0)e^{-kt}$.

11. It is found that when quantities a and b of two liquids are boiling in the same container the ratio of the amounts of the two liquids being vaporized at any instant is proportional to the ratio of the amounts remaining. Show that, if x and y are the remaining amounts of a and b respectively at time t, $y(dx/dt) = kx(dy/dt)$ and hence that $b^k x = ay^k$.

12. A rope passes round a cylindrical surface with coefficient of friction μ. Show that the change in tension δT in a distance in which the direction of pull changes by $\delta\theta$ is $\mu T \delta\theta$. Hence show that if the tension at the slacker end where the rope meets the surface is T_0, the tension where the direction of pull has changed by θ is $T_0 e^{\mu\theta}$.

13. A condenser of capacitance C farads with voltage v_0 is discharged through a resistance of R ohms. Show that if q coulombs is the charge on the condenser, i ampères the current and v the voltage at time t,

$$q = Cv, \quad v = Ri \quad \text{and} \quad i = -(dq/dt).$$

   Hence show that $v = v_0 e^{-t/RC}$.

14. A circuit consists of a resistance R ohms and a condenser of capacity C farads connected to a constant e.m.f. E. If q/C is the voltage of the condenser at time t after closing the circuit, show that $q/C = E - Ri$, and hence that the voltage at time t is $E(1 - e^{-t/RC})$.

15. A weight W falls from a height h into a cylinder in which it is brought to rest by compressing the air in the cylinder. When it has travelled a distance x in the cylinder the total force on it is $kaW/(a-x)$ opposing motion, a being the length of the cylinder. Show that the weight comes to rest after falling a distance $a(1 - e^{-h/ka})$ in the cylinder.

16. Stokes' law states that the resistance to motion of a sphere of radius a moving with speed v in a fluid of viscosity μ is $6\pi\mu av$. Show that when a sphere of mass m falls under gravity in a fluid through a distance x from rest we have $v\,dv/dx = \lambda(v_0-v)$, where v_0 is the terminal velocity and $\lambda = 6\pi\mu a/m$. Show also that the sphere reaches a velocity which is half of its terminal velocity after falling a distance $v_0(\log_e 2-\frac{1}{2})/\lambda$.

2.4 The linear first order differential equation

The general form for a linear differential equation of the first order is

$$\frac{dy}{dx}+Py = Q, \tag{1}$$

where P and Q are functions of x. This equation can be solved by two quadratures, that is by performing two integrations.

First consider the differential equation

$$\frac{dy}{dx}+Py = 0.$$

Since P is a function of x only, the variables are separable and we have

$$\frac{dy}{y}+P\,dx = 0,$$

giving

$$\log_e y+\int P\,dx = A,$$

that is

$$ye^{\int P\,dx} = B, \quad (B \text{ constant}). \tag{2}$$

Thus if

$$P = 1/x, \qquad \int P\,dx = \log_e x, \qquad e^{\int P\,dx} = x,$$

and the solution is

$$xy = B, \quad (B \text{ constant}).$$

If we differentiate the equation (2) we have

$$\frac{d}{dx}\left\{ye^{\int P\,dx}\right\} = e^{\int P\,dx}\left(\frac{dy}{dx}\right)+e^{\int P\,dx}Py = 0,$$

and it is clear that if the equation (1) is multiplied by $e^{\int P\,dx}$ the left-hand side of the equation is the exact differential of $ye^{\int P\,dx}$. The expression $e^{\int P\,dx}$ is called the *integrating factor* of the differential equation (1) and can often be found without much difficulty. Multiplying (1) by the integrating factor we have

$$e^{\int P\,dx}\frac{dy}{dx}+e^{\int P\,dx}Py = Qe^{\int P\,dx}.$$

This can be written

$$\frac{d}{dx}\left\{ye^{\int P\,dx}\right\} = Qe^{\int P\,dx},$$

giving on integration

$$ye^{\int P\,dx} = \int Qe^{\int P\,dx}\,dx+A,$$

and hence

$$y = e^{-\int P\,dx}\int Qe^{\int P\,dx}\,dx+Ae^{-\int P\,dx}.$$

This is the complete solution of the first order linear equation with its one arbitrary constant. An arbitrary constant is, of course, involved in the indefinite integral of P with respect to x, but this cancels out in the first part of the solution and only changes the value of A in $Ae^{-\int P\,dx}$. We may note that if P and Q are constants or P is a constant multiple of Q, the variables in the linear equation are separable and a solution can be found by the method of §2.2.

In view of later work on linear equations of higher order than the first it is pointed out here that the solution of a linear equation has two parts. The first is called the *complementary function* and this is the solution found by putting $Q = 0$ in the differential equation, that is, $y = Ae^{-\int P\,dx}$. The second is called the *particular integral* and is a solution of the equation as it stands. This is $e^{-\int P\,dx}\int Qe^{\int P\,dx}\,dx$ and, since the integral of $Qe^{\int P\,dx}$ is an indefinite one, may include a multiple of the complementary function. The general solution is then the sum of the complementary function, which contains the arbitrary constant, and the particular integral which may be any particular solution of the differential equation.

Example 6. *Solve the equation* $\dfrac{dy}{dx}+\dfrac{1}{x}y = x^2-1$.

Here

$$P = 1/x, \quad \int P\,dx = \log_e x$$

and the integrating factor is $e^{\log_e x} = x$. Multiplying by the integrating factor we have

$$x\frac{dy}{dx}+y = x^3-x,$$

that is

$$\frac{d}{dx}(xy) = x^3-x$$

giving

$$xy = \frac{x^4}{4}-\frac{x^2}{2}+A,$$

and hence

$$y = \frac{x^3}{4}-\frac{x}{2}+\frac{A}{x}.$$

Example 7. *Solve the equation* $\cos^2 x\dfrac{dy}{dx}+y = \tan x$.

Writing the equation in the standard form we have

$$\frac{dy}{dx}+\sec^2 x . y = \sec^2 x \tan x,$$

and

$$P = \sec^2 x, \quad \int P\,dx = \tan x, \quad e^{\int P\,dx} = e^{\tan x}.$$

Multiplying by the integrating factor

$$\frac{d}{dx}\{y e^{\tan x}\} = e^{\tan x}\sec^2 x \tan x.$$

This gives on integration

$$\begin{aligned} y e^{\tan x} &= \int e^{\tan x}\sec^2 x \tan x\,dx + A, \\ &= \int e^{\tan x}\tan x\, d(\tan x)+A, \\ &= e^{\tan x}(\tan x-1)+A, \end{aligned}$$

so that

$$y = \tan x-1+Ae^{-\tan x}.$$

2.5 Bernoulli's equation

The differential equation

$$\frac{dy}{dx}+Py = Qy^n, \tag{1}$$

is known as *Bernoulli's equation*. Although this equation is not of the linear form considered in §2.4 it can be reduced to that form by a simple substitution and then solved by the method of the previous paragraph.

Dividing by y^n we have

$$\frac{1}{y^n}\frac{dy}{dx}+P\frac{1}{y^{n-1}} = Q.$$

Now change the independent variable by writing $u = 1/y^{n-1}$. Then

$$\frac{du}{dx} = (1-n)\frac{1}{y^n}\frac{dy}{dx},$$

and

$$\frac{du}{dx}+(1-n)Pu = (1-n)Q.$$

This equation is of the linear form and can be solved for u by the method of §2.4, and hence a solution relating y to x can be obtained.

Example 8. *Solve the equation* $\frac{dy}{dx}+\frac{1}{x}y = x^3y^3$.

Dividing by y^3 we have

$$\frac{1}{y^3}\frac{dy}{dx}+\frac{1}{x}\frac{1}{y^2} = x^3.$$

Writing

$$u = \frac{1}{y^2}, \qquad \frac{du}{dx} = -\frac{2}{y^3}\frac{dy}{dx},$$

and we have

$$-\frac{1}{2}\frac{du}{dx}+\frac{u}{x} = x^3,$$

that is

$$\frac{du}{dx}-\frac{2}{x}u = -2x^3.$$

The integrating factor is $e^{-2\log_e x} = 1/x^2$. Then

$$\frac{1}{x^2}\frac{du}{dx}-\frac{2}{x^3}u = -2x,$$

giving

$$\frac{d}{dx}\left(\frac{u}{x^2}\right) = -2x.$$

Integrating we have

$$\frac{u}{x^2} = -x^2+A,$$

$$u = -x^4+Ax^2,$$

and the solution is

$$\frac{1}{y^2} = -x^4+Ax^2.$$

Exercises 2(*c*)

Solve the differential equations:

1. $(x+1)\frac{dy}{dx}+2y = x.$

2. $(x+1)\dfrac{dy}{dx}-2y = x.$

3. $x\dfrac{dy}{dx}+2y = e^x.$

4. $(x-1)\dfrac{dy}{dx}+3y = x^2.$

5. $(x+1)\dfrac{dy}{dx}+(2x-1)y = e^{-2x}.$

6. $x\dfrac{dy}{dx}+y = x\sin 2x.$

7. $x^2\dfrac{dy}{dx}+3xy = 1.$

8. $(x^2+1)\dfrac{dy}{dx}+xy = x.$

9. $\dfrac{dy}{dx}+y\cot x = \operatorname{cosec} x.$

10. $x(x-1)\dfrac{dy}{dx}+y = x(x-1)^2.$

11. $(1+x^2)\dfrac{dy}{dx}-xy = x(1+x^2).$

12. $(1+x^2)^{3/2}\dfrac{dy}{dx}+x(1+x^2)^{1/2}y = 1.$

13. $\dfrac{dy}{dx}+y\cos x = \sin 2x$, with $y = 0$ when $x = 0$.

14. $x(x-1)\cos y\dfrac{dy}{dx}+\sin y = x(x-1)^2.$

15. $2x(x-1)\dfrac{dy}{dx}+(2x-1)y = 1.$

16. $(e^{-2y}-2xy+x)\dfrac{dy}{dx} = y+1.$

17. $(1-3xy)\dfrac{dy}{dx}+y^2 = 0.$

18. $2x\dfrac{dy}{dx} = y-4y^3.$

19. $2y-3x\dfrac{dy}{dx} = e^x y^4.$

20. $3x^2\dfrac{dy}{dx}-6xy+2y^{5/2} = 0.$

21. $5(1+x^2)\dfrac{dy}{dx}-xy+x(1+x^2)y^6 = 0.$

2.6 Applications of the linear first order equation

Some physical problems leading to linear first order differential equations are considered in the following examples and exercises. These again illustrate the two stages in the solution of this type of problem.

Example 9. *A circuit consisting of a resistance of R ohms and an inductance L henries is connected to an e.m.f. of voltage E cos ωt. Find the current i ampères at time t after the circuit is closed.*

As in Example 5 (page 24) we equate the voltage supplied to the drop in voltage Ri due to the resistance and the drop $L\,(di/dt)$ due to the inductance and we have

$$L\frac{di}{dt}+Ri = E\cos\omega t.$$

This is a linear differential equation and the integrating factor is $e^{Rt/L}$. We have

$$e^{Rt/L}\left(\frac{di}{dt}+\frac{R}{L}i\right) = \frac{E}{L}e^{Rt/L}\cos\omega t,$$

giving

$$i\,e^{Rt/L} = \frac{E}{L}\int e^{Rt/L}\cos\omega t\,dt+A.$$

Now

$$\int e^{at}\cos bt\,dt = e^{at}(a\cos bt+b\sin bt)/(a^2+b^2),$$

therefore

$$i\,e^{Rt/L} = \frac{e^{Rt/L}E}{R^2+L^2\omega^2}(R\cos\omega t+L\omega\sin\omega t)+A,$$

and since $i = 0$ when $t = 0$.

$$A = -RE/(R^2+L^2\omega^2).$$

Writing

$$\tan\alpha = L\omega/R$$

we have

$$R\cos\omega t+L\omega\sin\omega t = (R^2+L^2\omega^2)^{1/2}\cos(\omega t-\alpha),$$

and

$$i = \frac{E}{(R^2+L^2\omega^2)^{1/2}}\cos(\omega t-\alpha)+A\,e^{-Rt/L}$$

$$= \frac{E}{(R^2+L^2\omega^2)^{1/2}}\{\cos(\omega t-\alpha)-e^{-Rt/L}\cos\alpha\}.$$

The quantity involving $e^{-Rt/L}$ is a transient component which quickly dies away leaving the steady current

$$i = \frac{E}{(R^2+L^2\omega^2)^{1/2}}\cos(\omega t-\alpha).$$

$(R^2+L^2\omega^2)^{1/2}$ is called the *impedance* and α is called the *phase lag*. Thus in the steady state the flow of current is the same as if the circuit contained merely a resistance $(R^2+L^2\omega^2)^{1/2}$ with a delay due to the phase lag.

Example 10. *In successive chemical reactions A → B → C . . . , the velocities of the first two reactions are given by the equations*

$$\frac{dx}{dt} = -kx,\qquad \frac{dy}{dt} = 2k(b+a-x-y),$$

where x is the concentration of A and y the concentration of C; initially $x = a$ and $y = 0$. Solve these equations and prove that at time t the concentration of B is $a\,e^{-kt}-(a-b)\,e^{-2kt}$.

The solution of the equation $\dot{x} = -kx$ is $x = Ae^{-kt}$ and, since $x = a$ when $t = 0$, we have

$$x = a\,e^{-kt}.$$

Substituting for x in the second equation, we have

$$\frac{dy}{dt}+2ky = 2kb+2ka-2ka\,e^{-kt}.$$

The integrating factor is e^{2kt} and we have

$$\frac{d}{dt}\{e^{2kt}y\} = 2k(a+b)\,e^{2kt}-2ka\,e^{kt},$$

giving

$$e^{2kt}y = (a+b)\,e^{2kt}-2a\,e^{kt}+A,$$

and, since $y = 0$ initially,

$$e^{2kt}y = (a+b)\,e^{2kt}-2a\,e^{kt}+(a-b).$$

Hence,

$$y = (a+b)-2a\,e^{-kt}+(a-b)\,e^{-2kt}.$$

From the form of the equations it is evident that the concentration of B at time t is $b+a-x-y$, that is

$$b+a-a\,e^{-kt}-(a+b)+2a\,e^{-kt}-(a-b)\,e^{-2kt} = a\,e^{-kt}-(a-b)\,e^{-2kt}.$$

Example 11. *Find the equation of the trajectory of a shell projected with velocity V at an inclination α to the horizontal, assuming that the retardation due to air resistance is kv^2 opposing motion, where v is the velocity.*

Let v be the velocity at a point of the trajectory where the direction of motion is inclined at an angle ψ to the horizontal. Then (Fig. 2) the retardation has

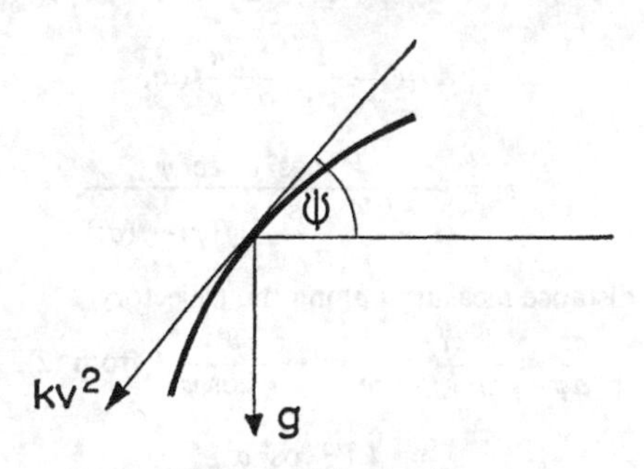

FIG. 2

components $kv^2+g\sin\psi$ and $g\cos\psi$ along and perpendicular to the direction of motion. General expressions for the acceleration along and perpendicular to the direction of motion are $\dot{v}$ and $v\dot{\psi}$. Therefore

$$\dot{v} = -kv^2-g\sin\psi, \tag{1}$$

$$v\dot{\psi} = -g\cos\psi. \tag{2}$$

Hence

$$\frac{\dot{v}}{\dot{\psi}} = \frac{dv}{d\psi} = \frac{kv^3}{g\cos\psi} + v\tan\psi,$$

so that

$$\cos\psi\frac{dv}{d\psi} - v\sin\psi = \frac{k}{g}v^3. \tag{3}$$

This is a differential equation of Bernoulli's type which is reduced to a linear equation by the substitution $u = 1/v^2$, giving

$$\frac{du}{d\psi} + (2\tan\psi)u = -\frac{2k}{g}\sec\psi. \tag{4}$$

Now

$$\int 2\tan\psi\, d\psi = \log_e \sec^2\psi$$

and the integrating factor for (4) is $\sec^2\psi$. We have

$$\sec^2\psi\frac{du}{d\psi} + (2\sec^2\psi\tan\psi)u = -\frac{2k}{g}\sec^3\psi,$$

which can be written

$$\frac{d}{d\psi}\{u\sec^2\psi\} = -\frac{2k}{g}\sec^3\psi.$$

This gives

$$u\sec^2\psi = -\frac{2k}{g}\int \sec^3\psi\, d\psi + A$$

$$= -\frac{k}{g}\{\sec\psi\tan\psi + \log_e(\sec\psi + \tan\psi)\} + A,$$

that is

$$\frac{1}{v^2\cos^2\psi} = -\frac{k}{g}f(\psi) + A,$$

where

$$f(\psi) = \sec\psi\tan\psi + \log_e(\sec\psi + \tan\psi).$$

Hence

$$A = \frac{1}{V^2\cos^2\alpha} + \frac{k}{g}f(\alpha),$$

and we have

$$v^2 = \frac{V^2\cos^2\alpha\sec^2\psi}{1 - \dfrac{kV^2\cos^2\alpha}{g}\{f(\psi) - f(\alpha)\}}.$$

Now if s is the distance measured along the trajectory

$$\frac{ds}{d\psi} = \frac{ds}{dt}\frac{dt}{d\psi} = \frac{v}{\dot{\psi}} = -\frac{v^2}{g\cos\psi}, \quad \text{from (2).}$$

Therefore

$$ds = -\frac{1}{2k}\frac{\dfrac{kV^2\cos^2\alpha}{g}(2\sec^3\psi)\, d\psi}{1 - \dfrac{kV^2\cos^2\alpha}{g}\{f(\psi) - f(\alpha)\}},$$

and remembering that $df(\psi) = 2\sec^3\psi\, d\psi$, we have

$$s = \frac{1}{2k}\log_e[1 - (kV^2/g)\cos^2\alpha\{f(\psi) - f(\alpha)\}].$$

This is one form of the equation of the trajectory.

Exercises 2(*d*)

1. A particle of unit mass is attracted to a fixed point O by a force μx when it is distant x from O in a medium which offers a resistance kv^2 to its motion where v is its velocity. Show that the equation of motion is $v(dv/dx)+kv^2+\mu x = 0$. If the particle is initially at rest at distance a from O show that it reaches O with velocity
$$\{(\mu/2k^2)(1-e^{2ka}+2ka\,e^{2ka})\}^{1/2}.$$

2. A particle of unit mass in a medium in which the resistance to motion is k times its velocity is acted on by a force of magnitude $a \cos pt$ in a fixed direction. Prove that if v is the velocity of the particle $(dv/dt)+kv = a \cos pt$. Prove also that, if the particle is initially at rest, its velocity at time t is $a(k^2+p^2)^{-1}(k \cos pt+p \sin pt-k e^{-kt})$.

3. A chain of mass m per unit length is coiled on the edge of a smooth table and begins to fall over the edge. Show that when a length x has fallen the momentum of the chain is $mx\dot{x}$ and hence that the equation of motion is $xv(dv/dx)+v^2 = gx$, v being the velocity. Hence show that $3v^2 = 2gx$.

4. A rocket has mass m when full and after burning for time t its mass is $m-bt$; the relative backward velocity of the gases is u. If the rocket is ignited and moves vertically upwards from rest against air resistance kbv, where v is the velocity, show that the equation of motion is
$$(m-bt)\frac{dv}{dt}-bu = -(m-bt)g-kbv.$$
Hence show that
$$v = \frac{u}{k}+\frac{g(m-bt)}{b(1-k)}-\left\{\frac{mg}{b(1-k)}+\frac{u}{k}\right\}\left(1-\frac{bt}{m}\right)^k.$$

5. A circuit with resistance R ohms and a condenser of capacitance C farads is charged by a voltage $E \cos \omega t$. Show that if q is the charge on the condenser at time t, $CR(dq/dt)+q = CE \cos \omega t$. Hence show that if $q = 0$ when $t = 0$, $q = CE \sin \beta \sin (\omega t+\beta)-CE \sin^2 \beta e^{-t/CR}$, where $\cot \beta = CR\omega$. Show also that when the transient component has disappeared the flow of current has impedance $R \sec \beta$ and phase lead β.

6. A circuit consists of a resistance R ohms and an inductance L henries. The arrangement is shunted by an equal resistance R and a condenser of capacitance C farads. An e.m.f. $E \sin \omega t$ produces currents i_1 and i_2 in the two branches. Show that
$$L\frac{di_1}{dt}+Ri_1 = E \sin \omega t,$$
$$R\frac{di_2}{dt}+\frac{i_2}{C} = \omega E \cos \omega t.$$
Show also that if $CR^2 = L$, the total current i_1+i_2 will be $(E/R) \sin \omega t$.

7. A tank, at time t, contains a volume v of a solution at a concentration C of a given reagent and a solution of constant concentration C_0 flows into the tank at a variable rate q (litres per min.), while solution is removed by an outlet pipe at a constant rate q_0. The concentration is uniform throughout the tank except in a very small region near the inlet. Prove that

$$\frac{d}{dt}\{(C-C_0)v\}+(C-C_0)q_0 = 0,$$

and show how to obtain the concentration at any instant from data giving v as a function of t. If initially $C = 0$ and if $v = v_0+q_0t$, find the concentration at time t. (L.U.)

8. The amounts x, y, z of three substances A, B, C satisfy the equations

$$\frac{dx}{dt} = -px, \qquad \frac{dy}{dt} = px-qy, \qquad \frac{dz}{dt} = qy,$$

and initially at $t = 0$, $x = a$ and $y = z = 0$. If $p \neq q$, show that y is a maximum when $(p-q)t = \log_e p-\log_e q$, and find the amounts of A, B and C at that instant. (L.U.)

9. By Newton's law of cooling the surface temperature T at time t of a sphere in surroundings at temperature T_0 is determined by the equation $(dT/dt) = -k(T-T_0)$, where k is a constant. Show that if

$$T_0 = a+b\cos pt \text{ and } T = T_1 \text{ at time } t = 0,$$

$$T = e^{-kt}(T_1-a)+a+\{kb/(k^2+p^2)\}(k\cos pt+p\sin pt-ke^{-kt}).$$

10. A body is heated so that its temperature T increases at a rate ct after time t, while at the same time its temperature decreases at a rate $k(T-T_0)$, where T_0 is the temperature of the surrounding medium. Show that $(dT/dt)+k(T-T_0) = ct$. Show further that, if $T = T_1$ when $t = 0$,

$$T-T_0 = (T_1-T_0+c/k^2)e^{-kt}+(c/k^2)(kt-1).$$

11. The radial compressive stress p and the hoop stress f at distance r from the axis of a cylindrical gun tube subjected to internal pressure p_0 and external pressure p_1 are given by $r(dp/dr)+p = f$ and $f+ap = b$, where a and b are constants. If r_0 and r_1 are respectively the inner and outer radii of the tube and the stress p_1 is negligible by comparison with p_0, show that $r_1 = r_0\{1-(a+1)p_0/b\}^{1/(a+1)}$.

12. A cable of weight w per unit length passes over a fixed pulley of radius a, whose axis is horizontal, and is on the point of slipping, the coefficient of friction between cable and pulley being μ. Show that if θ is the inclination to the vertical of the cable at a point, and $\theta = 0$ where the slacker end leaves the pulley, the tension T at the point is given by $(dT/d\theta)-\mu T = wa(\cos\theta+\mu\sin\theta)$. Show also that if T_0 and T_1 are the tensions at $\theta = 0$ and $\theta = \pi$ respectively,

$$(T_1-T_0e^{\mu\pi})(1+\mu^2) = 2\mu wa(1+e^{\mu\pi}).$$

13. A belt of mass w per unit length runs with speed v on a wheel of radius r being in contact over a length $r\alpha$ of the circumference. Show that if the belt is on the point of slipping on the wheel, the coefficient of friction being μ, the tension T at distance $r\theta$ from the slacker end is given by $(dT/d\theta)-\mu T = -\mu wv^2/g$. Hence show that if T_0 and T_1 are the tensions where $\theta = 0$ and $\theta = \alpha$ respectively, $T_1 = T_0\, e^{\mu\alpha}-wv^2(e^{\mu\alpha}-1)/g$.

2.7 Homogeneous equations

A function of two variables $f(x, y)$ is said to be a *homogeneous function of degree zero* if, when λx is substituted for x and λy is substituted for y, the function has the same value. Thus $(x^2+2y^2)/(x^2-y^2)$ is a homogeneous function of degree zero since

$$\frac{(\lambda x)^2+2(\lambda y)^2}{(\lambda x)^2-(\lambda y)^2} = \frac{x^2+2y^2}{x^2-y^2}.$$

The general differential equation of the first order and first degree

$$\frac{dy}{dx} = f(x, y) \tag{1}$$

can be solved when $f(x, y)$ is a homogeneous function of degree zero and such an equation is then called a *homogeneous equation.*

Since $f(x, y) = f(\lambda x, \lambda y)$ we may take $\lambda = 1/x$ and we have $f(x, y) = f(1, y/x)$. Thus $f(x, y)$ is a function of y/x and we may write a homogeneous equation in the form

$$\frac{dy}{dx} = \phi\left(\frac{y}{x}\right). \tag{2}$$

The form of the equation (2) suggests the substitution $z = y/x$. Making this substitution we have

$$y = zx,$$

$$\frac{dy}{dx} = x\frac{dz}{dx}+z,$$

and hence

$$x\frac{dz}{dx} = \phi(z)-z.$$

This is an equation in which the variables can be separated and the solution is

$$\int\frac{dx}{x} = \int\frac{dz}{\phi(z)-z}+A.$$

Having performed the integrations we replace z by y/x and thus obtain the solution giving y in terms of x.

Example 12. *Solve the equation* $x\frac{dy}{dx} = y+(x^2+y^2)^{1/2}$.

We have

$$\frac{dy}{dx} = \frac{y}{x}+\left(1+\frac{y^2}{x^2}\right)^{1/2}.$$

Putting $y = zx$, we have

$$x\frac{dz}{dx}+z = z+(1+z^2)^{1/2};$$

separating the variables,

$$\int\frac{dx}{x} = \int\frac{dz}{(1+z^2)^{1/2}}$$

giving

$$\log_e x = \log_e\{z+(1+z^2)^{1/2}\}+A,$$

that is

$$Bx = z+(1+z^2)^{1/2},$$

and, since $xz = y$,

$$Bx^2 = y+(x^2+y^2)^{1/2},$$

B being an arbitrary constant.

Example 13. *Solve the equation* $\frac{dy}{dx} = \frac{y(y-2x)}{x(x-2y)}$.

The equation can be written in the form

$$\frac{dy}{dx} = \frac{(y/x)(y/x-2)}{1-2y/x}.$$

Putting $y = zx$, we have

$$x\frac{dz}{dx}+z = \frac{z(z-2)}{1-2z},$$

that is

$$x\frac{dz}{dx} = \frac{3(z^2-z)}{1-2z}.$$

We then have

$$\frac{1-2z}{z^2-z}\,dz = 3\frac{dx}{x},$$

giving

$$-\log_e(z^2-z) = 3\log_e x+A,$$

that is

$$x^3(z^2-z) = B,$$

and, since $xz = y$,

$$xy^2-x^2y = B, \quad \text{where } B \text{ is a constant.}$$

2.8 Reducible equations

Certain equations which are not homogeneous can be made so by a change of variables. The simplest case is that of the equation

$$\frac{dy}{dx} = \frac{a_1x+b_1y+c_1}{a_2x+b_2y+c_2}. \tag{1}$$

Let (α, β) be the coordinates of the point of intersection of the straight lines $a_1x+b_1y+c_1 = 0$ and $a_2x+b_2y+c_2 = 0$. Then since (α, β) lies on each of the lines

$$a_1x+b_1y+c_1 \equiv a_1(x-\alpha)+b_1(y-\beta),$$

$$a_2x+b_2y+c_2 \equiv a_2(x-\alpha)+b_2(y-\beta).$$

Let $x-\alpha = X$ and $y-\beta = Y$. Then $dy/dx = dY/dX$ and we have

$$\frac{dY}{dX} = \frac{a_1X+b_1Y}{a_2X+b_2Y}.$$

This is a homogeneous equation which can be solved by the method of §2.7.

A difficulty arises when the two straight lines are parallel, so that there is no point of intersection (α, β). In this case we have $a_1/b_1 = a_2/b_2$ and $a_2x+b_2y = (a_2/a_1)(a_1x+b_1y)$ so that the function on the right-hand side in (1) is a function of a_1x+b_1y. This equation can then be solved by the method of §2.2 by substituting $z = a_1x+b_1y$, giving an equation of the form

$$\frac{dz}{dx} = a_1+b_1\phi(z)$$

in which the variables are separable.

Equations in which the right-hand side in (1) is any function of $(a_1x+b_1y+c_1)/(a_2x+b_2y+c_2)$ may be dealt with in the same way.

Example 14. *Solve the equation* $(x-y+3)\dfrac{dy}{dx} = (3x-y-1)$.

The lines $x-y+3 = 0$ and $3x-y-1 = 0$ intersect at the point (2, 5) and $x-y+3 = (x-2)-(y-5)$, $3x-y-1 = 3(x-2)-(y-5)$. Putting $X = x-2$, $Y = y-5$, we have

$$\frac{dY}{dX} = \frac{3X-Y}{X-Y}.$$

Putting $Y = ZX$ this can be written

$$X\frac{dZ}{dX} = \frac{3-Z}{1-Z}-Z,$$

$$= \frac{Z^2-2Z+3}{1-Z}.$$

Hence

$$\frac{(Z-1)\,dZ}{Z^2-2Z+3}+\frac{dX}{X} = 0,$$

and, on integration

$$X^2(Z^2-2Z+3) = A,$$

that is

$$Y^2-2XY+3X^2 = A,$$

or

$$3x^2-2xy+y^2-2x-6y = B, \quad (B \text{ constant}).$$

Exercises 2(*e*)

Solve the differential equations:

1. $(x-y)\dfrac{dy}{dx} = x+y.$

2. $(x+2y)\dfrac{dy}{dx} = 2x-y.$

3. $(x+y)\dfrac{dy}{dx} = y-2x.$

4. $(x^2+y^2)\dfrac{dy}{dx} = xy.$

5. $x^2\dfrac{dy}{dx}+xy-y^2 = 0.$

6. $(y^3-3xy^2)\dfrac{dy}{dx} = x^3+y^3.$

7. $(2x-y)\dfrac{dy}{dx} = 2y-x.$

8. $2x^2\dfrac{dy}{dx}-xy-y^2 = 0.$

9. $x(2y^4-x^4)\dfrac{dy}{dx} = y(y^4-x^4).$

10. $\dfrac{dy}{dx} = \sin(y/x)+(y/x).$

11. $(4x-10y)\dfrac{dy}{dx} = 2x-5y+2.$

12. $(3x+y-5)\dfrac{dy}{dx} = 2x+2y-2.$

13. $(x^2+y^2-2x-4y+5)\dfrac{dy}{dx} = xy-2x-y+2.$

14. $(x-y+1)^2\dfrac{dy}{dx} = (x-y-2)^2.$

15. A particle of unit mass moves in a straight line, being attracted to a fixed point O of the line by a force μx where x is its distance from O. There is also a frictional force kv opposing motion, v being the velocity. Show that the equation of motion is $v(dv/dx)+kv+\mu x = 0$. Hence show that $\log_e(v^2+kvx+\mu x^2)-(k/\omega)\tan^{-1}\{(v/\omega x)+(k/2\omega)\} = \text{constant}$, where $\omega^2 = \mu-k^2/4$.

2.9 Non-linear equations of the first order

We now consider some non-linear equations of the first order for which solutions can be easily obtained. Such equations are of the form

$f(x, y, p) = 0$, where p is conveniently written for (dy/dx). Typical examples are

$$(p-y)^2-(x-y)^2 = 0, \tag{1}$$

$$y = px+p^2x^2, \tag{2}$$

$$x = \frac{y}{p^2}-\frac{1}{p}. \tag{3}$$

If the equation is such that p can be obtained explicitly in terms of x and y, as in (1), the solution can often be obtained by the method of §1.6(*b*).

Example 15. *Solve the equation* $(p-y)^2-(x-y)^2 = 0$.

We have

$$p = y\pm(x-y),$$

that is

$$p = x, \quad \text{or} \quad p = 2y-x.$$

Taking $p = x$ we easily obtain the solution

$$2y-x^2 = A. \tag{4}$$

Taking $p = 2y-x$ and putting $z = 2y-x$ we have

$$\frac{dz}{dx} = 2p-1$$

$$= 2z-1.$$

Hence

$$2z-1 = A\,e^{2x},$$

that is

$$4y-2x-1 = A\,e^{2x}. \tag{5}$$

Thus either (4) or (5) is a solution of the differential equation and these solutions can be combined in the single relation

$$(2y-x^2-A)(4y-2x-1-A\,e^{2x}) = 0.$$

A standard procedure, which will often lead to a solution of the differential equation, is adopted when y can be obtained explicitly in terms of x and p as in (2). The equation is differentiated with respect to x, and, since $(dy/dx) = p$, an equation involving p, x and dp/dx is obtained.

Example 16. *Solve the equation* $y = px+p^2x^2$.

Differentiating the given equation, we have

$$p = \frac{dy}{dx} = p+x\frac{dp}{dx}+2px^2\frac{dp}{dx}+2xp^2,$$

and hence

$$(1+2px)\frac{dp}{dx}+2p^2 = 0.$$

If p is taken as the independent variable in this equation, we have

$$\frac{dx}{dp}+\frac{x}{p} = -\frac{1}{2p^2}.$$

Multiplying by p, which is the integrating factor, we have

$$\frac{d(xp)}{dp} = -\frac{1}{2p},$$

and hence

$$x = -\frac{1}{2p}\log_e p+\frac{A}{p}, \quad (A \text{ constant}). \tag{6}$$

Substituting this value of x in the original equation we have

$$y = -\tfrac{1}{2}\log_e p+A+(-\tfrac{1}{2}\log_e p+A)^2. \tag{7}$$

Equations (6) and (7) together form a complete solution of the differential equation since x and y are given in terms of a parameter p. A relation between x and y can, at least in theory, be found by eliminating p between the two equations.

If x is given explicitly in terms of y and p, the same process may be adopted (differentiating this time with respect to y) to obtain an equation in y, p and dp/dy.

Example 17. *Solve the equation* $x = y/p^2-1/p$.

Differentiating with respect to y, we have

$$\frac{1}{p} = \frac{1}{p^2}-\frac{2y}{p^3}\frac{dp}{dy}+\frac{1}{p^2}\frac{dp}{dy}.$$

This gives

$$p^2-p = (p-2y)\frac{dp}{dy},$$

that is

$$\frac{dy}{dp}+\frac{2}{p^2-p}y = \frac{1}{p-1}.$$

This is a linear equation with integrating factor $(1-1/p)^2$ and we have

$$\frac{d}{dp}\left\{\left(1-\frac{1}{p}\right)^2 y\right\} = \frac{1}{p}-\frac{1}{p^2}.$$

Integrating, we have

$$\left(1-\frac{1}{p}\right)^2 y = \frac{1}{p}+\log_e p+A,$$

hence

$$y = \frac{p^2}{(p-1)^2}\left\{\frac{1}{p}+\log_e p+A\right\},$$

and, from the original equation,

$$x = \frac{1}{(p-1)^2}\left\{\frac{1}{p}+\log_e p+A\right\}-\frac{1}{p}.$$

2.10 Clairaut's equation

The non-linear differential equation

$$y = px+f(p), \tag{1}$$

where $f(p)$ is some function of p, is known as *Clairaut's equation* and has the very simple solution

$$y = Ax+f(A), \tag{2}$$

where A is an arbitrary constant.

To show this we use the method of §2.9 and differentiate equation (1) with respect to x. This gives

$$p = p+x\frac{dp}{dx}+f'(p)\frac{dp}{dx},$$

that is

$$\frac{dp}{dx}\{x+f'(p)\} = 0. \tag{3}$$

A solution is obtained by taking $dp/dx = 0$, and this gives

$$p = A, \quad (A \text{ constant}).$$

Hence, (2) is a solution of the differential equation.

We also have the possibility of taking $x+f'(p) = 0$ in (3), and this leads to the solution

$$x = -f'(p),$$
$$y = f(p)-pf'(p), \quad \text{from (1)}.$$

This solution of Clairaut's equation contains no arbitrary constant. It is not, in general, derivable from the general solution (2) by giving a particular value to A and is therefore a singular solution of the equation (1) as defined in §1.5.

Example 18. *Solve the equation* $y = px-p^3$.

This equation is of Clairaut's form and we may write down the solution as

$$y = Ax-A^3. \tag{4}$$

The solution obtained by writing $x+f'(p) = 0$ is

$$x = 3p^2,$$
$$y = 2p^3.$$

In these equations we regard p as an arbitrary parameter, and it is clear that for all values of p we have $y = px-p^3$. But if we eliminate p between the expressions for x and y we have $4x^3 = 27y^2$. This singular solution is the equation of a semicubical parabola and is not one of the family of straight lines given by (4).

2.11 Singular solutions

We have defined (§1.5) a singular solution as being a solution of a differential equation which is not derivable from the general solution by giving a particular value to the arbitrary constant. A singular solution will usually represent the envelope of the family of curves which is the general solution of the differential equation. For if (x, y) be the co-ordinates and p the slope at the point of contact of any one of the family of curves with the envelope, the values of x, y and p will satisfy the differential equation. The envelope is the locus of such points and therefore its equation will be a solution of the differential equation.

We know from differential geometry that the envelope of the family of curves $f(x, y, A) = 0$, where A has an arbitrary value, is obtained by eliminating A between the equations

$$f(x, y, A) = 0 \quad \text{and} \quad \frac{\partial}{\partial A} f(x, y, A) = 0.$$

Thus the solution of Clairaut's equation $y = px + f(p)$ is the family of curves $y = Ax + f(A)$. For the envelope of the family we have therefore the equations

$$y = Ax + f(A),$$

and

$$0 = x + f'(A),$$

and the result of eliminating A between these equations is the envelope.

In Example 18,

$$f(A) = -A^3, \qquad f'(A) = -3A^2,$$

and hence for the envelope we eliminate A between the equations

$$x = 3A^2, \qquad y = 2A^3,$$

to obtain the equation

$$4x^3 = 27y^2.$$

Example 19. *Find the general solution and the singular solution of the equation* $y^2(1+p^2) = r^2$, (r *constant*).

We have

$$p = \pm \frac{(r^2 - y^2)^{1/2}}{y},$$

that is

$$\int \frac{y\,dy}{(r^2 - y^2)^{1/2}} = \pm \int dx + A.$$

Thus

$$-(r^2 - y^2)^{1/2} = \pm x + A,$$

and hence

$$(x \pm A)^2 + y^2 = r^2. \tag{1}$$

Thus the general solution represents a family of circles of radius r with centres on the x-axis. The envelope is found by differentiating (1) with respect to A, giving $x \pm A = 0$, and hence from (1) $y^2 = r^2$. The lines $y = \pm r$ are therefore the envelopes of the family of circles and it is easily seen that $y^2 = r^2$ gives $p = 0$ and hence is a singular solution of the differential equation.

Exercises 2(f)

Solve the differential equations:

1. $p^2+2xp-8x^2 = 0$.
2. $p^2+4p\cosh 2x+4 = 0$.
3. $x+y = p^3$.
4. $y = 4p+4p^3$.
5. $p^2-py+x = 0$.
6. $3p^4-py+1 = 0$.
7. $p^2+x = 2y+1$.

Find the general and singular solutions of:

8. $y = px+1/p$.
9. $y = px+(b^2+a^2p^2)^{1/2}$.
10. $y = px+p^n$.
11. $y = px-p/(1+p^2)^{1/2}$.
12. $y = px-\log_e p$.

2.12 Orthogonal trajectories

Given the equation of a family of curves $f(x, y, A) = 0$, we can find the equation of another family of curves $F(x, y, B) = 0$ which intersect the curves of the original family at right angles. The curves $F(x, y, B) = 0$ are called the *orthogonal trajectories* of the curves $f(x, y, A) = 0$.

The problem of finding orthogonal trajectories has many practical applications. Thus in the steady flow of a liquid the stream lines intersect the lines of equal velocity potential at right angles and, in two-dimensional problems of electrostatics, the sections of the equipotential surfaces are orthogonal to the lines of force.

To find the orthogonal trajectories of the family $f(x, y, A) = 0$, we first find the differential equation of the family. We have on differentiating

$$\frac{\partial f}{\partial x}+\frac{\partial f}{\partial y}\frac{dy}{dx} = 0. \tag{1}$$

Eliminating the constant A between the equation $f(x, y, A) = 0$ and equation (1) we obtain a differential equation of the form

$$\phi(x, y, p) = 0. \tag{2}$$

On the orthogonal trajectories the slope at the point (x, y) will be p' where $pp' = -1$, therefore these curves satisfy the differential equation

$$\phi(x, y, -1/p) = 0. \tag{3}$$

The solution of the differential equation (3) is the family of orthogonal trajectories.

Example 20. *The stream lines due to a doublet in two-dimensional flow are the circles* $x^2+(y-a)^2 = a^2$. *Find the lines of equal velocity potential.*

We have for the stream lines

$$x+(y-a)\frac{dy}{dx} = 0,$$

and hence

$$a = y+\frac{x}{p}.$$

Substituting for a we find the differential equation of the stream lines to be

$$x^2+y^2-2y\left(y+\frac{x}{p}\right) = 0,$$

that is

$$p(x^2-y^2)-2xy = 0.$$

The differential equation for the orthogonal lines of equal velocity potential is therefore

$$x^2-y^2+2pxy = 0,$$

that is

$$\frac{dy}{dx}+\frac{x^2-y^2}{2xy} = 0.$$

Putting $y = zx$, we have

$$x\frac{dz}{dx}+z+\frac{1-z^2}{2z} = 0.$$

Hence

$$\int\frac{2z\,dz}{z^2+1}+\int\frac{dx}{x} = 0,$$

giving

$$x(z^2+1) = A,$$

that is

$$x^2+y^2 = Ax.$$

This equation represents a family of circles whose centres lie on the x-axis and which pass through the origin.

2.13 Riccati's equation

The first order non-linear equation

$$\frac{dy}{dx}+y^2+Py+Q = 0, \tag{1}$$

where P and Q are functions of x, is known as *Riccati's equation.* This very general type of equation is not easily solved except in some special cases. Its importance lies in its relation to the general second order linear equation

$$\frac{d^2z}{dx^2}+P\frac{dz}{dx}+Qz = 0. \tag{2}$$

The tranformation

$$y = \frac{1}{z}\frac{dz}{dx}$$

transforms equation (1) into equation (2), for we have

$$\frac{dy}{dx} + y^2 = \frac{1}{z}\frac{d^2z}{dx^2},$$

and hence

$$\frac{dy}{dx} + y^2 + P\,y + Q = \frac{1}{z}\left(\frac{d^2z}{dx^2} + P\frac{dz}{dx} + Qz\right).$$

Thus the equations (1) and (2) are equivalent, and if y is a solution of (1), the solution of (2) is

$$z = A\,e^{\int y\,dx}, \quad (A \text{ constant}).$$

A particular solution of Riccati's equation can sometimes be found by inspection and if a particular solution is known the general solution of the equation can be found.

Let y_1 be a particular solution of (1) and let the general solution be $y = y_1 + 1/v$. Substituting in (1) we have

$$\frac{dy_1}{dx} - \frac{1}{v^2}\frac{dv}{dx} + \left(y_1 + \frac{1}{v}\right)^2 + P\left(y_1 + \frac{1}{v}\right) + Q = 0,$$

and, since y_1 is a solution,

$$-\frac{1}{v^2}\frac{dv}{dx} + \frac{2y_1}{v} + \frac{1}{v^2} + \frac{P}{v} = 0,$$

that is

$$\frac{dv}{dx} - (P + 2y_1)v = 1.$$

Now let

$$R = \int (P + 2y_1)\,dx.$$

Then e^{-R} is an integrating factor and we have

$$\frac{d}{dx}(e^{-R}v) = e^{-R},$$

and

$$v = e^{R}\left(\int e^{-R}\,dx + A\right), \quad (A \text{ constant}).$$

Example 21. *Find the solution of the equation* $\dot{y}+y^2+2ky+\mu = 0$, *k and μ being constants, and deduce the solution of the equation* $\ddot{x}+2k\dot{x}+\mu x = 0$.

The Riccati equation has variables separable and we have, writing

$$\mu = k^2+\omega^2,$$

$$\int \frac{dy}{(y+k)^2+\omega^2}+\int dt = 0,$$

that is

$$\tan^{-1}\left(\frac{y+k}{\omega}\right)+\omega t = \varepsilon, \quad (\varepsilon \text{ constant}).$$

Hence

$$y = -k-\omega \tan(\omega t-\varepsilon).$$

The transformation

$$y = \frac{1}{x}\frac{dx}{dt}$$

transforms the Riccati equation into the second order equation and we have

$$x = A e^{\int y dt}$$

$$= A e^{-kt+\log_e \cos(\omega t-\varepsilon)}$$

that is

$$x = A e^{-kt} \cos(\omega t-\varepsilon).$$

Exercises 2(g)

1. Find the orthogonal trajectories of the family of curves $xy = A$.
2. Find the orthogonal trajectories of the family of ellipses $4y^2+x^2 = A$.
3. Prove that, if $f\{r, \theta, r(d\theta/dr)\} = 0$ is the differential equation of a family of curves in polar coordinates, $f\{r, \theta, (-1/r)(dr/d\theta)\} = 0$ is the differential equation of the orthogonal trajectories. Find the orthogonal trajectories of the family of spirals $\log_e r = A\theta$.
4. Find the differential equation of the family of conics
$$x^2/(a^2+\lambda)+y^2/(b^2+\lambda) = 1.$$
Show that the family is self-orthogonal.
5. For a source in an infinite stream the stream lines are the curves $cy+\tan^{-1}(y/x) = A$, where A is the parameter of the family. Find the equation of the lines of equal velocity potential.
6. The stream lines for a doublet in an infinite stream are the curves $cy+y/(x^2+y^2) = A$, where A is the parameter of the family. Find the equation of the lines of equal velocity potential.
7. Show that $y = 1/x$ is a solution of the equation
$$(dy/dx)+y^2+y/x-1/x^2 = 0,$$
and find the general solution of the equation.
8. Use the result of Exercise 7 to find the general solution of the equation
$$\frac{d^2z}{dx^2}+\frac{1}{x}\frac{dz}{dx}-\frac{z}{x^2} = 0.$$

9. Show that $y = 3/x$ is a solution of the equation

$$x^2(x-2)(dy/dx)+x^2(x-2)y^2-2x(2x-3)y+6(x-1) = 0$$

and that the general solution is $y = (3Ax^2-2x+1)/(Ax^3-x^2+x)$, A being constant. Hence find the solution of the equation

$$x^2(x-2)(d^2z/dx^2)-2x(2x-3)(dz/dx)+6(x-1)y = 0.$$

10. Show that $y = -3/x$ is a solution of the equation

$$(dy/dx)+y^2+7y/x+9/x^2 = 0$$

and find the general solution. Deduce the general solution of the equation

$$(d^2z/dx^2)+(7/x)(dz/dx)+9z/x^2 = 0.$$

CHAPTER 3

LINEAR DIFFERENTIAL EQUATIONS WITH CONSTANT COEFFICIENTS

3.1 Introductory

The general form of the nth order linear differential equation with constant coefficients is

$$\frac{d^n y}{dx^n}+p_1\frac{d^{n-1}y}{dx^{n-1}}+\ldots+p_{n-1}\frac{dy}{dx}+p_n y=f(x), \tag{1}$$

where $p_1, p_2, \ldots$ are constants and $f(x)$ is some function of x. The solution of this equation involves first finding the solution of the equation with $f(x)$ replaced by zero, that is

$$\frac{d^n y}{dx^n}+p_1\frac{d^{n-1}y}{dx^{n-1}}+\ldots+p_{n-1}\frac{dy}{dx}+p_n y=0. \tag{2}$$

Equation (2) is called the *reduced equation* or the *homogeneous equation* corresponding to (1).

Many of the differential equations of this type which arise in physical problems are of the second order and we shall consider in detail the solution of the equation

$$\frac{d^2 y}{dx^2}+p\frac{dy}{dx}+qy=f(x), \tag{3}$$

where p and q are constants.

It is convenient in dealing with linear equations to use the operator notation, that is

$$Dy\equiv\frac{dy}{dx},\quad D^2y\equiv\frac{d^2y}{dx^2},\quad D^3y\equiv\frac{d^3y}{dx^3},\quad \text{and so on.}$$

Here D is an operator and it must not be thought that D^2, D^3, etc., are squares and cubes in the ordinary sense. They can however be handled as if they were, obeying many of the ordinary rules of algebra. Thus, using the ordinary rules of the differential calculus,

$$D^2(y+z)=D^2y+D^2z,$$

$$D^2(D^3y)=D^5y=D^3(D^2y),$$

$$(D-\alpha)(D-\beta)y=(D-\beta)(D-\alpha)y$$

$$=\{D^2-(\alpha+\beta)D+\alpha\beta\}y,$$

when α and β are constants. The symbol D^n is thus defined for positive integral values of n and is seen from the foregoing examples to obey the associative, distributive and commutative laws of algebra. It is also possible to use the symbols $D^{-1}y$ to represent $\int y\,dx$, $D^{-2}y$ to represent $\int dx\left\{\int y\,dx\right\}$, and so on. A meaning can also be found for such expressions as

$$\frac{1}{1+D}y = (1-D+D^2-D^3+\ldots)y,$$

$$e^D y = \left(1+D+\frac{1}{2!}D^2+\frac{1}{3!}D^3+\ldots\right)y,$$

$$\log_e(1+D)y = \left(D-\frac{1}{2}D^2+\frac{1}{3}D^3-\frac{1}{4}D^4+\ldots\right)y,$$

and techniques based on such forms can be used to find solutions of differential equations expeditiously. Some of these techniques will be considered in this chapter.

The expression

$$D^n+p_1D^{n-1}+p_2D^{n-2}+\ldots+p_{n-1}D+p_n, \tag{4}$$

is called a *linear differential operator* of order n. Denoting this operator by the symbol L, the equations (1) and (2) can be written respectively in the forms

$$L(y) = f(x), \tag{5}$$

and

$$L(y) = 0. \tag{6}$$

3.2 Nature of the solutions of linear equations

We now establish some results for linear equations in general. Consider the second order reduced equation

$$(D^2+PD+Q)y = 0, \tag{1}$$

where P and Q are functions of x or constants. Let y_1 be a solution of (1); then, A being any constant, since $DAy_1 = ADy_1$ and $D^2Ay_1 = AD^2y_1$, we have

$$(D^2+PD+Q)Ay_1 = A(D^2+PD+Q)y_1 = 0,$$

and hence Ay_1 is also a solution of equation (1). Also, if y_1 and y_2 are independent solutions of (1) and A and B are any constants,

$$D(Ay_1+By_2) = ADy_1+BDy_2,$$
$$D^2(Ay_1+By_2) = AD^2y_1+BD^2y_2,$$

and hence

$$(D^2+PD+Q)(Ay_1+By_2) = A(D^2+PD+Q)y_1+B(D^2+PD+Q)y_2 = 0.$$

Thus $y = Ay_1+By_2$ is a solution of equation (1) and, since it contains two arbitrary constants, it must be the general solution. We have assumed that y_1 and y_2 are linearly independent solutions, that is,that y_2 is not merely a constant multiple of y_1; if $y_2 = cy_1$, where c is a constant, we have

$$y = Ay_1+By_2 = (A+Bc)y_1 = Cy_1,$$

where

$$C = A+Bc,$$

and this is not the general solution as it contains only one arbitrary constant.

Similarly, if $y_1, y_2, \ldots, y_n$ are n linearly independent solutions of the nth order linear equation $L(y) = 0$, the general solution is

$$y = A_1y_1+A_2y_2+ \ldots +A_ny_n, \tag{2}$$

where $A_1, A_2, \ldots, A_n$ are arbitrary constants.

Now consider the second order equation

$$(D^2+PD+Q)y = f(x), \tag{3}$$

and suppose that we have found a solution y_0 which satisfies this equation. Then if y_1 and y_2 are independent solutions of the reduced equation (1), a solution of equation (3) is

$$y = Ay_1+By_2+y_0, \tag{4}$$

where A and B are arbitrary constants. This follows easily since

$$(D^2+PD+Q)y$$
$$= A(D^2+PD+Q)y_1+B(D^2+PD+Q)y_2+(D^2+PD+Q)y_0 = f(x).$$

Further, since the solution y has two arbitrary constants it is the general solution of equation (3). The solution of the reduced equation, Ay_1+By_2, is called the *complementary function* and y_0 is called a *particular integral.* Thus the general solution of (3) is the sum of the complementary function (C.F.) and the particular integral (P.I.).

As an example, consider the equation

$$(D^2+1)y = x.$$

A particular integral easily obtained by inspection is $y = x$. It can be verified that $y_1 = \cos x$ and $y_2 = \sin x$ are independent solutions of the reduced equation

$$(D^2+1)y = 0.$$

Then the complementary function is

$$A\cos x+B\sin x,$$

and the general solution is

$$y=A\cos x+B\sin x+x.$$

Similarly, the complementary function for the nth order linear equation

$$L(y)=f(x)$$

is the solution (2) of the reduced equation

$$L(y)=0,$$

and if y_0 is a particular integral, so that

$$L(y_0)=f(x),$$

the general solution is

$$y=A_1y_1+A_2y_2+\ \dots\ +A_ny_n+y_0.$$

3.3 Solution of reduced second order equations

If $y=e^{mx}$ is a solution of the reduced second order equation with constant coefficients

$$\frac{d^2y}{dx^2}+p\frac{dy}{dx}+qy=0, \tag{1}$$

we have, on substituting for y in the equation,

$$e^{mx}(m^2+pm+q)=0.$$

Therefore $y=e^{mx}$ is a solution of (1) if

$$m^2+pm+q=0. \tag{2}$$

Equation (2) is called the *auxiliary equation.* It gives, in general, two values of m, for which $y=e^{mx}$ is a solution of (1) and, therefore, two distinct solutions of (1). We now distinguish between the cases in which the roots of the auxiliary equation are real and distinct, complex or coincident.

(i) *Real and distinct roots*

Let the roots of the auxiliary equation (2) be

$$m=\alpha \quad\text{and}\quad m=\beta,\quad \alpha\neq\beta;$$

then we have two independent solutions $y_1=e^{\alpha x}$ and $y_2=e^{\beta x}$, and the general solution of (1) is

$$y=A\,e^{\alpha x}+B\,e^{\beta x}, \tag{3}$$

where A and B are constants.

Example 1. *Solve the equation* $(D^2-2D-3)y = 0$, *given that* $y = 1$ *and* $y' = 2$ *when* $x = 0$.

The auxiliary equation is

$$m^2-2m-3 = 0,$$

giving $m = -1$ or $m = 3$. Hence the general solution is

$$y = Ae^{-x}+Be^{3x}, \quad (A \text{ and } B \text{ constant}).$$

Since $y = 1$ when $x = 0$ we have

$$1 = A+B.$$

Now

$$y' = -Ae^{-x}+3Be^{3x},$$

and since $y' = 2$ when $x = 0$,

$$2 = -A+3B.$$

Hence,

$$B = \tfrac{3}{4}, \qquad A = \tfrac{1}{4}$$

and the solution is

$$4y = e^{-x}+3e^{3x}.$$

(ii) *Complex roots*

If p and q are real and $p^2 < 4q$, the roots of the auxiliary equation will be of the form

$$m = \lambda \pm i\mu,$$

where λ and μ are real and i is the square root of -1. Then the general solution of equation (1) is

$$\begin{aligned} y &= A\,e^{(\lambda+i\mu)x}+B\,e^{(\lambda-i\mu)x} \\ &= e^{\lambda x}(A\,e^{i\mu x}+B\,e^{-i\mu x}). \end{aligned}$$

Remembering that $e^{\pm i\mu x} = \cos\mu x \pm i\sin\mu x$, this can be rewritten in the more convenient form

$$\begin{aligned} y &= e^{\lambda x}\{(A+B)\cos\mu x+i(A-B)\sin\mu x\} \\ &= e^{\lambda x}(E\cos\mu x+F\sin\mu x), \end{aligned} \tag{4}$$

where E and F are arbitrary constants. It is emphasized that E and F are completely arbitrary and need not involve imaginary quantities; nor need A and B be purely real.

Example 2. *Solve the equation* $(D^2+4D+13)y = 0$, *given that* $y = 1$ *and* $y' = 2$ *when* $x = 0$.

The auxiliary equation is

$$m^2+4m+13 = 0,$$

and hence

$$\begin{aligned} m &= -2\pm\sqrt{(-9)} \\ &= -2\pm 3i. \end{aligned}$$

The general solution is therefore, from (4),

$$y = e^{-2x}(E\cos 3x+F\sin 3x).$$

Since $y = 1$ when $x = 0$, we have $1 = E$.

Now

$$y' = -2e^{-2x}(E\cos 3x + F\sin 3x) + 3e^{-2x}(-E\sin 3x + F\cos 3x),$$

and since

$$y' = 2 \quad \text{when } x = 0,$$
$$2 = -2E + 3F.$$

Hence $F = \frac{4}{3}$ and the solution is

$$3y = e^{-2x}(3\cos 3x + 4\sin 3x).$$

The solution (4) may be expressed in a slightly different and often more useful form by writing

$$F/E = \tan\varepsilon.$$

Then

$$\begin{aligned} y &= (E^2 + F^2)^{\frac{1}{2}} e^{\lambda x}(\cos\varepsilon\cos\mu x + \sin\varepsilon\sin\mu x) \\ &= C\, e^{\lambda x}\cos(\mu x - \varepsilon), \end{aligned} \tag{5}$$

where C and ε are arbitrary constants. Similarly $y = C'\, e^{\lambda x}\sin(\mu x - \varepsilon')$ is a solution.

(iii) *Coincident roots*

When $p^2 = 4q$, the auxiliary equation yields only one value for m, namely $m = \alpha = -\frac{1}{2}p$, and hence the solution $y = Ae^{\alpha x}$. This is not the general solution as it does not contain the necessary two arbitrary constants and in this case we proceed as follows. Assume that $y = ve^{\alpha x}$, where v is a function of x to be deterined.

Then

$$\begin{aligned} y &= v\, e^{\alpha x}, \\ y' &= v'\, e^{\alpha x} + \alpha v\, e^{\alpha x}, \\ y'' &= v''\, e^{\alpha x} + 2\alpha v'\, e^{\alpha x} + \alpha^2 v\, e^{\alpha x}. \end{aligned}$$

Substituting for y, y' and y'' in the differential equation we have

$$e^{\alpha x}\{v'' + 2\alpha v' + \alpha^2 v + p(v' + \alpha v) + qv\} = 0,$$

and hence

$$v'' + v'(p + 2\alpha) + v(\alpha^2 + p\alpha + q) = 0.$$

Now $\alpha^2 + p\alpha + q = 0$ and $p + 2\alpha = 0$, so that $v'' = 0$. Hence, integrating,

$$v = Ax + B,$$

where A and B are arbitrary constants, and the general solution of equation (1) is

$$y = (Ax + B)\, e^{\alpha x}. \tag{6}$$

Example 3. *Solve the equation* $(D^2-4D+4)y = 0$ *given that* $y = 1$ *and* $y' = 3$ *when* $x = 0$.

The auxiliary equation

$$m^2-4m+4 = 0$$

has the one root $m = 2$. The general solution is therefore, from (6),

$$y = (Ax+B)e^{2x}.$$

Since $y = 1$ when $x = 0$ we have $1 = B$. Now

$$y' = 2(Ax+B)e^{2x}+Ae^{2x},$$

and since $y' = 3$ when $x = 0$,

$$3 = 2B+A.$$

Hence $A = 1$ and the solution is

$$y = (x+1)e^{2x}.$$

Exercises 3(*a*)

Solve the differential equations:

1. $\ddot{y}+5\dot{y}+4y = 0$.
2. $\ddot{y}+4\dot{y}-21y = 0$.
3. $4y''+20y'+25y = 0$.
4. $y''+4y'+13y = 0$.
5. $4\ddot{y}+4\dot{y}+5y = 0$.
6. $4\ddot{y}+24\dot{y}+37y = 0$.
7. $\ddot{y}+4y = 0$, given that $y = 4$ and $\dot{y} = 0$ when $t = 0$.
8. $l\ddot{\theta}+g\theta = 0$, given that $\theta = a$ and $\dot{\theta} = 0$ when $t = 0$.
9. $y''-4y = 0$, given that $y = 1$ and $y' = 2$ when $x = 0$.
10. $y''+10y'+25y = 0$, given that $y = y' = 1$ when $x = 0$.
11. $\ddot{y}+8\dot{y}+15y = 0$, given that $y = 1$ and $\dot{y} = 3$ when $t = 0$.
12. $y''+2y'-15y = 0$, given that $y = 1$ when $x = 0$ and $y \to 0$ as $x \to \infty$.
13. $\ddot{y}+2\dot{y}+5y = 0$, given that $y = 2$ and $\dot{y} = 0$ when $t = 0$.
14. $\ddot{y}+2a\dot{y}+10a^2y = 0$, given that $y = 0$ and $\dot{y} = 3a$ when $t = 0$.
15. $\ddot{x}+2k\dot{x}+(k^2+p^2)x = 0$, given that $x = a$ and $\dot{x} = v$ when $t = 0$.
16. $L\ddot{q}+R\dot{q}+q/C = 0$, given that $CR^2 = 4L$, and $q = Q, \dot{q} = 0$ when $t = 0$

3.4 Solution of reduced *n*th order equations

The reduced *n*th order equation

$$(D^n+p_1D^{n-1}+\ldots+p_{n-1}D+p_n)y = 0 \tag{1}$$

is solved in much the same way as the second order equation. Assuming a solution $y = e^{mx}$, we derive the auxiliary equation

$$m^n+p_1m^{n-1}+\ldots+p_{n-1}m+p_n = 0. \tag{2}$$

If the n roots of the auxiliary equation are distinct, whether real or imaginary, the general solution is

$$y = Ae^{\alpha x} + Be^{\beta x} + \ldots + Ke^{\kappa x},$$

and this contains the requisite n arbitrary constants. In taking the general solution in this form it is to be understood that the part of the solution corresponding to a pair of complex roots $\lambda \pm i\mu$ can, if desired, be rewritten in the form $e^{\lambda x}(E\cos\mu x + F\sin\mu x)$.

A root $m = \alpha$ repeated r times in the auxiliary equation will yield as its contribution to the general solution the expression

$$y = e^{\alpha x}(A + Bx + Cx^2 + \ldots + Kx^{r-1}), \tag{3}$$

containing r arbitrary constants. Similarly imaginary roots $\lambda \pm i\mu$ repeated r times gives the expression

$$e^{\lambda x}\{(E\cos\mu x + F\sin\mu x) + x(E_1\cos\mu x + F_1\sin\mu x) + \ldots + x^{r-1}(E_{r-1}\cos\mu x + F_{r-1}\sin\mu x)\}, \tag{4}$$

containing $2r$ arbitrary constants.

To establish the contribution (3) to the solution for repeated roots, let

$$y = e^{\alpha x}(A + Bx + Cx^2 + \ldots + Kx^{r-1}). \tag{5}$$

Then

$$e^{-\alpha x}y = A + Bx + Cx^2 + \ldots + Kx^{r-1}. \tag{6}$$

Now

$$\frac{d}{dx}(e^{-\alpha x}y) = e^{-\alpha x}\frac{dy}{dx} - \alpha e^{-\alpha x}y$$

$$= e^{-\alpha x}(D-\alpha)y.$$

Also

$$\frac{d^2}{dx^2}(e^{-\alpha x}y) = e^{-\alpha x}\frac{d^2y}{dx^2} - 2\alpha e^{-\alpha x}\frac{dy}{dx} + \alpha^2 e^{-\alpha x}y$$

$$= e^{-\alpha x}(D-\alpha)^2 y.$$

Similarly

$$D^r(e^{-\alpha x}y) = e^{-\alpha x}(D-\alpha)^r y.$$

Hence, differentiating equation (6) r times we have

$$D^r(e^{-\alpha x}y) = e^{-\alpha x}(D-\alpha)^r y = 0.$$

Thus the function y as defined by equation (5) is a solution of the equation $(D-\alpha)^r y = 0$, and is therefore also a solution of the equation $f(D)(D-\alpha)^r y = 0$, where $f(D)$ is any polynomial in D. A similar argument applies to the solution (4) corresponding to repeated imaginary roots.

Example 4. *Solve the equation* $\dfrac{d^4y}{dx^4}-\alpha^4 y = 0.$

The auxiliary equation is

$$m^4-\alpha^4 = 0,$$

and its roots are $m = \pm\alpha$ and $\pm i\alpha$. Therefore the solution is

$$y = A\,e^{\alpha x}+B\,e^{-\alpha x}+C\cos\alpha x+D\sin\alpha x.$$

This solution can be written in the alternative forms

$$y = A_1\cosh\alpha x+B_1\sinh\alpha x+C\cos\alpha x+D\sin\alpha x,$$

or $$y = E\cosh(\alpha x+\varepsilon)+F\cos(\alpha x+\eta),$$

where E, F, ε and η are constants.

Example 5. *Solve the equation* $\dfrac{d^3y}{dx^3}-6\dfrac{d^2y}{dx^2}+12\dfrac{dy}{dx}-8y = 0$ *with the conditions that when* $x = 0$, $y = 1$, $y' = 0$ *and* $y'' = 0$.

The auxiliary equation is

$$m^3-6m^2+12m-8 = 0,$$

that is

$$(m-2)^3 = 0.$$

The root $m = 2$ is repeated three times and the general solution is

$$y = (A+Bx+Cx^2)e^{2x}.$$

The condition $y = 1$ when $x = 0$ gives $A = 1$. Now

$$y' = \{2(A+Bx+Cx^2)+(B+2Cx)\}e^{2x},$$

and the condition $y' = 0$ when $x = 0$ gives $2A+B = 0$, and hence $B = -2$. For y'' we have $e^{2x}\{4(A+Bx+Cx^2)+4(B+2Cx)+2C\}$ and the condition $y'' = 0$ when $x = 0$ gives $4A+4B+2C = 0$. Hence $C = 2$ and the solution is

$$y = (1-2x+2x^2)e^{2x}.$$

Example 6. *Solve the equation* $\dfrac{d^4y}{dx^4}+2\dfrac{d^2y}{dx^2}+y = 0$ *with the condition that when* $x = 0$, $y = 1$, $y' = y'' = y''' = 0$.

The auxiliary equation $m^4+2m^2+1 = 0$ has the roots $\pm i$ repeated twice. Hence the solution is

$$y = (A+Bx)\cos x+(E+Fx)\sin x.$$

We have

$$y' = (E+B+Fx)\cos x+(F-A-Bx)\sin x,$$
$$y'' = (2F-A-Bx)\cos x+(-2B-E-Fx)\sin x,$$
$$y''' = (-E-3B-Fx)\cos x+(-3F+A+Bx)\sin x.$$

The initial conditions give

$$1 = A,$$
$$0 = E+B,$$
$$0 = 2F-A,$$
$$0 = E+3B.$$

Hence $A = 1$, $B = 0$, $E = 0$, $F = \frac{1}{2}$, and the solution is

$$y = \cos x+\tfrac{1}{2}x\sin x.$$

Exercises 3(*b*)

Solve the differential equations:

1. $(D^3+9D^2+31D+39)y = 0$, $D \equiv d/dx$.
2. $(D^3+2D^2+4D+8)y = 0$, $D \equiv d/dx$.
3. $(D^3+6D^2-91D)y = 0$, $D \equiv d/dx$.
4. $(D^4-16)y = 0$, $D \equiv d/dx$.
5. $(D^4+4)y = 0$, $D \equiv d/dx$.
6. $(D^3+8)y = 0$, $D \equiv d/dx$.
7. $(D^2+2D+2)(D^2+2D+5)y = 0$, $D \equiv d/dt$.
8. $(D^4-2D^2+1)y = 0$, $D \equiv d/dt$.
9. $(D^4+6D^2+25)y = 0$, $D \equiv d/dt$.
10. $(D^4+4aD^3+6a^2D^2+4a^3D+a^4)y = 0$, $D \equiv d/dx$.

3.5 Particular integrals of second order equations

A particular integral of the second order equation

$$\frac{d^2y}{dx^2}+p\frac{dy}{dx}+qy = f(x) \tag{1}$$

can be obtained by integration. Let the auxiliary equation

$$m^2+pm+q = 0$$

have distinct roots α and β, which may be real or complex. Then, a particular integral of equation (1) is

$$y_0 = \frac{1}{\alpha-\beta}\left\{e^{\alpha x}\int e^{-\alpha x}f(x)\,dx - e^{\beta x}\int e^{-\beta x}f(x)\,dx\right\}. \tag{2}$$

To establish this formula, differentiate equation (2) twice, giving

$$Dy_0 = \frac{\alpha}{\alpha-\beta}e^{\alpha x}\int e^{-\alpha x}f(x)\,dx - \frac{\beta}{\alpha-\beta}e^{\beta x}\int e^{-\beta x}f(x)\,dx,$$

$$D^2y_0 = \frac{\alpha^2}{\alpha-\beta}e^{\alpha x}\int e^{-\alpha x}f(x)\,dx - \frac{\beta^2}{\alpha-\beta}e^{\beta x}\int e^{-\beta x}f(x)\,dx + f(x).$$

Therefore

$$(D^2+pD+q)y_0 = \frac{(\alpha^2+p\alpha+q)}{\alpha-\beta}e^{\alpha x}\int e^{-\alpha x}f(x)\,dx$$

$$-\frac{(\beta^2+p\beta+q)}{\alpha-\beta}e^{\beta x}\int e^{-\beta x}f(x)\,dx + f(x) = f(x),$$

since $\alpha^2+p\alpha+q = \beta^2+p\beta+q = 0$. Therefore y_0 is a particular integral of equation (1).

Example 7. *Solve the equation* $y''-3y'+2y = \cos 3x$.

The auxiliary equation $m^2-3m+2 = 0$ has roots $m = 1$, $m = 2$ and the complementary function is therefore

$$y = Ae^x+Be^{2x}.$$

Now

$$\int e^{ax}\cos bx\,dx = \frac{e^{ax}}{a^2+b^2}(a\cos bx+b\sin bx),$$

therefore

$$e^x\int e^{-x}\cos 3x\,dx = \tfrac{1}{10}(-\cos 3x+3\sin 3x),$$

and

$$e^{2x}\int e^{-2x}\cos 3x\,dx = \tfrac{1}{13}(-2\cos 3x+3\sin 3x).$$

Hence the particular integral given by equation (2) with $\alpha = 2$ and $\beta = 1$ is

$$y_0 = -\tfrac{1}{130}(7\cos 3x+9\sin 3x),$$

and the general solution is

$$y = Ae^x+Be^{2x}-(\tfrac{1}{130})(7\cos 3x+9\sin 3x).$$

If the auxiliary equation $m^2+pm+q = 0$ has equal roots $m = \alpha$, a particular integral of equation (1) is

$$y_0 = e^{\alpha x}\int\left\{\int e^{-\alpha x}f(x)\,dx\right\}dx. \tag{3}$$

To establish this formula differentiate $y_0e^{-\alpha x}$ twice. Then

$$D(y_0e^{-\alpha x}) = e^{-\alpha x}(y_0'-\alpha y_0) = \int e^{-\alpha x}f(x)\,dx,$$

$$D^2(y_0e^{-\alpha x}) = e^{-\alpha x}(y_0''-2\alpha y_0'+\alpha^2y_0) = e^{-\alpha x}f(x).$$

Hence, since $p = -2\alpha$ and $q = \alpha^2$, we have

$$y_0''+py_0'+qy_0 = f(x).$$

Therefore y_0 is a particular integral of equation (1).

Example 8. *Solve the equation* $y''+4y'+4y = \cos x$.

The auxiliary equation $m^2+4m+4 = 0$ has equal roots $m = -2$. Now

$$\int e^{2x}\cos x\,dx = \frac{e^{2x}}{5}(2\cos x+\sin x),$$

and

$$\int\frac{e^{2x}}{5}(2\cos x+\sin x)\,dx = \frac{e^{2x}}{25}(3\cos x+4\sin x).$$

Hence, the particular integral given by (3) is

$$y_0 = (\tfrac{1}{25})(3\cos x+4\sin x),$$

and the general solution of the equation is

$$y = (A+Bx)e^{-2x}+(\tfrac{1}{25})(3\cos x+4\sin x).$$

3.6 Particular integrals of equations of higher order

For equations of higher order than the second, particular integrals may be obtained in a similar way by integration. Consider the third order equation

$$F(D)y \equiv (D^3+pD^2+qD+r)y = f(x). \tag{1}$$

Let α, β and γ be the roots of the auxiliary equation

$$F(m) \equiv m^3+pm^2+qm+r = 0$$

and let these roots be distinct. Equation (1) can be written in the form

$$F(D)y \equiv (D-\alpha)(D-\beta)(D-\gamma)y = f(x).$$

We can obtain a formal expression for $1/F(m)$ in partial fractions, namely

$$\frac{1}{F(m)} = \frac{1}{F'(\alpha)}\frac{1}{m-\alpha}+\frac{1}{F'(\beta)}\frac{1}{m-\beta}+\frac{1}{F'(\gamma)}\frac{1}{m-\gamma},$$

and hence, multiplying by $F(m)$,

$$1 = \frac{(m-\beta)(m-\gamma)}{F'(\alpha)}+\frac{(m-\gamma)(m-\alpha)}{F'(\beta)}+\frac{(m-\alpha)(m-\beta)}{F'(\gamma)}. \tag{2}$$

Now let

$$y_1 = \frac{1}{F'(\alpha)}e^{\alpha x}\int e^{-\alpha x}f(x)\,dx,$$

$$y_2 = \frac{1}{F'(\beta)}e^{\beta x}\int e^{-\beta x}f(x)\,dx,$$

$$y_3 = \frac{1}{F'(\gamma)}e^{\gamma x}\int e^{-\gamma x}f(x)\,dx.$$

Then

$$\frac{d}{dx}(e^{-\alpha x}y_1) = e^{-\alpha x}(D-\alpha)y_1 = \frac{e^{-\alpha x}}{F'(\alpha)}f(x),$$

that is

$$(D-\alpha)y_1 = \frac{1}{F'(\alpha)}f(x),$$

and

$$F(D)y_1 = \frac{1}{F'(\alpha)}(D-\beta)(D-\gamma)f(x).$$

With similar expressions for $F(D)y_2$ and $F(D)y_3$, we have

$$F(D)(y_1+y_2+y_3) = \left\{\frac{(D-\beta)(D-\gamma)}{F'(\alpha)}+\frac{(D-\gamma)(D-\alpha)}{F'(\beta)}+\frac{(D-\alpha)(D-\beta)}{F'(\gamma)}\right\}f(x).$$

The expression on the right-hand side in brackets is unity by virtue of (2) with m replaced by D, and hence

$$F(D)(y_1+y_2+y_3) = f(x).$$

Thus a particular integral of (1) is

$$y_0 = \frac{e^{\alpha x}}{F'(\alpha)}\int e^{-\alpha x}f(x)\,dx+\frac{e^{\beta x}}{F'(\beta)}\int e^{-\beta x}f(x)\,dx+\frac{e^{\gamma x}}{F'(\gamma)}\int e^{-\gamma x}f(x)\,dx. \qquad (3)$$

The formula (3) is easily generalized for an equation of the nth order $F(D)y = f(x)$ whose auxiliary equation $F(m) = 0$ has n distinct roots $\alpha_1, \alpha_2, \ldots, \alpha_n$. By the ordinary theory of partial fractions

$$\frac{1}{F(m)} = \sum_{r=1}^{n}\frac{1}{F'(\alpha_r)}\cdot\frac{1}{m-\alpha_r},$$

and hence a particular integral of $F(D)y = f(x)$ is

$$y_0 = \sum_{r=1}^{n}\frac{e^{\alpha_r x}}{F'(\alpha_r)}\int e^{-\alpha_r x}f(x)\,dx. \qquad (4)$$

If the auxiliary equation has coincident roots, repeated integrations are necessary. Again $1/F(m)$ must be expanded in partial fractions giving an expression such as

$$\frac{1}{F(m)} = \frac{a}{(m-\alpha_1)^2}+\frac{b}{m-\alpha_1}+\frac{c}{m-\alpha_2}+\ldots,$$

if α_1 is a once-repeated root. Then the particular integral will be

$$y_0 = a\,e^{\alpha_1 x}\int dx\left\{\int e^{-\alpha_1 x}f(x)\,dx\right\}+b\,e^{\alpha_1 x}\int e^{-\alpha_1 x}f(x)\,dx$$
$$+c\,e^{\alpha_2 x}\int e^{-\alpha_2 x}f(x)\,dx+\ldots.$$

Example 9. *Solve the equation $y'''+6y''+11y'+6y = x$.*

The auxiliary equation $F(m) \equiv m^3+6m^2+11m+6 = 0$, that is

$$(m+1)(m+2)(m+3) = 0,$$

has roots -1, -2 and -3. We have

$$F'(m) = 3m^2+12m+11,$$

and hence

$$F'(-1) = 2, \qquad F'(-2) = -1, \qquad F'(-3) = 2.$$

Also

$$e^{-x}\int e^{x}x\,dx = x-1,$$

$$e^{-2x}\int e^{2x}x\,dx = (2x-1)/4,$$

$$e^{-3x}\int e^{3x}x\,dx = (3x-1)/9.$$

Hence, the particular integral given by (3) is

$$y_0 = \tfrac{1}{2}(x-1)-\tfrac{1}{4}(2x-1)+\tfrac{1}{18}(3x-1)$$
$$= (6x-11)/36,$$

and the general solution is

$$y = Ae^{-x}+Be^{-2x}+Ce^{-3x}+(6x-11)/36.$$

Exercises 3(*c*)

Solve the differential equations:

1. $y''-3y'+2y = 2e^{3x}$.
2. $y''-3y'+2y = e^{2x}$.
3. $y''-3y'+2y = 2x-1$.
4. $y''-3y'+2y = x^2$.
5. $y''+4y'+4y = 2e^{-2x}$.
6. $y''+2y'+y = e^{-2x}$.
7. $y''+6y'+9y = 18x$.
8. $y''+2y'+5y = 10\cos x$.
9. $\ddot{y}+\omega^2 y = \cos pt$.
10. $y'''-3y''+3y'-y = e^{x}$.
11. $y'''-2y''-y'+2y = 2e^{x}$.
12. $y'''-2y''-y'+2y = 4x^2$.

3.7 Use of operators to find particular integrals

Symbolic methods involving the linear operator $F(D)$ are a powerful means of finding a particular integral of the equation $F(D)y = f(x)$. Here $F(D) \equiv D^n+p_1D^{n-1}+\ldots+p_{n-1}D+p_n$, and is a polynomial of degree n in D. The operator can be factorized, it being understood that the factors such as $(D-\alpha)$ are operators.

We require three preliminary lemmas, namely

$$(a)\ F(D)e^{ax} = F(a)e^{ax},$$
$$(b)\ F(D)e^{ax}u(x) = e^{ax}F(D+a)u(x),$$
$$(c)\ F(D^2)\cos ax = F(-a^2)\cos ax,$$
$$F(D^2)\sin ax = F(-a^2)\sin ax.$$

The first lemma is obvious since $D^r(e^{ax}) = a^r e^{ax}$. Lemma (*b*) is proved by differentiating the product $e^{ax}u$, r times using Leibnitz' theorem. We have

$$D^r(e^{ax}u) = uD^r(e^{ax})+rDu \,.\, D^{r-1}(e^{ax})+\frac{r(r-1)}{2!}D^2u \,.\, D^{r-2}(e^{ax})+\,..$$

$$= e^{ax}\left\{a^ru+ra^{r-1}Du+\frac{r(r-1)}{2!}a^{r-2}D^2u+\ldots\right\}$$

$$= e^{ax}(D+a)^ru.$$

Therefore

$$\Sigma p_rD^r(e^{ax}u) = e^{ax}\Sigma p_r(D+a)^ru,$$

and hence

$$F(D)\,e^{ax}u = e^{ax}F(D+a)u.$$

In lemma (c) we suppose that only even indices are involved in the operator, and since $D^2(\cos ax) = (-a^2)\cos ax$, the result follows easily. Similarly $F(D^2)\sin ax = F(-a^2)\sin ax$.

We now consider some cases in which the particular integral can be quickly found.

(i) *The exponential case*

Let $f(x) = e^{ax}$; then a particular integral of the equation $F(D)y = f(x)$ is

$$y_0 = \frac{1}{F(a)}e^{ax}, \quad F(a) \neq 0. \tag{1}$$

This follows by lemma (a) from which we see that $F(D)y_0 = F(a)y_0 = e^{ax}$.

If $(D-a)$ is a factor of $F(D)$, $F(a) = 0$ and the solution (1) breaks down. In this case if the factor $(D-a)$ is repeated r times in $F(D)$, so that $F(D) = (D-a)^rG(D)$ where $G(a) \neq 0$, a particular integral is

$$y_0 = \frac{1}{G(a)}e^{ax}\frac{x^r}{r!}. \tag{2}$$

To verify this we have

$$F(D)y_0 = \frac{G(D)}{G(a)}(D-a)^r e^{ax}\frac{x^r}{r!}$$

$$= e^{ax}\frac{G(D+a)}{G(a)}D^r\frac{x^r}{r!}, \quad \text{from lemma } (b)$$

$$= e^{ax}\frac{G(D+a)}{G(a)}.1.$$

Now $G(D+a)$ operating on 1 gives $G(a)$, since all derivatives are zero, therefore $F(D)y_0 = e^{ax}$.

It will be seen that the particular integral (1) could be obtained by writing

$$F(D)y_0 = e^{ax},$$

$$y_0 = \frac{1}{F(D)} e^{ax}$$

$$= \frac{1}{F(a)} e^{ax}, \quad \text{from lemma } (a).$$

Again, the particular integral (2) could be obtained by writing

$$y_0 = \frac{1}{(D-a)^r G(D)} e^{ax}$$

$$= e^{ax} \frac{1}{D^r G(D+a)} \,.\, 1, \quad \text{from lemma } (b)$$

and taking

$$\frac{1}{G(D+a)} \,.\, 1 = \frac{1}{G(a)},$$

$$y_0 = \frac{e^{ax}}{G(a)} D^{-r} \,.\, 1$$

$$= \frac{e^{ax}}{G(a)} \frac{x^r}{r!},$$

$D^{-r}.1$ being interpreted as the rth integral of 1. This is a formal use of operators which would require careful justification but which is useful for writing down the particular integral quickly.

Example 10. *Solve the equation* $y''+4y'+3y = 10e^{-2x}$.

Here

$$F(D) \equiv D^2+4D+3,$$

and

$$F(-2) = 4-8+3 = -1,$$

hence

$$y_0 = \frac{10}{(-1)} e^{-2x}$$

is a particular integral. The roots of the auxiliary equation being -1 and -3 the general solution is

$$y = Ae^{-x}+Be^{-3x}-10e^{-2x}.$$

Example 11. *Solve the equation* $y'''-8y = 60e^{2x}$.

Here

$$F(D) \equiv (D^3-8)$$

$$= (D-2)(D^2+2D+4) = (D-2)G(D).$$

The critical factor $(D-2)$ occurs once only and $G(2) = 12$; therefore the particular integral is $(\frac{60}{12})e^{2x}x$. The auxiliary equation has roots 2 and $-1 \pm i\sqrt{3}$, and hence the general solution is

$$y = Ae^{2x} + e^{-x}(B \cos \sqrt{3}x + C \sin \sqrt{3}x) + 5xe^{2x}.$$

(ii) *The trigonometrical case*

When $f(x) = \cos ax$ or $\sin ax$ one method is to write

$$f(x) = e^{iax} = \cos ax + i \sin ax,$$

and find a particular integral for this exponential. The particular integrals for $\cos ax$ and $\sin ax$ are then respectively the real and imaginary parts of this particular integral.

A useful simplification in the working is obtained by replacing D^2 by $-a^2$, D^3 by $-a^2 D$, etc., in $F(D)$ by virtue of lemma (*c*). Thus we can write formally

$$\frac{1}{D^2+b^2} \cos ax = \frac{1}{-a^2+b^2} \cos ax,$$

and also

$$\frac{1}{D^3+b^3} \cos ax = \frac{1}{b^3 - a^2 D} \cos ax$$

$$= \frac{b^3 + a^2 D}{b^6 - a^4 D^2} \cos ax$$

$$= \frac{b^3 + a^2 D}{b^6 + a^6} \cos ax$$

$$= \frac{1}{b^6 - a^6}(b^3 \cos ax - a^3 \sin ax).$$

Example 12. *Solve the equation* $y'' + 4y = 15 \cos 3x$.

Here, $F(D) \equiv D^2 + 4$, and we have the particular integral

$$y_0 = \frac{1}{D^2+4} . 15 \cos 3x$$

$$= \frac{1}{4-9} . 15 \cos 3x = -3 \cos 3x,$$

and the general solution is

$$y = A \cos 2x + B \sin 2x - 3 \cos 3x.$$

Example 13. *Solve the equation* $y'' + 4y = 8 \cos 2x$.

Here the method of Example 12 breaks down and we find a particular integral for $8e^{2ix}$. We have

$$(D-2i)(D+2i)y_0 = 8e^{2ix}.$$

Noting the critical factor $(D-2i)$ we have the particular integral

$$y_0 = \frac{8}{(2i+2i)} e^{2ix} x$$
$$= -2ix(\cos 2x + i \sin 2x).$$

The real part of y_0 yields the particular integral $2x \sin 2x$ and the general solution is

$$y = A \cos 2x + B \sin 2x + 2x \sin 2x.$$

Example 14. *Find a particular integral of the equation* $y''' + 3y'' + 4y' + 2y = 170 \cos 3x$.

For the particular integral y_0 we have formally

$$y_0 = \frac{170}{D^3 + 3D^2 + 4D + 2} \cos 3x,$$

and, putting $D^2 = -9$,

$$= \frac{170}{-9D - 27 + 4D + 2} \cos 3x$$
$$= -\frac{34}{D+5} \cos 3x.$$

This is formally,

$$y_0 = -\frac{34(D-5)}{D^2 - 25} \cos 3x$$
$$= (D-5) \cos 3x,$$

and so

$$y_0 = -(3 \sin 3x + 5 \cos 3x).$$

Exercises 3(*d*)

Solve the differential equations:

1. $y'' + 6y' + 8y = 70e^{3x}$.
2. $y'' + 6y' + 8y = 8e^{-2x}$.
3. $y'' - 4y' + 4y = 4e^{2x}$.
4. $y'' - 4y = 12 \cosh 2x$.
5. $y'' + 4y' + 13y = 27e^{-2x}$.
6. $y''' + 6y'' + 9y' = 12e^{-3x}$.
7. $y''' - 9y' = 108 \sinh 3x$.
8. $y^{\text{iv}} - 8y'' + 16y = 64 \cosh 2x$.
9. $y'' + 9y = 15 \cos 2x$.
10. $y'' + 2y' + 5y = 85 \cos 2x$.
11. $y'' + 4y = 8 \cos 2x$.
12. $y'' + 4y' + 13y = 80 \cos 3x$.
13. $\ddot{y} + 2k\dot{y} + p^2 y = \cos pt$, $(p > k)$.
14. $L\ddot{q} + R\dot{q} + q/C = E \cos \omega t$, where $CR^2 = 4L(1 - LC\omega^2)$.
15. $y'' + 2y' + 2y = 650e^{2x} \cos 3x$.
16. $y'' + 4y' = 195e^{2x} \sin 3x$.

17. $4\ddot{y}+4\dot{y}+37y = 99\cos t-12\sin t$, given that $y = \dot{y} = 1$ when $t = 0$. (C.U.)

18. $y''+4y'+13y = 5\cos 3x$, if $y = \frac{1}{4}$ and $y' = 2$ when $x = 0$. (C.U.)

(iii) *Particular integral of a power of x*

Finding a particular integral by the use of the operator D of the equation $F(D)y = x^m$ requires the formal expansion of $1/F(D)$ in ascending powers of D as far as the term in D^m. Thus if $F(D) = 1-D$, we write

$$\frac{1}{F(D)} = \frac{1}{1-D} = 1+D+D^2+\ldots+D^m+\ldots.$$

Then a particular integral is, since $D^{m+r}x^m = 0$,

$$y_0 = \frac{1}{F(D)}x^m = (1+D+D^2+\ldots+D^m)x^m$$
$$= x^m+mx^{m-1}+m(m-1)x^{m-2}+\ldots+m!.$$

In general, if a formal expansion of $1/F(D)$ is

$$\frac{1}{F(D)} = a_0+a_1D+\ldots+a_mD^m+\ldots,$$

a particular integral is

$$y_0 = (a_0+a_1D+\ldots+a_mD^m)x^m.$$

Example 15. *Solve the equation* $y''+4y = 8x^3$.

We have formally for the particular integral

$$y_0 = \frac{8}{4+D^2}x^3$$
$$= 2(1+D^2/4)^{-1}x^3.$$

Expanding by the binomial theorem,

$$y_0 = (2-\tfrac{1}{2}D^2+\ldots)x^3$$
$$= 2x^3-3x.$$

Hence the general solution is

$$y = A\cos 2x+B\sin 2x+2x^3-3x.$$

Justification of this method of finding a particular integral can be found by comparing the method with that given in §3.6. It is shown there that when the auxiliary equation has distinct roots $\alpha_1, \alpha_2, \ldots, \alpha_n$, a particular integral is, with $f(x) = x^m$,

$$y_0 = \sum_{r=1}^{n} \frac{1}{F'(\alpha_r)} e^{\alpha_r x}\int e^{-\alpha_r x}x^m\,dx,$$

that is, on integrating by parts,

$$y_0 = \sum_{r=1}^{n} \frac{1}{F'(\alpha_r)}\left(-\frac{x^m}{\alpha_r} - \frac{mx^{m-1}}{\alpha_r^{\,2}} - \ldots - \frac{m!}{\alpha_r^{\,m+1}}\right)$$

$$= \sum_r \frac{-1/\alpha_r}{F'(\alpha_r)}\left(1+\frac{D}{\alpha_r}+\frac{D^2}{a_r^{\,2}}+\ldots+\frac{D^m}{\alpha_r^{\,m}}\right)x^m.$$

This gives formally, since for $r > m$, $D^r(x^m) = 0$,

$$y_0 = \sum_r \frac{1}{F'(\alpha_r)} \cdot \frac{1}{D-\alpha_r} \cdot x^m$$

$$= \frac{1}{F(D)}x^m.$$

Example 16. *Solve the equation $y''+4y = 48x \cos 2x$.*

We have for a particular integral when $\cos 2x$ is replaced by e^{2ix}

$$y_0 = \frac{48}{D^2+4} x\, e^{2ix}$$

$$= e^{2ix} \frac{48}{(D+2i)^2+4} x, \quad \text{using lemma } (b).$$

Hence

$$y_0 = -12i\, e^{2ix} \frac{1}{D(1+D/4i)} x$$

$$= -12i\, e^{2ix} \frac{1}{D}\left(1-\frac{D}{4i}\right)x,$$

that is

$$y_0 = -12i\, e^{2ix} \frac{1}{D}\left(x+\frac{i}{4}\right)$$

$$= -12i\, e^{2ix}\left(\frac{x^2}{2}+\frac{ix}{4}\right)$$

$$= e^{2ix}(3x-6ix^2).$$

The real part of y_0 is $3x \cos 2x+6x^2 \sin 2x$, and hence the general solution is

$$y = A \cos 2x+B \sin 2x+3x \cos 2x+6x^2 \sin 2x.$$

Exercises 3(*e*)

Solve the differential equations:

1. $y''+4y = 8x^2$.
2. $y''-4y'+3y = 27x^2$.
3. $y''-4y'+4y = 16x^3$.
4. $y''+6y'+13y = 13x^2-x+22$.
5. $y'''+4y' = 48x^2$.

6. $y^{iv}-4y'' = 96x^2$.
7. $y''-4y'+3y = e^{2x}x^2$.
8. $y''+4y'+13y = 27e^{-2x}x^3$.
9. $y''+y = 4x\sin x$.
10. $y''+4y'+8y = 1000x^2\cos 2x$.
11. $y''+4y = 96x^2\cos 2x$.
12. $y''+6y'+18y = 108x^2e^{-3x}\cos 3x$.

3.8 Other methods of finding particular integrals

Differential equations of the type considered in this chapter can be solved by the use of the *Laplace transform*. This method gives the complete solution (complementary function and particular integral) and is particularly useful when a solution satisfying given initial conditions is required. Details of this method will be found in Chapter 8.

Methods of trial and error will often yield results as quickly as any other method when the form that the particular integral will take is known. Thus for the equation

$$(D^2-5D+3)y = e^x x^2,$$

knowing that the particular integral will be of the form

$$y_0 = e^x(ax^2+bx+c),$$

we can substitute for y_0 in the differential equation and, by comparing coefficients of x^2, x and the constant term, determine a, b and c. Thus

$$\begin{aligned}(D^2-5D+3)y_0 &= e^x\{-ax^2-(b+6a)x+(2a-c-3b)\}\\ &= e^x x^2.\end{aligned}$$

Hence $a = -1$, $b = 6$, $c = -20$, and

$$y_0 = e^x(-x^2+6x-20).$$

It would be necessary in the case where $F(D)y = e^{ax}x^m$ to see whether $D-a$ is a factor of $F(D)$, and, if this factor is repeated r times, to expect y_0 to be of the form $e^{ax}(bx^{m+r}+cx^{m+r-1}+\ldots)$.

Another method, sometimes called the *symbolic method*, is used to find a particular integral of the equation $F(D)y = f(x)$ when $f(x)$ is of the form $a\sin(\omega x+\varepsilon)$. We have

$$\begin{aligned}(b+cD)a\sin(\omega x+\varepsilon) &= a\{b\sin(\omega x+\varepsilon)+c\omega\cos(\omega x+\varepsilon)\}\\ &= a(b^2+c^2\omega^2)^{\frac{1}{2}}\sin(\omega x+\varepsilon+\alpha),\end{aligned}$$

where $\tan\alpha = c\omega/b$.

Thus the effect of $(b+cD)$ operating on $a\sin(\omega x+\varepsilon)$ is to multiply its amplitude a by $(b^2+c^2\omega^2)^{\frac{1}{2}}$ and to increase the phase ε by α. It follows that the effect of the inverse operator $(b+cD)^{-1}$ operating on $a\sin(\omega x+\varepsilon)$ is to divide its amplitude by $(b^2+c^2\omega^2)^{\frac{1}{2}}$ and diminish the phase ε by α. That is

$$\frac{1}{b+cD}a\sin(\omega x+\varepsilon)=\frac{a}{(b^2+c^2\omega^2)^{\frac{1}{2}}}\sin(\omega x+\varepsilon-\alpha),$$

where $\tan\alpha = c\omega/b$.

Example 17. *Solve the equation* $y'''+3y''+4y'+2y = 6\sin x$.

For the particular integral we have

$$y_0 = \frac{6}{D^3+3D^2+4D+2}\sin x,$$

and, replacing D^2 by -1,

$$y_0 = \frac{6}{-1+3D}\sin x$$

$$= \frac{6}{\sqrt{(10)}}\sin(x+\tan^{-1}3).$$

The roots of the auxiliary equation being -1 and $-1\pm i$, the general solution is

$$y = e^{-x}(A+B\cos x+C\sin x)+(3\sqrt{10}/5)\sin(x+\tan^{-1}3).$$

Exercises 3(*f*)

Solve the differential equations:

1. $y''-3y'+2y = 4x^2$.
2. $y''+y'-2y = 2e^{x}(3x+1)$.
3. $y''+4y = 8x\cos 2x+2\sin 2x$.
4. $y'''+3y''+4y'+2y = 2x^2$.
5. $EIy''+Py = -\frac{1}{2}wlx+\frac{1}{2}wx^2$, given that $y=0$ when $x=0$ and when $x=l$.
6. $y''+2y'+5y = 17\cos 2x$.
7. $y''+2y'+5y = 4\cos 2x-3\sin 2x$.
8. $y'''-2y''+6y'-5y = 5\cos 2x$.
9. Show that a particular integral of the equation $\ddot{y}+2k\dot{y}+(k^2+p^2)y=f(t)$ is
$$y = \frac{1}{p}\int^{t} f(u)e^{-k(t-u)}\sin p(t-u)\,du.$$
10. One end A of a uniform bar is maintained at constant temperature T_1 while the other end and the surroundings are at temperature T_0 $(<T_1)$. In the steady state the temperature T at distance x from A satisfies the differential equation $(d^2T/dx^2)-a^2T = -a^2T_0$. Show that
$$T-T_0 = (T_1-T_0)\sinh a(l-x)/\sinh al.$$

3.9 The Euler linear equation

The linear equation

$$x^n\frac{d^ny}{dx^n}+p_1x^{n-1}\frac{d^{n-1}y}{dx^{n-1}}+\ldots+p_{n-1}x\frac{dy}{dx}+p_ny=f(x), \quad (1)$$

in which the derivative of the rth order is multiplied by x^r and by a constant, is known as *Euler's linear equation*. The substitution $x = e^t$ reduces the equation to a linear equation with constant coefficients with t as the independent variable. If $x = e^t$ it follows that $dx/dt = x$. Then

$$\frac{dy}{dx}=\frac{dy}{dt}\frac{dt}{dx}, \quad \text{and} \quad x\frac{dy}{dx}=\frac{dy}{dt}.$$

Again

$$x\frac{d}{dx}\left(x\frac{dy}{dx}\right)=\frac{d^2y}{dt^2},$$

and hence

$$x^2\frac{d^2y}{dx^2}=\frac{d^2y}{dt^2}-\frac{dy}{dt}=\frac{d}{dt}\left(\frac{d}{dt}-1\right)y.$$

Similarly

$$x^3\frac{d^3y}{dx^3}=\frac{d}{dt}\left(\frac{d}{dt}-1\right)\left(\frac{d}{dt}-2\right)y,$$

and

$$x^n\frac{d^ny}{dx^n}=\frac{d}{dt}\left(\frac{d}{dt}-1\right)\ldots\left(\frac{d}{dt}-n+1\right)y.$$

Substituting for $x^r\dfrac{d^ry}{dx^r}$ in (1) the equation transforms into

$$\frac{d^ny}{dt^n}+q_1\frac{d^{n-1}y}{dt^{n-1}}+\ldots+q_{n-1}\frac{dy}{dt}+q_ny=f(e^t), \quad (2)$$

in which $q_1, q_2 \ldots,$ are constants.

Example 18. *Solve the equation* $x^2\dfrac{d^2y}{dx^2}+6x\dfrac{dy}{dx}+6y=\dfrac{1}{x^2}.$

Putting $x = e^t$,

$$x\frac{dy}{dx}=\frac{dy}{dt}, \quad x^2\frac{d^2y}{dx^2}=\frac{d^2y}{dt^2}-\frac{dy}{dt},$$

and, substituting in the equation,

$$\frac{d^2y}{dt^2}+5\frac{dy}{dt}+6y=e^{-2t}.$$

The auxiliary equation $m^2+5m+6 = 0$ has roots $m = -2$ and $m = -3$, and a particular integral is

$$y_0 = \frac{1}{(D+2)(D+3)}e^{-2t} = te^{-2t}.$$

The general solution is

$$y = Ae^{-2t}+Be^{-3t}+te^{-2t},$$

that is

$$y = \frac{A}{x^2}+\frac{B}{x^3}+\frac{\log_e x}{x^2}.$$

Exercises 3(*g*)

Solve the differential equations:

1. $x^2y''-xy'+y = 0.$
2. $x^2y''+6xy'+4y = 0.$
3. $4x^2y''+24xy'+25y = 0.$
4. $x^2y''+xy'-9y = 0.$
5. $x^2y''-2xy'+2y = 12x^4.$
6. $x^3y'''+3x^2y''+xy' = 27x^3.$
7. $x^2y''-xy'+2y = x\log_e x.$
8. $x^2y''-3xy'+4y = 2x^2.$
9. $x^3y''+5x^2y'+4xy = x^2+1.$
10. $x^2y''-5xy'+9y = 27\log_e x.$
11. $(x-1)^2y''-4(x-1)y'+4y = x^2.$
12. $x^4y^{\text{iv}}+6x^3y'''+2x^2y''-4xy'+4y = 40x^3.$

3.10 The vibration equation

The differential equation

$$\frac{d^2x}{dt^2}+2k\frac{dx}{dt}+\omega^2x = E\cos qt \tag{1}$$

is known as the *vibration equation* and is of great importance in engineering theory. Equation (1) is the equation of motion of a particle moving in a straight line. The term ω^2x is an acceleration directed towards the origin and proportional to distance from the origin. The term $2k\dot{x}$ is a retardation due to frictional damping; this is always a retardation since $2k\dot{x}$ changes sign with $\dot{x}$. The term on the right-hand side, $E\cos qt$ (or $E\sin qt$), is the acceleration caused by a periodic applied force, and when this is present the vibration is called a *forced oscillation*.

The vibration equation can be solved by the methods of the previous sections and the solution is discussed below.

(i) *Undamped vibration without forcing*

In this case equation (1) becomes

$$\ddot{x}+\omega^2 x = 0, \tag{2}$$

the equation of simple harmonic motion, and the solution is

$$x = A\cos\omega t + B\sin\omega t$$

or

$$x = C\cos(\omega t-\varepsilon),$$

where A and B or C and ε are constants depending on the initial conditions. If at time $t = 0$, $x = a$ and $\dot{x} = u$, the constants are easily found and the solution is

$$x = a\cos\omega t + (u/\omega)\sin\omega t,$$

and in the alternative form $C = (a^2+u^2/\omega^2)^{\frac{1}{2}}$ and $\tan\varepsilon = u/a\omega$. The motion is thus periodic with period $2\pi/\omega$ and amplitude $(a^2+u^2/\omega^2)^{\frac{1}{2}}$.

(ii) *Undamped forced vibrations*

In this case equation (1) becomes

$$\ddot{x}+\omega^2 x = E\cos qt. \tag{3}$$

A particular integral is

$$x = \frac{E}{D^2+\omega^2}\cos qt = \frac{E}{\omega^2-q^2}\cos qt, \quad \text{if} \quad q \neq \omega.$$

If $q = \omega$, a particular integral is the real part of

$$\frac{E}{D^2+\omega^2}e^{i\omega t} = E\,e^{i\omega t}\frac{1}{D(D+2i\omega)}\,.\,1$$

$$= \frac{E}{2i\omega}t\,e^{i\omega t}.$$

Thus a particular integral is $x = (E/2\omega)t\sin\omega t$. If $q \neq \omega$, the solution of (3) is

$$x = C\cos(\omega t-\varepsilon)+\frac{E}{\omega^2-q^2}\cos qt, \tag{4}$$

and, if $q = \omega$, the solution is

$$x = C\cos(\omega t-\varepsilon)+\frac{E}{2\omega}t\sin\omega t. \tag{5}$$

The motion, therefore, consists of the free oscillation with the natural period of vibration of the system and the forced oscillation with the period of the applied force. If q is nearly equal to ω the amplitude $E/(\omega^2-q^2)$ will be very large and we have the phenomenon of *resonance* when the amplitude becomes indefinitely great. The manner in which the amplitude builds up with time is shown by the solution (5).

(iii) *Damped vibration without forcing*

In this case equation (1) becomes

$$\ddot{x}+2k\dot{x}+\omega^2x = 0. \tag{6}$$

The roots of the auxiliary equation $m^2+2km+\omega^2 = 0$ are

$$m = -k\pm(k^2-\omega^2)^{\frac{1}{2}},$$

and if $\omega \leqq k$ the solution is of the form

$$x = Ae^{-\alpha t}+Be^{-\beta t}, \quad \text{or} \quad x = (A+Bt)\,e^{-\alpha t},$$

α and β being positive. Because of the exponential terms, x diminishes rapidly without oscillation. Oscillations only occur, therefore, when $\omega > k$ and it is convenient to write $\omega^2 = k^2+p^2$, where p is real and positive. The auxiliary equation then has roots $m = -k\pm ip$ and the solution of equation (6) is

$$x = e^{-kt}(A\cos pt+B\sin pt),$$

or

$$x = Ce^{-kt}\cos(pt+\varepsilon),$$

where A and B, C and ε are constants depending on the initial conditions. The motion has period $2\pi/p$ but the amplitude diminishes exponentially to zero. The graph of displacement against time is shown in Fig. 3.

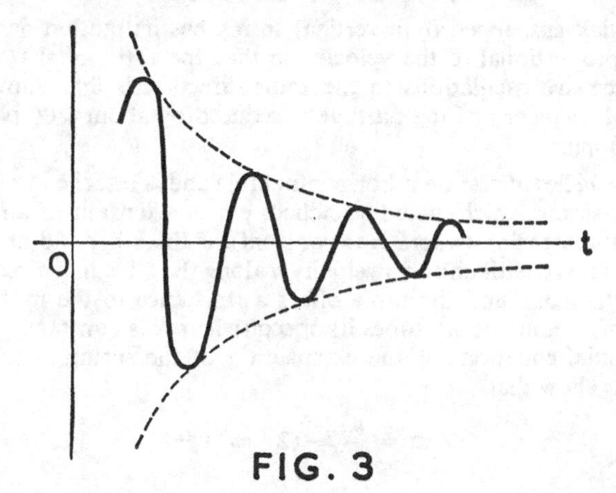

FIG. 3

(iv) *Damped and forced vibrations*

In this case equation (1) may be written as

$$\ddot{x}+2k\dot{x}+(k^2+p^2)x = E\cos qt, \tag{7}$$

and has a particular integral

$$\frac{E}{D^2+2kD+k^2+p^2}\cos qt = \frac{E}{k^2+p^2-q^2+2kD}\cos qt,$$

and, using the symbolic method of §3.8, this is

$$\frac{E}{\{(\omega^2-q^2)^2+4k^2q^2\}^{\frac{1}{2}}}\cos(qt-\alpha),$$

where $\tan\alpha = 2kq/(\omega^2-q^2)$. This is the forced oscillation, the free oscillation being as before $Ce^{-kt}\cos(pt+\varepsilon)$. When the period of the forced oscillation is the natural period of vibration, that is $q = \omega$, the forced oscillation is

$$x = (E/2k\omega)\sin\omega t,$$

and the amplitude, although it may be large, does not increase indefinitely.

Exercises 3(*h*)

1. A particle of mass m moves in a straight line in a medium whose resistance is $5mnv$, where v is the velocity, and is attracted towards a fixed point O in the line by a force $4mn^2x$, where x is the distance from O. It is projected towards O with velocity u from a point at a distance a from O, at time $t = 0$. Find x in terms of t. Discuss the motion
   (i) if $u \leqq 4na$, (ii) if $u > 4na$. (L.U.)
2. A particle suspended by a vertical spring has its motion damped by a force proportional to the velocity so that the ratio of the amplitudes of successive oscillations in the same direction is 0·8. Show that the natural frequency of the particle is reduced by about 0·06 per cent. by the damping.
3. A mass m lies at rest on a horizontal table and is attached to one end of a light spring which, when stretched, exerts a tension of amount $m\omega^2$ times its extension (where ω is constant). If the other end of the spring is now moved with uniform velocity u along the table in a direction away from the mass, and the table offers a resistance to the motion of the mass of an amount mk times its speed (where k is constant), obtain the differential equation for the extension x of the spring after time t. If $k = 2\omega$ show that
   $$x = \frac{u}{\omega}\{2-(2+\omega t)e^{-\omega t}\}. \qquad \text{(L.U.)}$$

4. A 4-lb. mass hangs at rest on a spring producing in the spring an extension of 1 ft. The upper end of the spring is now made to execute a vertical simple harmonic oscillation $x = \sin 4t$, x being measured vertically downwards in feet. If the mass is subject to a frictional resistance whose magnitude in lb.wt. is one-quarter of its velocity in ft. per sec., obtain the differential equation for the motion of the mass and find the expression for its displacement at time t, when t is large. (L.U.)

5. A mass m is supported on a horizontal platform to which it is attached by a spring of stiffness λ, and its vibration is damped by a damper which applies a force kv when the velocity of the mass relative to the platform is v. If the platform oscillates horizontally, its displacement at any time being y, while the displacement of the mass at the same time is x, show that $m\ddot{x}+k\dot{x}+\lambda x = k\dot{y}+\lambda y$. If $y = a \sin pt$, determine the amplitude of the steady oscillation of the mass and show that it attains its maximum value when
$$p^2 = \lambda^2\{(1+2k^2/m\lambda)^{1/2}-1\}/k^2. \qquad \text{(L.U.)}$$

6. A body of mass M performs oscillations controlled by a spring of stiffness λ and subject to a frictional force of constant magnitude F, and displacement is measured from the position in which the tension in the spring is zero. If the body be released from rest with a displacement a, greater than F/λ, show that it next comes to rest with displacement $-(a-2F/\lambda)$. Show that whatever the initial displacement, the body comes to rest in a finite time with a displacement numerically less than F/λ. (L.U.)

7. A rigid pendulum has moment of inertia Mk^2 about its axis of rotation and its centre of gravity G is distant h from its axis at O. The axis is now made to oscillate horizontally so that its distance from O at time t is $a \cos \omega t$. Show that the equation of motion of the pendulum is $Mk^2\ddot{\theta} = Mah\omega^2 \cos \theta \cos \omega t - Mgh \sin \theta$, where θ is the inclination of OG to the downward vertical. Hence show that if θ is small a solution is of the form $\theta = A \cos \{(gh/k^2)^{1/2}t+\varepsilon\}+(ah\omega^2 \cos \omega t)/(gh-k^2\omega^2)$.

8. A uniform rod of length l is supported by its upper end and is free to swing in a vertical plane. If the upper end is given a horizontal reciprocal motion, the displacement x from the mean position being given by $x = a \sin 2\pi nt$, obtain an expression for the angular movement of the rod after the motion has become steady, on the assumption that the oscillations generated are small in amplitude. (C.U.)

9. A horizontal beam is mounted on a vertical axis passing through its centre of gravity and its moment of inertia about the axis is I. The beam is made to perform angular oscillations under the influence of an alternating couple $L \cos 2\pi nt$ acting in a horizontal plane. If the friction at the pivot introduces a couple resisting motion of magnitude $\mu \dfrac{d\theta}{dt}$, where θ is the angular displacement of the beam, show that when the motion has become steady
$$\theta = \frac{-L \cos (2\pi nt+\phi)}{2\pi n(\mu^2+4\pi^2n^2I^2)^{1/2}}, \quad \text{where } \tan \phi = \frac{\mu}{2\pi nI}.$$

Show that if $\theta = 0$ and $\dfrac{d\theta}{dt} = 0$ when $t = 0$, the earlier stage of the motion is given by

$$\theta = L\left\{\frac{Ie^{-\mu t/I}}{\mu^2+4\pi^2n^2I^2} - \frac{\cos(2\pi nt+\phi)}{2\pi n(\mu^2+4\pi^2n^2I^2)^{1/2}}\right\}. \quad \text{(C.U.)}$$

10. A weight is hung on a spring, the upper end of which is given a vertical simple harmonic motion of frequency equal to the natural undamped frequency of oscillation of the weight. The oscillation of the weight is restricted by means of a dashpot which provides a damping force proportional to the velocity, so that its amplitude is equal to that of the upper end of the spring. Show that if the weight is allowed to oscillate freely under the same damping conditions the ratio of successive displacements in the same direction will be about 38:1. (C.U.)

3.11 Applications to the theory of structures

Linear equations with constant coefficients are met with in the study of the theory of struts, the vibration of beams, the deflexions of beams on elastic foundations, the whirling of shafts and many other structural problems. Some of the methods used in the formation and solution of these equations are illustrated in the following examples and exercises

Example 19. *A long strut of length l, of uniform cross-section, flexural rigidity EI and negligible weight is designed to support a load which acts along its length. Find the crippling load for the strut (i) when the ends are fixed in position and direction, (ii) when they are fixed in position only.*

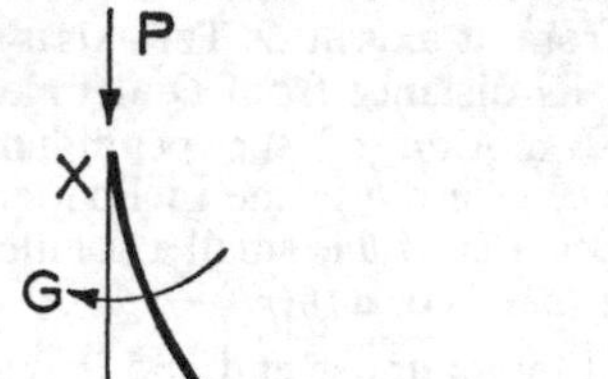

FIG. 4

Let P be the load and G the couple required to keep the ends fixed in direction. Then if y is the deflexion at distance x from one end (Fig. 4) the bending moment at $Q(x, y)$ is $-Py+G$. Thus, since the general expression for the bending moment is $EI(d^2y/dx^2)$, we have the deflexion equation

$$EI\frac{d^2y}{dx^2} = -Py+G.$$

Writing $P = EIn^2$, this becomes

$$\frac{d^2y}{dx^2}+n^2y = \frac{n^2G}{P}.$$

A particular integral is $y = G/P$, and the general solution is

$$y = A\cos nx+B\sin nx+G/P.$$

Since $y = dy/dx = 0$ when $x = 0$, $A = -G/P$ and $B = 0$, so that

$$y = \frac{G}{P}(1-\cos nx).$$

Also, $y = 0$ when $x = l$, so that $\cos nl = 1$ and hence $nl = 2\pi$. We have therefore $P = 4\pi^2 EI/l^2$. Here P is the load that will hold the strut in a deflected position and is called *Euler's crippling load* for the strut.

If the ends are not fixed in direction, $G = 0$ and hence

$$y = A \cos nx + B \sin nx.$$

The end conditions $y = 0$ for $x = 0$ and for $x = l$ give $A = 0$ and $\sin nl = 0$. Hence $nl = \pi$ and $P = \pi^2 EI/l^2$.

Example 20. *A uniform shaft of length l, mass ml and flexural rigidity EI runs in bearings which do not restrain the directions of its ends. Find the critical whirling speed of the shaft.*

Let ω be the angular velocity of the shaft. Then if y be the deflexion (Fig. 5) of an element δx of the shaft, the centrifugal force on the element is $m\,\delta x\,\omega^2 y/g$, so that the centrifugal force is equivalent to a load of intensity $m\omega^2 y/g$. Equating this to the general expression $EI(d^4y/dx^4)$ for the loading, we have the deflexion equation

$$EI\frac{d^4y}{dx^4} = \frac{m\omega^2 y}{g},$$

that is, writing $EIg\alpha^4 = m\omega^2$,

$$\frac{d^4y}{dx^4} - \alpha^4 y = 0.$$

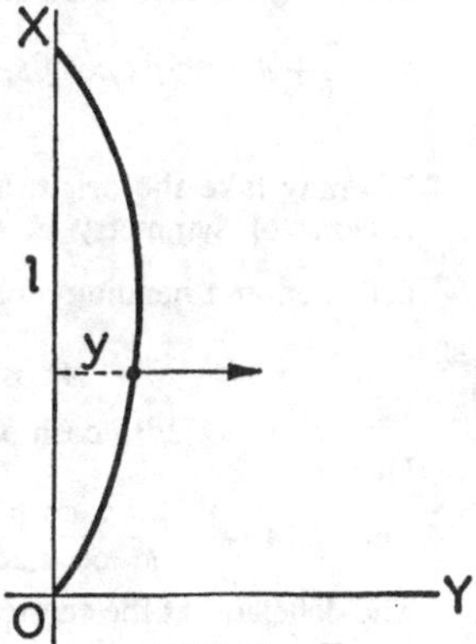

To solve this equation we note that the auxiliary equation $m^4 - \alpha^4 = 0$ has roots $m = \pm\alpha$ and $m = \pm i\alpha$, so that the general solution is

$$y = A\cosh \alpha x + B \sinh \alpha x + C\cos \alpha x + D \sin \alpha x.$$

FIG. 5

The end conditions are zero deflexion and bending moment at the ends, that is $y = d^2y/dx^2 = 0$ for $x = 0$ and $x = l$. The conditions at $x = 0$ give

$$A + C = 0 = \alpha^2 A - \alpha^2 C,$$

and hence $A = C = 0$.
The conditions at $x = l$ give

$$B \sinh \alpha l + D \sin \alpha l = 0,$$

and

$$\alpha^2 B \sinh \alpha l - \alpha^2 D \sin \alpha l = 0.$$

Hence $B = 0$ and, if D is not to be zero, $\sin \alpha l = 0$ so that $\alpha l = n\pi$, where n is an integer. This gives, substituting for α,

$$\omega = \left(\frac{gEI}{m}\right)^{1/2} \frac{n^2\pi^2}{l^2}.$$

The critical whirling speed, which is the lowest speed at which the shaft can hold a deflexion, is given by taking $n = 1$. At higher speeds, critical values of ω are found by taking $n = 2, 3, \ldots$.

Example 21. *A uniform beam of length 2l and flexural rigidity EI is freely supported at its ends at the same level and carries a load of uniform intensity w. The beam rests on an elastic foundation of modulus k. Find the deflexion at the centre of the beam.*

If y is the deflexion at any point, the elastic foundation exerts an upthrust ky per unit length at this point. Hence the deflexion equation is

$$EI\frac{d^4y}{dx^4} = w-ky.$$

Writing $k = 4EI\beta^4$, we have

$$\frac{d^4y}{dx^4}+4\beta^4y = \frac{w}{EI}.$$

A particular integral of this equation is $y = w/(4EI\beta^4) = w/k$ and the auxiliary equation is $m^4+4\beta^4 = 0$ which has roots $m = \beta(\pm 1\pm i)$. This gives the complementary function

$$y = e^{\beta x}(P\cos\beta x+Q\sin\beta x)+e^{-\beta x}(R\cos\beta x+S\sin\beta x),$$

and the general solution can be written as

$$y = \frac{w}{k}+A\cosh\beta x\cos\beta x+B\sinh\beta x\sin\beta x+C\cosh\beta x\sin\beta x$$
$$+D\sinh\beta x\cos\beta x.$$

We may take the origin at the mid-point of the undeflected beam and, from reasons of symmetry y is an even function of x so that $C = D = 0$. Zero deflexion and bending moment at the ends gives $y = \frac{d^2y}{dx^2} = 0$ for $x = l$. Thus

$$A\cosh\beta l\cos\beta l+B\sinh\beta l\sin\beta l = -w/k,$$
$$2\beta^2B\cosh\beta l\cos\beta l-2\beta^2A\sinh\beta l\sin\beta l = 0.$$

Hence

$$A = -\frac{2w}{k}.\frac{\cosh\beta l\cos\beta l}{\cosh 2\beta l+\cos 2\beta l},\qquad B = -\frac{2w}{k}.\frac{\sinh\beta l\sin\beta l}{\cosh 2\beta l+\cos 2\beta l}.$$

The deflexion at the centre is $w/k+A$.

Exercises 3(i)

1. A light strut of length l and modulus of rigidity EI is placed in a vertical position with its lower end clamped and its upper end free. The strut supports a load P. Show that the deflexion equation is $y''+n^2y = n^2a$, where a is the deflexion of the upper end and $P = EIn^2$. Hence show that the crippling load for the strut is $\pi^2EI/4l^2$.
2. A light strut of length l and modulus of rigidity EI has its ends fixed in position. A thrust P is applied to the ends parallel to the strut but acting at a distance e from the centre line of the strut. Show that the deflexion equation is $y''+n^2y+n^2e = 0$, where $P = EIn^2$, and that the greatest bending moment in the strut is $Pe\sec(nl/2)$.
3. A uniform tie bar of length l and flexural rigidity EI is subjected to a tensile force P at its ends and carries a uniform lateral load of intensity w. Show that the deflexion equation is $y''-n^2y = wn^2x(x-l)/2P$, where $P = EIn^2$. Hence show that the greatest bending moment in the bar is $w(1-\text{sech}\,\frac{1}{2}nl)/n^2$.
4. The lower end of a uniform light cantilever of length l and flexural rigidity EI is clamped at an angle α to the vertical. A vertical load W is applied to the upper end. Show that the deflexion y from the unde-

flected position at x from the lower end is given by the equation $y''+n^2y = n^2a+n^2(l-x)\tan\alpha$, where $EIn^2 = W\cos\alpha$ and a is the end deflexion. Show that at the free end the value of dy/dx is $\tan\alpha\,(\sec nl-1)$. (L.U.)

5. A light uniform pole of length l and flexural rigidity EI is fixed vertically in the ground and its upper end is acted upon by a force T which makes an angle α with the downward vertical. The small horizontal deflexion of the upper end is a. Show that $y''+n^2y = n^2a+n^2(l-x)\tan\alpha$, x being measured upwards from the ground and $EIn^2 = T\cos\alpha$. Hence show that $na = \tan\alpha(\tan nl-nl)$. (L.U.)

6. A uniform thin lath of length l and flexural rigidity EI is clamped vertically at its lower end and its upper end carries a small light bracket of length a fixed perpendicularly to the lath. When a load W is hung from the bracket it deflects a small horizontal distance b and a negligible vertical distance. State the bending moment at a point on the lath distant x vertically and y horizontally from the clamped end. Find b and the bending moment at the clamp in terms of the other quantities given. Evaluate W when $b = a$. (L.U.)

7. A light uniform flexible lath of length l is clamped vertically at its upper end and carries a small mass m at its lower end. Show that the time period of its small flexural oscillations is equal to that of a simple pendulum of length $(nl-\tanh nl)/n$, where EI is the flexural rigidity of the lath and $EIn^2 = mg$. (L.U.)

8. A uniform light strut of length l and flexural rigidity EI has an initial curvature so that its shape is that of a sine curve with its centre at a distance a from the line joining its ends. The strut is subjected to a compressive force P at its ends, which are fixed in position only. Show that the deflexion y at distance x from one end is given by the equation $y''+n^2y = -(a\pi^2/l^2)\sin\pi x/l$, where $EIn^2 = P$. Show that the greatest bending moment is $-Pa\pi^2/(\pi^2-n^2l^2)$.

9. A uniform beam of length $2l$, flexural rigidity EI and weight w per unit length rests on an elastic foundation giving an upthrust ky per unit length where y is the deflexion. Show that if y is the deflexion at a horizontal distance x from the centre of the undeflected beam

$$(d^4y/dx^4)+4\beta^4y = w/EI, \quad \text{where } 4EI\beta^4 = k.$$

Show that if the ends of the beam are clamped horizontally at the same level, the deflexion at the centre is $A+w/k$, where

$$A = -(w/k)(\cosh\beta l\sin\beta l+\sinh\beta l\cos\beta l)/(\cosh\beta l\sinh\beta l+\cos\beta l\sin\beta l).$$

10. A uniform shaft of length l, flexural rigidity EI and mass m per unit length rotates in bearings with angular velocity ω. Show that the deflexion equation is $(d^4y/dx^4)-a^4y = 0$, where $gEIa^4 = m\omega^2$. Show that, if a non-zero solution exists giving $y = dy/dx = 0$ for $x = 0$ and $x = l$, $\cosh al\cos al = 1$, and that a solution of this equation is $al = 1{\cdot}506\pi$ approximately. Hence find the least whirling speed of the shaft.

3.12 Application to electric circuits

The flow of current in an electric circuit and the charge on a condenser are found as the solutions of linear differential equations with constant coefficients.

The following notation is used:

i = current in *ampères*,
q = charge on a condenser in *coulombs*,
v = voltage drop or potential difference across the plates of a condenser,
C = capacitance of a condenser in *farads*,
L = coefficient of self-inductance in *henries*,
R = resistance in *ohms*,
e = electromotive force in *volts*.

An inductance, a resistance and a condenser each causes a drop in voltage according to the following laws:

$L\frac{di}{dt}$ = the voltage drop across the inductance,

Ri = the voltage drop across the resistance,

$\frac{q}{C} = v$ = the voltage drop across the plates of the condenser.

When an electromotive force is connected to a circuit containing a condenser, the current is the rate of change of the charge on the condenser, that is $i = dq/dt$. When, however, a condenser is discharging through a circuit $i = -dq/dt$.

The differential equation of a circuit connected to an electromotive force is formed by equating the total voltage drop due to inductances, resistances and condensers to the voltage supplied by the electromotive force. When a condenser discharges through a circuit the voltage on the condenser is equated to the drop due to inductances and resistances.

In the following circuits for which the differential equations are given, the electromotive force e is usually a constant E or alternating and equal to $E \cos \omega t$ or $E \sin \omega t$.

Circuits with electromotive force

(i) Circuit with resistance and inductance,

$$L\frac{di}{dt}+Ri = e. \tag{1}$$

(ii) Circuit with resistance and condenser,

$$Ri+\frac{q}{C} = e \quad \text{and} \quad i = \frac{dq}{dt},$$

so that

$$R\frac{dq}{dt}+\frac{q}{C} = e, \tag{2}$$

or

$$R\frac{dv}{dt}+\frac{v}{C} = \frac{e}{C}, \tag{3}$$

or

$$R\frac{di}{dt}+\frac{i}{C} = \frac{de}{dt}. \tag{4}$$

(iii) Circuit with resistance, inductance and condenser,

$$L\frac{di}{dt}+Ri+\frac{q}{C} = e \quad \text{and} \quad i = \frac{dq}{dt},$$

so that

$$L\frac{d^2q}{dt^2}+R\frac{dq}{dt}+\frac{q}{C} = e, \tag{5}$$

or

$$L\frac{d^2v}{dt^2}+R\frac{dv}{dt}+\frac{v}{C} = \frac{e}{C}, \tag{6}$$

or

$$L\frac{d^2i}{dt^2}+R\frac{di}{dt}+\frac{i}{C} = \frac{de}{dt}. \tag{7}$$

Discharge of a condenser

(iv) Circuit with resistance only,

$$Ri = \frac{q}{C} \quad \text{and} \quad i = -\frac{dq}{dt},$$

so that

$$R\frac{dq}{dt}+\frac{q}{C} = 0, \tag{8}$$

or

$$R\frac{dv}{dt}+\frac{v}{C} = 0, \tag{9}$$

or

$$R\frac{di}{dt}+\frac{i}{C} = 0. \tag{10}$$

(v) Circuit with resistance and inductance,

$$L\frac{di}{dt}+Ri=\frac{q}{C} \quad \text{and} \quad i=-\frac{dq}{dt},$$

so that

$$L\frac{d^2q}{dt^2}+R\frac{dq}{dt}+\frac{q}{C}=0, \tag{11}$$

or

$$L\frac{d^2v}{dt^2}+R\frac{dv}{dt}+\frac{v}{C}=0, \tag{12}$$

or

$$L\frac{d^2i}{dt^2}+R\frac{di}{dt}+\frac{i}{C}=0. \tag{13}$$

Example 22. *A circuit with inductance L, resistance R and a condenser of capacitance C is connected to an electromotive force E* $\sin \omega t$. *Find the steady state current at time t after the circuit is closed.*

The differential equation is equation (7), that is

$$L\frac{d^2i}{dt^2}+R\frac{di}{dt}+\frac{i}{C}=\omega E\cos\omega t.$$

The auxiliary equation $m^2+(R/L)m+1/(LC)=0$ has roots

$$m=-R/2L\pm(R^2/4L^2-1/LC)^{1/2}.$$

Thus the current given by the complementary function may be oscillatory or not according as $R^2 <$ or $\geqq 4L/C$, but because of the factor $e^{-Rt/2L}$ it is transient and diminishes to zero. The current given by the particular integral is called the steady state current, and is, using §3.8,

$$i=\frac{E\omega}{LD^2+RD+1/C}\cos\omega t=\frac{E}{\{1/(\omega C)-L\omega\}+(R/\omega)D}\cos\omega t$$

$$=\frac{E}{[\{1/(\omega C)-L\omega\}^2+R^2]^{1/2}}\cos(\omega t-\alpha),$$

where $\{1/(\omega C)-L\omega\}\tan\alpha=R$. Thus the impedance is $[\{1/(\omega C)-L\omega\}^2+R^2]^{1/2}$ and the phase lag is α. It is easily seen that the amplitude has its maximum value E/R when $\omega=1/(LC)^{1/2}$, and this state is analogous to that of resonance in mechanical systems.

Exercises 3(*j*)

1. A condenser of capacity C discharges through a circuit of resistance R and self-inductance L. Find the condition that the discharge be just non-oscillatory. Obtain also the formulae for the time variation of charge and current in this case, when the initial voltage is E. (L.U.)

2. A condenser of capacity C and initial charge Q_0 is discharged through a resistance R and an inductance L in series. Prove that if $R^2C<4L$ the current at time t is $Q_0e^{-ht}(k+h^2/k)\sin kt$, where $-h\pm ik$ are the roots of the equation $CLx^2+CRx+1=0$. (L.U.)

3. A circuit consists of inductance L and capacity C in series. An alternating e.m.f. $E \sin nt$ is applied to the circuit commencing at time $t = 0$, the initial current and charge on the condenser being zero. Prove that the current at time t is given by $I = \dfrac{nE}{L(n^2-\omega^2)}(\cos \omega t - \cos nt)$ where $CL\omega^2 = 1$. (L.U.)

4. A condenser of capacitance C is charged so that the potential difference of the plates is v_0. The plates are connected by a wire of resistance R and inductance L. If $R = 100$ ohms, $C = 10^{-5}$ farads, $L = 5 \times 10^{-2}$ henries and $v_0 = 800$ volts, find the potential difference of the plates t seconds after the circuit is closed.

5. An uncharged condenser of capacity C is charged by an e.m.f. of $E \sin \{t/(LC)^{1/2}\}$ through leads of self-inductance L and negligible resistance. Prove that at time t the charge on one of the plates is

$$\tfrac{1}{2}CE\{\sin [t/(LC)^{1/2}] - [t/(LC)^{1/2}] \cos [t/(LC)^{1/2}]\}.$$

If, in addition, there is a small resistance, in what respect is the mathematical form of the above result altered?

6. An uncharged condenser of capacity C is charged by applying an e.m.f. $E \sin \pi t$ through leads of self-inductance L and small resistance R. After a time $2T$, where T is a large integer, the e.m.f. remains zero. Find the charge on the condenser at time t, where $t > 2T$. (L.U.)

7. An e.m.f. $E \cos pt$ is switched across an electric circuit consisting of a coil of inductance L henries, resistance R ohms, in series with a condenser of capacity C farads. State the differential equation for the charge q on the condenser at any time t and the form of the general solution when the resistance is just sufficient to prevent natural oscillations.
If $L = 0{\cdot}0001$, $R = 2$, find the value of C for this condition to hold and calculate the amplitude of the steady current for an imposed peak voltage of 100 at 50 cycles per second. (L.U.)

8. A voltage $E \sin \omega t$ is applied to a circuit containing an inductance L henries, a resistance R ohms and a capacitance C farads. Write down the differential equation for the charge q on a plate of the condenser and the current i flowing into this plate at time t. Find a formula for i in the steady state and sketch a graph of the amplitude of i for different values of ω. (L.U.)

9. An alternating e.m.f. $E \sin \omega t$ is supplied to a circuit containing inductance L, resistance R and capacitance C. Obtain the differential equation satisfied by the current i.
Find the resistance if it is just large enough to prevent natural oscillations. For this value of R and $LC\omega^2 = 1$ prove that

$$i = E(\sin \omega t - \omega t e^{-\omega t})/2k,$$

where $k^2 = L/C$, when the current and the charge on the condenser are both zero at time $t = 0$. (L.U.)

10. An electric circuit consists of an inductance L, resistance R and capacitance C in series. A constant e.m.f. E is applied in series with the circuit

at time $t = 0$ when the current i and the potential v across the condenser are zero. Obtain the differential equation for v. Find the values of v and i at time t, given that $5CL\omega^2 = 1$ and $5CR\omega = 2$, and prove that the greatest value of v is $E(1+e^{-\pi/2})$. (L.U.)

3.13 Application to servomechanisms

A servomechanism is an automatic control mechanism designed to ensure that a certain variable quantity θ_0, called the output, is in agreement with a given variable quantity θ_i, called the input. If at any time θ_0 is not equal to θ_i, a force proportional to the error, that is $\theta_i - \theta_0$, is brought into play to correct the value of θ_0.

For example, if θ_0 and θ_i are the angular displacements of two shafts, the shafts may be coupled to a differential gear which will give an angular displacement $\theta_i - \theta_0$ to a third shaft. This third shaft moves the control of a potentiometer and this supplies a voltage proportional to $\theta_i - \theta_0$ to a motor coupled to the output shaft. Thus the acceleration of this shaft is proportional to $\theta_i - \theta_0$ and there will also be frictional damping proportional to $\dot{\theta}_0$. The differential equation of the motion is therefore of the form

$$\frac{d^2\theta_0}{dt^2} + 2\zeta\omega\frac{d\theta_0}{dt} = \omega^2(\theta_i - \theta_0),$$

or, writing D for d/dt,

$$(D^2 + 2\zeta\omega D + \omega^2)\theta_0 = \omega^2\theta_i. \qquad (1)$$

This equation is often called the *instrument equation*, ω is called the *undamped natural frequency* of the mechanism and ζ is called the *damping ratio*.

In other mechanisms the acceleration of the output shaft may vary also with the derivative of $\theta_i - \theta_0$, leading to a differential equation of the form

$$(D^2 + aD + b)\theta_0 = (cD + b)\theta_i. \qquad (2)$$

Equations of higher order than the second arise with more complicated mechanisms.

(i) *Solution of the differential equation*

The auxiliary equation for equation (1) is $m^2 + 2\zeta\omega m + \omega^2 = 0$, with roots $m = -\zeta\omega \pm \omega(\zeta^2 - 1)^{\frac{1}{2}}$. The complementary function is

if $\zeta = 0$, $\quad \theta_0 = A\cos\omega t + B\sin\omega t$,

if $0 < \zeta < 1$, $\quad \theta_0 = e^{-\zeta\omega t}(A\cos\mu\omega t + B\sin\mu\omega t)$, $\quad \mu^2 = 1 - \zeta^2$,

if $\zeta = 1$, $\quad \theta_0 = e^{-\omega t}(A + Bt)$,

if $\zeta > 1$, $\quad \theta_0 = e^{-\zeta\omega t}(A\cosh\lambda\omega t + B\sinh\lambda\omega t)$, $\quad \lambda^2 = \zeta^2 - 1$.

Thus in all cases in which damping is present ($\zeta > 0$), the complementary function is transient and does not affect the ultimate value of θ_0. The complementary function is often termed the *transient response* and the following table shows its behaviour for various values of the damping ratio.

Damping ratio	*Transient response*
$\zeta = 0$	No damping; system oscillates with constant amplitude and frequency ω.
$0 < \zeta < 1$	Damped oscillations of frequency less than ω.
$\zeta = 1$	Critical damping; oscillations just cease.
$\zeta > 1$	Over-damped; no oscillations.

It is essential in a servomechanism that the complementary function should be transient, and for this it is necessary and sufficient that the roots of the auxiliary equation should all be negative or have negative real parts. A system for which this is the case is said to be *stable*. Criteria from the theory of algebraic equations are available for determining the stability of systems whose differential equations are of order higher than the second.

The important part of the solution is the particular integral which may be found when the input θ_i is known; this is called the *steady state* output. This integral is written in the form

$$\theta_0 = f(D)\theta_i,$$

and for equation (1)

$$f(D) = \frac{\omega^2}{D^2+2\zeta\omega D+\omega^2}.$$

The operator $f(D)$ is called the *transfer function* of the system. If the mechanism is linked to a second mechanism with transfer function $f_1(D)$ in series, the final relation of output to input is $\theta_0 = f_1(D)f(D)\theta_i$, that is, the transfer functions are multiplied together.

Example 23. *Find the output in the steady state given by the system of equation* (1) *if $\theta_i = \Omega t$, where Ω is constant.*

We have

$$\theta_0 = \frac{\omega^2}{D^2+2\zeta\omega D+\omega^2}\Omega t$$

$$= \left(1-\frac{2\zeta}{\omega}D+\ldots\right)\Omega$$

$$= \Omega\left(t-\frac{2\zeta}{\omega}\right), \quad \text{by §3.7 (iii}$$

Thus the output follows the input but with a time lag $2\zeta/\omega$. The angular lag $2\Omega\zeta/\omega$, being proportional to the angular velocity, is called the *velocity lag*.

(ii) *Harmonic response of a system*

The *response* of a stable system to an input θ_i is the output θ_0. Since the input can usually be expressed as a trigonometrical series, the general nature of the response may be studied by considering the output corresponding to the input $\theta_i = a \sin pt$. The steady state output due to this value of θ_i is called the *harmonic response* of the system.

The harmonic response of the system of equation (1) is

$$\theta_0 = \frac{a\omega^2}{D^2+2\zeta\omega D+\omega^2}\sin pt$$

$$= \frac{a\omega^2}{(\omega^2-p^2)+2\zeta\omega D}\sin pt$$

$$= am\sin(pt-\alpha), \quad \text{using §3.8},$$

where $m = \omega^2/\{(\omega^2-p^2)^2+4\zeta^2\omega^2p^2\}^{\frac{1}{2}}$ and $(\omega^2-p^2)\tan\alpha = 2\zeta\omega p$. Here α is the phase lag and m is called the *amplitude magnification factor*.

(iii) *Unit response of a system*

Another standard input, called the *unit step function*, is used to find the values of the constants in the complementary function in the output when the system begins to operate.

The step function $h(t)$, shown in Fig. 6, is a function of the time; it

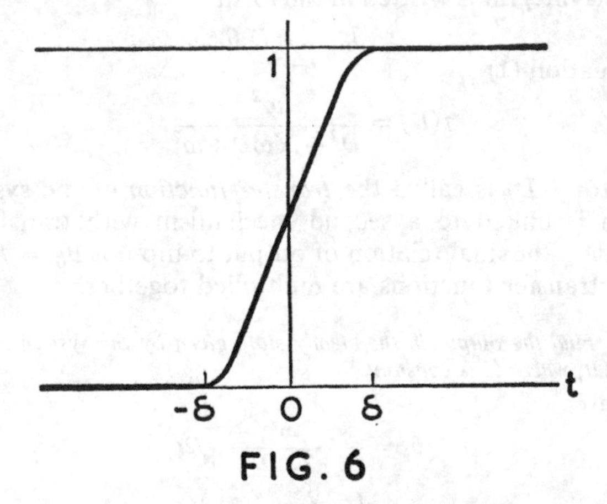

FIG. 6

is zero for $t < -\delta$ and unity for $t > \delta$, where δ is small, changing continuously from 0 to 1 in the interval $-\delta$ to δ. By finding the values of θ_0 and its derivatives for a value of t in the interval $-\delta$ to δ, and letting δ tend to zero we find the initial values of θ_0 and its derivatives, and hence the constants for the transient part of the output.

Example 24. *Find the unit response of the system whose transfer function is*

$$(5D+26)/(D^3+6D^2+21D+26).$$

If $\theta_i = h(t) = 1$ the particular integral is $\theta_0 = 1$. The complementary function is

$$\theta_0 = Ae^{-2t}+e^{-2t}(B\cos 3t+C\sin 3t).$$

Now

$$(D^3+6D^2+21D+26)\theta_0 = (5D+26)h(t),$$

and integrating this equation from $-\delta$ to t, where $-\delta \leqq t \leqq \delta$, we have

$$(D^2+6D+21)\theta_0+26\int_{-\delta}^{t}\theta_0\,dt = 5h(t)+26\int_{-\delta}^{t}h(t)\,dt.$$

The values of θ_0 and $h(t)$ being bounded in the interval, the integrals of these quantities tend to zero as δ tends to zero, and hence in the limit for $t = 0$ we have

$$D^2\theta_0+6D\theta_0+21\theta_0 = 5.$$

Integrating again twice and using the same limiting process after each integration we find for $t = 0$

$$D\theta_0+6\theta_0 = 0,$$

and

$$\theta_0 = 0.$$

Thus we find that initially on application of the input

$$\theta_0 = 0, \quad \dot{\theta}_0 = 0, \quad \ddot{\theta}_0 = 5.$$

The complete solution is

$$\theta_0 = e^{-2t}(A+B\cos 3t+C\sin 3t)+1$$

and

$$\dot{\theta}_0 = e^{-2t}\{-2A+(3C-2B)\cos 3t-(3B+2C)\sin 3t\},$$

$$\ddot{\theta}_0 = e^{-2t}\{4A-(12C+5B)\cos 3t+(12B-5C)\sin 3t\}.$$

Inserting the values found for θ_0, $\dot{\theta}_0$ and $\ddot{\theta}_0$ when $t = 0$ we find $A = -\frac{8}{9}$, $B = -\frac{1}{9}$, $C = -\frac{2}{3}$, and

$$\theta_0 = 1-(e^{-2t}/9)(8+\cos 3t+6\sin 3t)$$

is the unit response of the system.

The input and the response are shown graphically in Fig. 7.

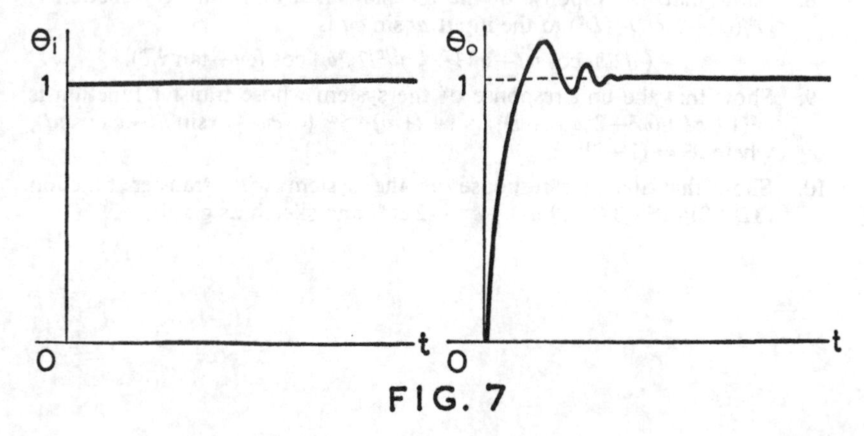

FIG. 7

Exercises 3(k)

1. Show that the response of the system whose transfer function is $(b+cD)/(b+aD+D^2)$ to the input Ωt has velocity lag $(a-c)/b$.

2. Show that the response of the mechanism whose transfer function is $\omega^2(\omega^2+2\zeta\omega D+D^2)^{-1}$ to the input Ωt^2 is $\Omega\{(t-2\zeta/\omega)^2+(4\zeta^2-2)/\omega^2\}$.

3. Show that the harmonic response of the system whose transfer function is $1/(1+aD+bD^2+cD^3)$ has amplitude magnification factor
$$\{(1-bp^2)^2+(a-cp^2)^2p^2\}^{-1/2}$$
and phase lag $\tan^{-1}\{(a-cp^2)p/(1-bp^2)\}$.

4. The acceleration $\ddot{\theta}_0$ of the output shaft of a mechanism is given by the equation $J\ddot{\theta}_0 = Ci-f\dot{\theta}_0$, where J is the moment of inertia of the armature, f a damping factor, C a constant and i the current supplied to the motor. The current is supplied from a source with controlled voltage $K(\theta_i-\theta_0)$ through leads of resistance R and self-inductance L. Show that the transfer function of the mechanism is
$$G/\{TJD^3+(J+fT)D^2+fD+G\}, \quad \text{where } G = CK/R \text{ and } T = L/R.$$

5. Given that the cubic equation $x^3+ax^2+bx+c = 0$, where a, b and c are positive, has roots whose real parts are negative if $ab-c > 0$, show that the mechanism described in Exercise 4 above is stable if
$$f(J+fT) > JTG.$$

6. Show that the harmonic response of the mechanism described in Exercise 4 above has phase lag $\tan^{-1}\{p(f-TJp^2)/(G-Jp^2-fTp^2)\}$ and amplitude magnification factor $G\{(G-Jp^2-fTp^2)^2+p^2(f-TJp^2)^2\}^{-1/2}$.

7. Show that the harmonic response of the system whose transfer function is $\omega^2(1+cD)/(\omega^2+2\zeta\omega D+D^2)$ has phase lag
$$\tan^{-1}\{2\zeta\omega p/(\omega^2-p^2)\}-\tan^{-1} cp,$$
and amplitude magnification factor
$$\omega^2(1+c^2p^2)^{1/2}\{(\omega^2-p^2)^2+4\zeta^2\omega^2p^2\}^{-1/2}.$$

8. Show that the response of the mechanism whose transfer function is $\omega^2/(\omega^2+2\zeta\omega D+D^2)$ to the input $at \sin \omega t$ is
$$-(a/2\zeta)t\cos\omega t+\{a(1+\zeta^2)^{1/2}/2\zeta^2\omega\}\cos(\omega t-\tan^{-1}\zeta).$$

9. Show that the unit response of the system whose transfer function is $\omega^2(1+cD)/(\omega^2+2\zeta\omega D+D^2)$ is $1+(1/\mu)e^{-\zeta\omega t}\{\omega(c\omega-\zeta)\sin\mu t-\mu\cos\mu t\}$, where $\mu^2 = (1-\zeta^2)\omega^2$.

10. Show that the unit response of the system with transfer function $(3D+2)/(D^2+3D+2)$ is $1+e^{-t}-2e^{-2t}$ and sketch its graph.

CHAPTER 4

SIMULTANEOUS EQUATIONS; REDUCIBLE EQUATIONS

4.1 Introductory

In this chapter we discuss the solutions of simultaneous differential equations with constant coefficients. These are equations in which there are two or more dependent variables and one independent variable, usually the time t, there being as many equations as there are dependent variables. The solution of such equations consists in expressing each variable separately in terms of the independent variable.

Simultaneous differential equations arise in many problems of dynamics and electric circuit theory, and some of these problems are considered here. Some of the most important applications are to the theory of vibrations of mechanical systems. The simultaneous equations which occur in this theory are of a particular type and we shall see how the normal modes of the vibrations can be quickly determined from the equations.

In the later part of the chapter we consider certain types of ordinary differential equations, both linear and non-linear, which can be solved by progressively reducing the order of the equation.

4.2 Simultaneous equations of the first order

Simultaneous differential equations are solved by eliminating all but one of the dependent variables so as to form a single differential equation for the remaining variable. The order of this equation will be less or equal to the sum of the orders of the simultaneous equations. Thus if we have three simultaneous equations involving variables x, y, z and their first derivatives the eliminant will be an equation of the third or lower order for x, y or z. The method of elimination is much the same as that used in solving simultaneous algebraic equations, operators being used as if they were multipliers. The solution will be a set of equations giving each of the dependent variables separately as a function of the independent variable. Care must be taken to ensure that the total number of arbitrary constants in the solution is not greater than the sum of the orders of the equations.

Example 1. *Solve the equations* $\dot{x}+2y+3x = 0$, $\dot{y}+3x-2y = 0$.

Writing D for d/dt, the equations are

$$(D+3)x+2y = 0, \tag{1}$$

$$3x+(D-2)y = 0. \tag{2}$$

Operating on equation (1) with the operator $(D-2)$ and multiplying equation (2) by 2 we have

$$(D-2)(D+3)x+2(D-2)y = 0,$$
$$6x+2(D-2)y = 0.$$

Subtracting,

$$(D^2+D-12)x = 0,$$

and the solution of this second order equation is, by the method of §3.3,

$$x = A\,e^{3t}+B\,e^{-4t}.$$

We can eliminate x by subtracting equation (1) multiplied by 3 from equation (2) operated on by $(D+3)$, giving

$$(D^2+D-12)y = 0,$$

and hence

$$y = E\,e^{3t}+F\,e^{-4t}.$$

There can be only two independent arbitrary constants in the solution, and hence E and F must be related to A and B. These relations can be found by substituting the solutions found for x and y in one of the original equations. Substituting in equation (1) we have, for all values of t,

$$(D+3)x+2y = e^{3t}(6A+2E)+e^{-4t}(-B+2F) = 0,$$

and hence $E = -3A$, $F = \frac{1}{2}B$. Thus the final solution is

$$x = A\,e^{3t}+B\,e^{-4t},$$
$$y = -3A\,e^{3t}+\tfrac{1}{2}B\,e^{-4t}.$$

In this case the expression for y could have been obtained directly from equation (1) by writing $y = -\frac{1}{2}(D+3)x$.

Example 2. *Solve the equations* $2\dot{x}+\dot{y}-2x-2y = 5e^t$, $\dot{x}+\dot{y}+4x+2y = 5e^{-t}$, *given that when* $t = 0$, $x = 2$, $y = 0$.

The equations are

$$2(D-1)x+(D-2)y = 5e^t,$$
$$(D+4)x+(D+2)y = 5e^{-t}.$$

Operating on the first equation with $(D+2)$ and on the second equation with $(D-2)$ and subtracting,

$$\{2(D+2)(D-1)-(D-2)(D+4)\}x = 5(D+2)e^t-5(D-2)e^{-t},$$

that is

$$(D^2+4)x = 15e^t+15e^{-t}.$$

The particular integral of this equation is $3e^t+3e^{-t} = 6\cosh t$, and

$$x = A\cos 2t+B\sin 2t+6\cosh t.$$

Similarly, eliminating x, we find the equation for y to be

$$\{2(D-1)(D+2)-(D+4)(D-2)\}y = -5(D+4)e^t+10(D-1)e^{-t}$$

that is

$$(D^2+4)y = -25e^t-20e^{-t}.$$

A particular integral is $-5e^t-4e^{-t} = -9\cosh t-\sinh t$, and

$$y = E\cos 2t+F\sin 2t-9\cosh t-\sinh t.$$

We now substitute the complementary functions of x and y in the first equation. We have

$$2(D-1)x = (4B-2A)\cos 2t-(4A+2B)\sin 2t,$$
$$(D-2)y = (2F-2E)\cos 2t-(2F+2E)\sin 2t,$$

and hence

$$F - E = A - 2B, \quad F + E = -2A - B.$$

From the initial conditions for x and y we have

$$2 = A + 6, \quad 0 = E - 9.$$

Hence $A = -4$, $B = 6$, $E = 9$, $F = -7$ and the solutions are

$$x = -4\cos 2t + 6\sin 2t + 6\cosh t, \quad y = 9\cos 2t - 7\sin 2t - 9\cosh t - \sinh t.$$

Exercises 4(*a*)

Solve the simultaneous differential equations:

1. $\dot{x} - 3x + 5y = 0$, $\dot{y} + 3y - 5x = 0$.
2. $\dot{x} + x + 5y = 0$, $2\dot{x} - \dot{y} + 12y = 0$.
3. $\dot{x} - x + 2y = 0$, $\dot{y} - 5x - 3y = 0$, $x = y = 1$ when $t = 0$. (L.U.)
4. $\dot{x} + 2x + y = 0$, $\dot{y} + x + 2y = 0$, $x = 1$ and $y = 0$ when $t = 0$. (L.U.)
5. $\dot{x} + 4x + 3y = 0$, $\dot{y} + 2x + 5y = e^t$, $x = y = 0$ when $t = 0$. (L.U.)
6. $3\dot{x} + 2x - y = t$, $2\dot{y} + y - x = 5e^{-t}$, $x = y = 0$ when $t = 0$. (L.U.)
7. $4\dot{x} + 5x - 3y = 4e^t$, $4\dot{y} - 5y + 3x = 4e^t$, $x = y = 0$ when $t = 0$. (L.U.)
8. $\dot{x} + 5x - 2y = t$, $\dot{y} + 2x + y = 0$, $x = y = 0$ when $t = 0$. (L.U.)
9. $\dot{x} = x + y$, $\dot{x} + \dot{y} = \cos t - x$, $x = 0$ when $t = 0$, $y = 0$ when $t = \pi/2$. (L.U.)
10. $\dot{x} + 3x - 2y = 1$, $\dot{y} - 2x + 3y = e^t$, $x = y = 0$ when $t = 0$. (L.U.)
11. $\dot{x} = y$, $\dot{y} = z$, $\dot{z} = 8x$, $x = 1$, $y = 2$, $z = 4$ when $t = 0$. Prove that independently of the above conditions $4x + 2y + z = Ae^{2t}$, where A is constant. (L.U.)
12. If $z(dx/dz) + 2y = \sin(\log_e z)$ and $z(dy/dz) + 3x - y = 0$, find and solve the differential equation for y and hence determine x. (L.U.)

4.3 Simultaneous equations of higher orders

Simultaneous differential equations in which derivatives of the second and higher orders occur are solved in the same way by eliminating all but one of the dependent variables. The solution of simultaneous algebraic equations is expressed conveniently by means of determinants. Thus the equations

$$\left.\begin{aligned} a_1x + b_1y + c_1z &= d_1 \\ a_2x + b_2y + c_2z &= d_2 \\ a_3x + b_3y + c_3z &= d_3 \end{aligned}\right\}, \tag{1}$$

have solution

$$\begin{vmatrix} a_1 & b_1 & c_1 \\ a_2 & b_2 & c_2 \\ a_3 & b_3 & c_3 \end{vmatrix} x = \begin{vmatrix} d_1 & b_1 & c_1 \\ d_2 & b_2 & c_2 \\ d_3 & b_3 & c_3 \end{vmatrix}, \tag{2}$$

with similar expressions for y and z, provided that the first determinant is not zero. Now if the quantities a, b, c are operators and d a function of t, equation (2) is the eliminant of the differential equations (1). It will be seen from this that when two of the variables have been eliminated the reduced equation is the same for x, y and z, and the complementary functions are therefore the same with different values of the arbitrary constants. It will also be seen that the order of the eliminant equation is not necessarily the sum of the orders of the original equations. Thus if each a is a second order operator and each b and c a first order operator the order of the eliminant will be that of abc, that is, the eliminant will be the fourth order, and the number of arbitrary constants in the solution will be four.

Example 3. *Solve the equations* $\ddot{x}+5x-4y = 2t^2$, $x+\ddot{y} = 4t$, *given that* $x = 1$, $y = 2$, *and* $\dot{x} = \dot{y} = 0$ when $t = 0$.

We have

$$(D^2+5)x-4y = 2t^2,$$
$$x+D^2y = 4t.$$

Operating on the first equation with D^2, multiplying the second equation by 4 and adding, we have

$$(D^4+5D^2+4)x = 4+16t.$$

Alternatively, we can write

$$\begin{vmatrix} D^2+5 & -4 \\ 1 & D^2 \end{vmatrix} x = \begin{vmatrix} 2t^2 & -4 \\ 4t & D^2 \end{vmatrix} = \begin{vmatrix} 4 & 2t^2 \\ -D^2 & 4t \end{vmatrix},$$

leading to the same eliminant. The auxiliary equation has roots $\pm i$ and $\pm 2i$ and a particular integral is $1+4t$, so that

$$x = A\cos t+B\sin t+E\cos 2t+F\sin 2t+1+4t.$$

Since $x = 1$ when $t = 0$, $A+E = 0$, and since $\dot{x} = 0$ when $t = 0$, $B+2F+4=0$, so that

$$x = A(\cos t-\cos 2t)+(B\sin t-2\sin 2t-\tfrac{1}{2}B\sin 2t)+1+4t.$$

From the first equation we have $4y = (D^2+5)x-2t^2$, and hence

$$4y = A(4\cos t-\cos 2t)+B(4\sin t-\tfrac{1}{2}\sin 2t)-2\sin 2t+5+20t-2t^2.$$

Since

$$y = 2 \text{ when } t = 0, \quad 3A+5 = 8 \text{ and } A = 1.$$

Since

$$\dot{y} = 0 \text{ when } t = 0, \quad 3B-4+20 = 0 \text{ and } B = -16/3.$$

Hence

$$x = \cos t-\cos 2t+(-16\sin t+2\sin 2t)/3+1+4t,$$
$$y = \cos t-\tfrac{1}{4}\cos 2t+(-16\sin t+\tfrac{1}{2}\sin 2t)/3+(5+20t-2t^2)/4.$$

Exercises 4(*b*)

Solve the simultaneous differential equations:

1. $\dot{x}+2y = 6t$, $\ddot{x}-\dot{y} = 0$, if $x = y = 0$ when $t = 0$.
2. $\ddot{x}+\omega\dot{y} = \omega^2 c$, $\omega\dot{x}-\ddot{y} = 0$, if $x = y = \dot{x} = \dot{y} = 0$ when $t = 0$.
3. $\ddot{x}-4\dot{x}+4x-y = 0$, $\ddot{y}+4\dot{y}+4y-25x = 0$, if $x = 8$, $\dot{x} = y = \dot{y} = 0$ when $t = 0$,
4. $2\ddot{x}+x-y = 3\cos 3t$, $3\dot{y}-x+2y = 0$. (L.U.)

5. $\ddot{x}+x+y = 0$, $4\ddot{y}-x = 0$, given that when $t = 0$, $x = 2a$, $y = -a$, $\dot{x} = 2b$, $\dot{y} = -b$, and show that the solution is then purely periodic. (L.U.)

6. $\ddot{x}+15x+3y+30 = 0$, $\ddot{y}+2x+10y+4 = 0$. (L.U.)

7. $\ddot{x}+x = \dot{y}$, $4\dot{x}+2x = \dot{y}+2y$. If (x, y) are the cartesian coordinates of a point in a plane, and if, when $t = 0$, $x = 0$, $y = 1$, $\dot{x} = 2$, prove that the point lies on the parabola $(5x-2y)^2 = 4(y-2x)$. (L.U.)

8. $6\ddot{x}-3\dot{x}+6x+\dot{y} = t$, $\ddot{x}+8x+y = 0$; if when $t = 0$, $x = \frac{11}{36}$, $y = -\frac{22}{9}$, $\dot{x} = \frac{1}{6}$, show that the locus of the point (x, y) is a straight line through the origin. (L.U.)

9. $5\ddot{x}+96(x-y) = 0$, $2\ddot{y}-96(x-y) = 0$, given that when $t = 0$, $x = y = \dot{y} = 0$ and $\dot{x} = 4$.

10. $\ddot{x}+\ddot{y} = 4a\omega^2$, $\ddot{x}+\omega^2x-\ddot{y}-\omega^2y = -4a\omega^2$, given that when $t = 0$, $x = y = \dot{x} = \dot{y} = 0$.

4.4 Applications of simultaneous equations

Problems which involve the formation and solution of simultaneous differential equations are considered in the following examples and exercises. The equations involved can always be solved by the methods of the previous sections, but the work is often considerably shortened by solving for combinations of the variables instead of for the individual variables. This is done in Example 4 below.

Example 4. *Particles of masses m_1 and m_2 are connected by a light spring of stiffness s and rest on a smooth horizontal table with the spring just unstretched. The mass m_1 is projected away from m_2 with velocity v. Form and solve the equations of motion and find the distance each particle will have moved when the spring first regains its natural length.*

Let x and y be the displacements of m_1 and m_2 respectively from their initial positions. Then the extension of the spring is $x-y$ and the tension in the spring is $sg(x-y)$. The equations of motion are

$$m_1\ddot{x} = -sg(x-y), \tag{1}$$

and

$$m_2\ddot{y} = sg(x-y), \tag{2}$$

with the initial conditions that $x = y = \dot{y} = 0$ and $\dot{x} = v$ when $t = 0$.

Adding equations (1) and (2) we have

$$m_1\ddot{x}+m_2\ddot{y} = 0,$$

and hence, integrating twice,

$$m_1x+m_2y = A+Bt.$$

The initial conditions give $A = 0$ and $B = m_1v$, so that

$$m_1x+m_2y = m_1vt. \tag{3}$$

Subtracting equation (2) multiplied by m_1 from equation (1) multiplied by m_2 we have

$$m_1m_2(\ddot{x}-\ddot{y}) = -sg(m_1+m_2)(x-y).$$

This is an equation of simple harmonic motion for $x-y$ and has solution

$$x-y = A\cos\omega t+B\sin\omega t,$$

where $\omega^2 = sg(m_1+m_2)/m_1m_2$. Using the initial conditions we find $A = 0$ and $B = v/\omega$, so that

$$x-y = (v/\omega)\sin\omega t. \tag{4}$$

From (3) and (4)

$$(m_1+m_2)x = m_1vt+(m_2v/\omega)\sin\omega t,$$

$$(m_1+m_2)y = m_1vt-(m_1v/\omega)\sin\omega t.$$

The spring regains its natural length when $x-y = 0$, that is when $\omega t = \pi$, and for this value of t

$$x = y = \frac{m_1v\pi}{(m_1+m_2)\omega} = v\pi\left\{\frac{m_1^3m_2}{sg(m_1+m_2)^3}\right\}^{1/2}.$$

Example 5. *An electromotive force* $E\cos\omega t$ *is applied through a resistance S to two branches each containing a resistance R, one a condenser of capacitance C and the other a coil of inductance L. If* $LC\omega^2 = 1$, *find the total current supplied in the steady state.*

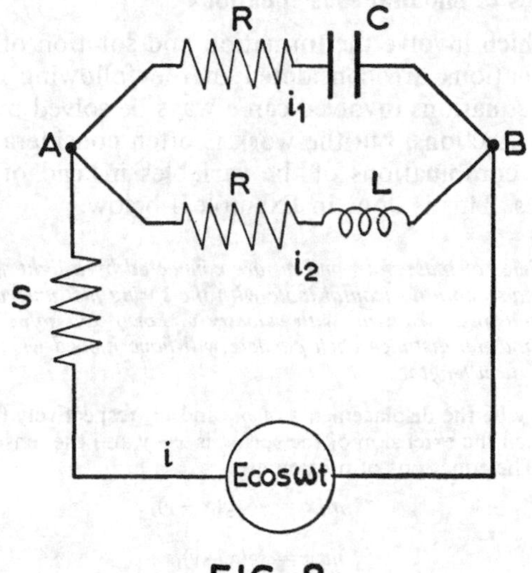

FIG. 8

Let i be the total current, i_1 and i_2 the currents in the two branches, (Fig. 8), so that $i = i_1+i_2$. The voltage drop between A and B is the same in each branch so that $L\dfrac{di_2}{dt}+Ri_2 = Ri_1+\dfrac{q_1}{C}$, where $i_1 = \dfrac{dq_1}{dt}$.

We have therefore

$$Si+Ri_1+\frac{q_1}{C} = E\cos\omega t,$$

$$Si+Ri_2+L\frac{di_2}{dt} = E\cos\omega t.$$

Differentiating the first equation and replacing i_1 by $i-i_2$,

$$(S+R)\frac{di}{dt}-R\frac{di_2}{dt}+\frac{i-i_2}{C} = -E\omega \sin \omega t.$$

The simultaneous equations for i and i_2 are therefore

$$\{(CS+CR)D+1\}i-(CRD+1)i_2 = -CE\omega \sin \omega t, \tag{5}$$

$$Si+(LD+R)i_2 = E\cos \omega t. \tag{6}$$

Eliminating i_2,

$$[(LD+R)\{(CS+CR)D+1\}+S(CRD+1)]i$$
$$= (CRD+1)E\cos\omega t-(LD+R)CE\omega \sin \omega t,$$

that is, since $LC\omega^2 = 1$,

$$\{LC(S+R)D^2+(CR^2+2SCR+L^2C\omega^2)D+(S+R)\}i = -2ECR\omega \sin \omega t.$$

The total current in the steady state is the particular integral of this equation, that is

$$i = \frac{-2ECR\omega}{-LC(S+R)\omega^2+(CR^2+2SCR+L^2C\omega^2)D+(S+R)}.\sin \omega t$$

$$= \frac{-2ER\omega}{(R^2+2SR+L^2\omega^2)D}.\sin \omega t$$

$$= \frac{2ER\cos \omega t}{R^2+2SR+L^2\omega^2}.$$

Example 6. *The primary circuit of a transformer consists of resistance R_1, inductance L_1 and a constant applied voltage E. The secondary circuit contains resistance R_2 and inductance L_2 only. Show that the current in the primary circuit is*

$$A\,e^{\alpha t}+B\,e^{\beta t}+E/R_1,$$

where A and B are constants and α and β are the roots of the equation

$$(L_1L_2-M^2)x^2+(L_1R_2+L_2R_1)x+R_1R_2 = 0,$$

M being the mutual inductance of the circuits.

This is an example of coupled circuits in which there is a mutual inductance M causing a voltage drop $M\,di/dt$ in each circuit, where i is the current in the other. The differential equation for each circuit thus involves the current in the other, and to obtain expressions for the separate currents a pair of simultaneous differential equations has to be solved. Let i_1 and i_2 be the currents in the primary and secondary circuits respectively. Then, for the primary circuit,

$$L_1\frac{di_1}{dt}+R_1i_1+M\frac{di_2}{dt} = E,$$

while for the secondary circuit,

$$L_2\frac{di_2}{dt}+R_2i_2+M\frac{di_1}{dt} = 0.$$

Eliminating i_2,

$$\{(L_2D+R_2)(L_1D+R_1)-M^2D^2\}i_1 = R_2E,$$

that is

$$\{(L_1L_2-M^2)D^2+(L_1R_2+L_2R_1)D+R_1R_2\}i_1 = R_2E.$$

Hence, a particular integral is E/R_1 and

$$i_1 = E/R_1+A\,e^{\alpha t}+B\,e^{\beta t},$$

where A, B are constants and α, β are the roo ts of the given auxiliary equation.

Exercises 4(*c*)

1. The motion of a particle, whose coordinates in a plane are (x, y), is given by the equations $\ddot{x} = a-n\dot{y}$, $\ddot{y} = -a+n\dot{x}$, where a and n are constants. Solve these equations and show that, if the particle starts from rest at the origin, its path is given by the equations
$$n^2(x-y) = 2a(1-\cos nt),$$
$$n^2(x+y) = 2a(nt-\sin nt). \qquad \text{(L.U.)}$$

2. Two weights, each of mass m, are connected by a light spring which exerts a force s for each unit length of extension. They are placed on a rough horizontal plane, the coefficient of friction being μ. One of the weights is projected along a line directly away from the other and at the instant when the second weight begins to move the first is travelling with velocity v. Show that during the subsequent motion the tension in the spring reaches a maximum value $mg(\mu^2+sv^2/2mg^2)^{1/2}$. (C.U.)

3. Two particles, A and B, of equal mass m, are attached to the ends of a light spring which exerts a tension of amount s per unit extension. Initially the particles are at rest on a smooth horizontal table with the spring just taut, and a constant force of magnitude sa is then applied to particle B in the direction AB. Obtain the differential equations for the displacements, x and y, of the particles A and B respectively at time t, and show that
$$y+x = \tfrac{1}{4}a\omega^2t^2,\ y-x = \tfrac{1}{2}a(1-\cos \omega t), \text{ where } m\omega^2 = 2s. \text{ (L.U.)}$$

4. A light spring has particles of masses m_1 and m_2 fixed to its ends and rests on a smooth horizontal table. The mass m_2 is placed against a fixed stop and the mass m_1 is moved towards m_2 until the spring is compressed by an amount h and then released. Show that the mass m_2 comes to rest at distances $2\pi nh\left\{\dfrac{m_1{}^2m_2}{(m_1+m_2)^3}\right\}^{1/2}$ from the stop, where $n=1, 2, 3, \ldots$, etc., and that the spring is then unstretched.

5. An electric circuit consists of two branches in parallel containing respectively a resistance R and an inductance L, and a resistance R and a capacitance C. Construct the differential equations for the currents in the branches due to an applied e.m.f. and solve them for a steady e.m.f. E suddenly applied at $t = 0$. Show that if $CR^2 = L$ the circuit behaves, as far as total current is concerned, as a pure resistance of magnitude R. (L.U.)

6. An electric cable has resistance R per unit length and the resistance per unit length of the insulation is n^2R, where n and R are constants. The voltage v and the current i at a distance x from one end satisfy the equations
$$\frac{dv}{dx}+Ri = 0 \quad\text{and}\quad \frac{di}{dx}+\frac{v}{n^2R} = 0.$$
If current is supplied at one end A of a cable, of length l, at a voltage V

and the cable is insulated at the other end, show that the current entering at A is

$$\frac{V}{nR}\tanh\frac{l}{n}. \qquad \text{(L.U.)}$$

7. Two points, A, B are joined by a wire of resistance R without self-induction; B is joined to a third point C by two wires each of resistance R, of which one is without self-induction and the other has a coefficient of self-induction L. If the ends A, C are kept at a potential difference $E\cos\omega t$, and if there are no mutual inductances, prove that the current in AB is

$$\frac{E}{R}\left\{\frac{4R^2+L^2\omega^2}{9R^2+4L^2\omega^2}\right\}^{1/2}\cos(\omega t-a) \quad \text{where } \tan a=\frac{RL\omega}{6R^2+2L^2\omega^2}.$$

Find also the difference of potential at B and C. (L.U.)

8. An alternating e.m.f. of amplitude E and frequency $\omega/2\pi$ is supplied to a coil of inductance L and resistance R. Write down the differential equation for the current in the coil and solve it, indicating the transient term. If the coil is shunted by a condenser of capacitance C and resistance S show that the circuit can be replaced (as far as permanent current is concerned) by a non-inductive resistance provided that

$$CR^2-L=\omega^2CL(CS^2-L). \qquad \text{(L.U.)}$$

9. Show that a combination of an inductance L and a resistance R in parallel with a resistance R and a capacitance C is equivalent to a simple resistance R for all applied e.m.f's if $R^2C=L$.

10. Two coupled circuits have resistances R_1, R_2, self-inductances L_1, L_2, e.m.f's e_1, e_2 and mutual inductance M. If $R_1=R_2=4$ ohms, $L_1=L_2=0{\cdot}1$ henry, $M=0{\cdot}6$ henry, $e_1=8$ volts and $e_2=0$, find expressions for the currents at time t, given that they are initially zero.

11. In a cyclic chemical reaction the concentrations x, y, z of three substances satisfy the simultaneous equations $\dot{x}=k_2y-k_1x$, $\dot{y}=k_4z-k_3y$, $x+y+z=a$, where a is constant. Show that if α and β are the roots of the equation $m^2-(k_1+k_3+k_4)m+k_1k_3+k_1k_4+k_2k_4=0$,

$$x=a(k_2k_4)/(k_2k_4+k_1k_3+k_1k_4)+Ae^{-\alpha t}+Be^{-\beta t},$$

$$y=a(k_1k_4)/(k_2k_4+k_1k_3+k_1k_4)+A(k_1-\alpha)e^{-\alpha t}/k_2+B(k_1-\beta)e^{-\beta t}/k_2.$$

4.5 Small oscillations and normal modes

Oscillations which involve small changes in coordinates $x, y, \theta, \ldots$, lead to dynamical equations in which linear combinations of $\ddot{x}, \ddot{y}, \ddot{\theta}, \ldots$, are equated to linear combinations of $x, y, \theta, \ldots$, there being as many equations as there are variables. As an example, consider the solution of the simultaneous equations

$$4\ddot{x}+\ddot{y}+21x+14y=0, \qquad (1)$$

$$2\ddot{x}-3\ddot{y}+3x-22y=0. \qquad (2)$$

Addition and subtraction of these equations leads to two new equations

$$6\ddot{x}-2\ddot{y}+24x-8y = 0,$$

$$2\ddot{x}+4\ddot{y}+18x+36y = 0,$$

and hence

$$D^2(3x-y)+4(3x-y) = 0,$$

$$D^2(x+2y)+9(x+2y) = 0.$$

These are equations of simple harmonic motion for the quantities $3x-y$ and $x+2y$ with solutions

$$(3x-y) = A\cos(2t+\varepsilon), \tag{3}$$

$$(x+2y) = B\cos(3t+\eta). \tag{4}$$

Hence

$$7x = 2A\cos(2t+\varepsilon)+B\cos(3t+\eta),$$

$$7y = -A\cos(2t+\varepsilon)+3B\cos(3t+\eta),$$

where A, B, ε and η are arbitrary constants to be determined by the initial conditions. Thus x and y are combinations of harmonic oscillations but $(3x-y)$ and $(x+2y)$ are simple harmonic oscillations of periods π and $2\pi/3$ respectively. The combinations of the variables which correspond to a single harmonic oscillation are called the *normal modes* of vibration.

(i) *Periods of normal modes.*

The periods of the normal modes are easily found from the equations of motion by assuming that $x = A\cos\omega t$, $y = B\cos\omega t, \ldots$, and hence that $\ddot{x} = -\omega^2 x$, $\ddot{y} = -\omega^2 y, \ldots$.

Thus in equations (1) and (2) substituting for $\ddot{x}$ and $\ddot{y}$ we have

$$x(21-4\omega^2)+y(14-\omega^2) = 0,$$

$$x(3-2\omega^2)+y(-22+3\omega^2) = 0.$$

Hence, eliminating x and y,

$$\begin{vmatrix} 21-4\omega^2 & 14-\omega^2 \\ 3-2\omega^2 & -22+3\omega^2 \end{vmatrix} = 0,$$

that is,

$$-14\omega^4+182\omega^2-504 = 0,$$

or,

$$\omega^4-13\omega^2+36 = 0.$$

Since ω is positive, the appropriate roots are $\omega_1 = 2$, $\omega_2 = 3$ and the periods of the normal modes are $2\pi/\omega_1$ and $2\pi/\omega_2$. The solutions of the differential equations are then

$$x = A\cos(\omega_1 t+\varepsilon)+B\cos(\omega_2 t+\eta),$$
$$y = E\cos(\omega_1 t+\varepsilon)+F\cos(\omega_2 t+\eta).$$

Substitution for x and y in one of the original equations gives E and F in terms of A and B and hence the solutions in the forms (3) and (4) may be derived. The method is easily generalized for three or more coordinates.

(ii) *Method of multipliers.*

When there are only two coordinates the normal modes and their periods can be found directly from the equations of motion by a method of multipliers.

Applying this method to equations (1) and (2) we multiply equation (1) by λ and equation (2) by μ and add, λ and μ being constants as yet undetermined. We have

$$(4\lambda+2\mu)\ddot{x}+(\lambda-3\mu)\ddot{y}+(21\lambda+3\mu)x+(14\lambda-22\mu)y = 0. \tag{5}$$

This will be the differential equation for one of the normal modes if the ratio of the coefficients of $\ddot{x}$ and $\ddot{y}$ is the same as the ratio of the coefficients of x and y, that is

$$\frac{4\lambda+2\mu}{\lambda-3\mu} = \frac{21\lambda+3\mu}{14\lambda-22\mu}.$$

Then

$$(4\lambda+2\mu)(14\lambda-22\mu)-(21\lambda+3\mu)(\lambda-3\mu) = 0,$$

giving $35\lambda^2-35\mu^2 = 0$. Thus $\lambda = \pm\mu$, and taking $\lambda = 1$, $\mu = 1$, equation (5) becomes

$$6\ddot{x}-2\ddot{y}+4(6x-2y) = 0.$$

Taking $\lambda = 1$, $\mu = -1$, we have

$$2\ddot{x}+4\ddot{y}+9(2x+4y) = 0.$$

From these equations the normal modes and their periods are seen at once.

Example 7. *Find the complete solution of the equations* $35\ddot{x}-6\ddot{y}+13x = 0$, $-6\ddot{x}+30\ddot{y}+13y = 0$, *given that when* $t = 0$, $x = y = a$, $\dot{x} = \dot{y} = 0$.

Multiplying the first equation by λ, the second by μ and adding,

$$(35\lambda-6\mu)\ddot{x}+(-6\lambda+30\mu)\ddot{y}+13\lambda x+13\mu y = 0.$$

This represents a normal mode if

$$\frac{35\lambda-6\mu}{-6\lambda+30\mu} = \frac{\lambda}{\mu},$$

that is,

$$6\lambda^2+5\lambda\mu-6\mu^2 = 0,$$

giving

$$(2\lambda+3\mu)(3\lambda-2\mu) = 0.$$

Taking $\lambda = 2$ and $\mu = 3$,

$$52\ddot{x}+78\ddot{y}+13(2x+3y) = 0,$$

that is

$$D^2(2x+3y)+\tfrac{1}{2}(2x+3y) = 0.$$

Taking $\lambda = 3$ and $\mu = -2$, we have

$$117\ddot{x}-78\ddot{y}+13(3x-2y) = 0,$$

or

$$D^2(3x-2y)+\tfrac{1}{3}(3x-2y) = 0.$$

The normal modes are therefore $2x+3y$ and $3x-2y$ with periods $2\pi\sqrt{(2)}$ and $2\pi\sqrt{(3)}$ respectively and we have

$$2x+3y = A\cos(t/\sqrt{2})+B\sin(t/\sqrt{2}),$$
$$3x-2y = E\cos(t/\sqrt{3})+F\sin(t/\sqrt{3}).$$

From the initial conditions it is seen that $A = 5a$, $E = a$ and $B = F = 0$, and hence

$$13x = 10a\cos(t/\sqrt{2})+3a\cos(t/\sqrt{3}),$$
$$13y = 15a\cos(t/\sqrt{2})-2a\cos(t/\sqrt{3}).$$

Example 8. *Find the periods of the normal modes of the vibration given by the equations*

$$29\ddot{x}+6\ddot{y}+6\ddot{z}+15x = 0, \quad 6\ddot{x}+4\ddot{y}+3y = 0, \quad 6\ddot{x}+4\ddot{z}+3z = 0.$$

Putting $\ddot{x} = -\omega^2 x$, $\ddot{y} = -\omega^2 y$, $\ddot{z} = -\omega^2 z$ in the equations and eliminating x, y and z,

$$\begin{vmatrix} 15-29\omega^2 & -6\omega^2 & -6\omega^2 \\ -6\omega^2 & 3-4\omega^2 & 0 \\ -6\omega^2 & 0 & 3-4\omega^2 \end{vmatrix} = 0,$$

that is,

$$(15-29\omega^2)(3-4\omega^2)^2-72\omega^4(3-4\omega^2) = 0,$$

or

$$(3-4\omega^2)(\omega^2-3)(44\omega^2-15) = 0.$$

Thus the periods of the normal modes are $4\pi/\sqrt{(3)}$, $2\pi/\sqrt{(3)}$ and $2\pi(44/15)^{1/2}$. By addition and subtraction of the given equations, the corresponding normal modes are found to be $y-z$, $5x-2y-2z$ and $4x+y+z$.

Example 9. *A uniform light shaft of torsional rigidity μ, free to rotate in bearings, carries three flywheels whose moments of inertia are I, $2I$ and I. The flywheels are equally spaced at distances l apart, the wheel of largest moment being central. Calculate the periods of free oscillation of the wheels about their mean position.*

Let θ, ϕ, ψ be the angular deviations of the flywheels from their mean positions; then the torque in the portions of the shaft between the wheels will be $\mu(\theta-\phi)/l$ and $\mu(\phi-\psi)/l$, and the equations of motion are

$$I\ddot{\theta} = -\frac{\mu}{l}(\theta-\phi),$$

$$2I\ddot{\phi} = -\frac{\mu}{l}(\phi-\theta)-\frac{\mu}{l}(\phi-\psi),$$

$$I\ddot{\psi} = -\frac{\mu}{l}(\psi-\phi).$$

Writing $\ddot{\theta} = -\omega^2\theta$, $\ddot{\phi} = -\omega^2\phi$, $\ddot{\psi} = -\omega^2\psi$ in these equations,

$$(lI\omega^2-\mu)\theta+\mu\phi = 0,$$
$$\mu\theta+2(lI\omega^2-\mu)\phi+\mu\psi = 0,$$
$$\mu\phi+(lI\omega^2-\mu)\psi = 0.$$

We now eliminate θ, ϕ, ψ and, writing $\lambda = lI\omega^2-\mu$, we have

$$\begin{vmatrix} \lambda & \mu & 0 \\ \mu & 2\lambda & \mu \\ 0 & \mu & \lambda \end{vmatrix} = 0,$$

that is

$$2\lambda^3-2\lambda\mu^2 = 0.$$

Thus $\lambda = 0$ or $\lambda = \pm\mu$, and $lI\omega^2 = 0$, μ or 2μ. There are therefore two periods of oscillation $2\pi(lI/\mu)^{1/2}$ and $2\pi(lI/2\mu)^{1/2}$. The absence of a third period is accounted for by the fact that $I\ddot{\theta}+2I\ddot{\phi}+I\ddot{\psi} = 0$ throughout the motion, showing that there are only two independent coordinates. The normal modes are easily seen to be $\theta+\psi$ and $\theta-2\phi+\psi$.

Exercises 4(*d*)

Solve the equations:

1. $2\ddot{x}+5x-3y = 0$, $2\ddot{y}-3x+5y = 0$, given that $x = y = \dot{x} = 1$ and $\dot{y} = 0$ when $t = 0$.
2. $5\ddot{x}+17x-6y = 0$, $5\ddot{y}-6x+8y = 0$, given that $x = y = 2$ and $\dot{x} = \dot{y} = 0$ when $t = 0$.
3. $5\ddot{x}-\ddot{y}+30x-19y = 0$, $4\ddot{x}+7\ddot{y}+6x+43y = 0$, given that $x = y = 1$, $\dot{x} = \dot{y} = 0$ when $t = 0$.
4. Two particles of equal mass m and distance a apart are attached to a taut string at equal distances ka from the fixed end points. Obtain the simultaneous differential equations for small transverse displacements x, y of the particles when the tension in the string has the constant value $kman^2$. Show that if the particles start from rest at $t = 0$ with $x = b$, $y = 0$ then x and y can be expressed in the forms $C(\cos nt+\cos \lambda nt)$, $C(\cos nt-\cos \lambda nt)$ respectively, and evaluate the constants λ, C. (L.U.)
5. A light string of length $3a$ is stretched between two fixed points and masses $8m$ and $3m$ are attached to the points of trisection. Gravity is neglected and the tension in the string is T. If both masses are drawn aside a small distance c and released show that the displacements of the masses at time t are $-(c/14)\cos 3\omega t+(15c/14)\cos\sqrt{2}\omega t$, and $(4c/14)\cos 3\omega t+(10c/14)\cos\sqrt{2}\omega t$, where $\omega^2 = T/12ma$.
6. A double pendulum is performing oscillations under gravity. The angles θ and ϕ made with the vertical at time t by the two parts of the pendulum satisfy the equations $5a\ddot{\theta}+12a\ddot{\phi}+6g\theta = 0$, $5a\ddot{\theta}+16a\ddot{\phi}+6g\phi = 0$. Find θ and ϕ in terms of t given that $\theta = 7a/4$, $\phi = a$, $\dot{\theta} = 0 = \dot{\phi}$ when $t = 0$. (L.U.)
7. A uniform rod AB of mass $2m$ and length $2a$ is free to turn about A. A second rod BC of mass $3m$ and length $2a$ is freely hinged to the first at B

and the system makes small oscillations in a vertical plane. Show that the equations of motion give $44a\ddot{\theta}+18a\ddot{\phi}+24g\theta = 0$, $6a\ddot{\theta}+4a\ddot{\phi}+3g\phi=0$, where θ and ϕ are the inclinations of the rods to the vertical. Find the normal modes and their periods.

8. A light inelastic string AB of length $7a$ is attached to a fixed point at A. A particle of mass m is attached at B and there is a similar particle at a point C of the string where $AC = 4a$. In small oscillations in a vertical plane the portions AC and CB of the string make angles θ and ϕ with the vertical. Show that $8a\ddot{\theta}+3a\ddot{\phi}+2g\theta = 0$, $4a\ddot{\theta}+3a\ddot{\phi}+g\phi = 0$ and that normal modes are $2\theta+\phi$ and $4\theta-3\phi$. Find the periods of these modes.

9. The equations of motion of a system oscillating with three degrees of freedom are

$$2\ddot{x}+3x-y-z = 0, \quad \ddot{y}+3x+7y+3z = 0, \quad 2\ddot{z}-7x-11y-3z = 0.$$

Find the normal modes and their periods.

10. Four flywheels, each of moment of inertia I, are equally spaced at distance l apart on a uniform shaft of torsional rigidity μ. Show that the squares of the principal periods of oscillation of the wheels about their mean positions are in the ratio $1:2-\sqrt{2}:2+\sqrt{2}$.

11. Two particles, each of mass m, rest on a smooth horizontal table connected by a spring of stiffness s. Each particle is joined to a fixed point by a spring of stiffness s_1 and all three springs are stretched. Show that if x and y are the small displacements of the particles along the line of the springs, the normal modes of vibration are $x+y$ and $x-y$ with periods $2\pi(m/gs_1)^{1/2}$ and $2\pi\{m/g(2s+s_1)\}^{1/2}$.

12. Two particles, each of mass m, rest on a smooth horizontal table connected by a spring of stiffness s and joined to points A and B respectively of the table by springs of stiffness s_1. The particles are initially at rest in the line AB and all the springs are taut. If the table is given a displacement z along the line AB, and x and y are the displacements of the masses in this direction, show that

$$m\ddot{x}+s(x-y)+s_1(x-z) = 0,$$
$$m\ddot{y}+s(y-x)+s_1(y-z) = 0.$$

Solve these equations when $z = a\cos\omega t$.

4.6 Reducible equations

Most of the general methods of solving differential equations considered so far in this book are applicable to equations of the second and higher orders only when the coefficients appearing in the equation are constants. When the coefficients are variable there are no *general* methods of obtaining a formal solution in closed form. Solutions can, however, be found as infinite series and methods of finding such solutions are given in the next chapter.

In certain cases, however, the order of a differential equation can be reduced by a change of the dependent or independent variable. This has obvious advantages when, for example, a second order equation can be reduced to a first order equation for which methods of obtaining a solution are available. In this section we deal with three specific cases in which the order of an equation can be reduced.

(i) *Dependent variable absent*

If y is explicitly absent from a differential equation, only appearing in dy/dx, d^2y/dx^2, etc., a change of the dependent variable from y to p, where $p = dy/dx$, has the effect of reducing the order of the equation.

Example 10. *Solve the equation* $\dfrac{d^2y}{dx^2}+\dfrac{1}{x}\dfrac{dy}{dx} = e^x(x+2)$.

Writing p for dy/dx, the equation becomes

$$\frac{dp}{dx}+\frac{p}{x} = e^x(x+2).$$

This is a first order linear equation for p and the integrating factor is easily found to be x. Hence

$$\frac{d}{dx}(xp) = e^x(x^2+2x),$$

and integrating

$$xp = e^x x^2+A,$$

or

$$p = e^x x+\frac{A}{x}.$$

Integrating once more,

$$y = e^x(x-1)+A\log_e x+B,$$

where A and B are constants.

(ii) *Independent variable absent*

If x is explicitly absent, only appearing in dy/dx, d^2y/dx^2, etc., the order of the equation can be reduced by writing $p = dy/dx$ and changing the independent variable from x to y. We then have

$$\frac{dy}{dx} = p,$$

$$\frac{d^2y}{dx^2} = \frac{dp}{dx} = \frac{dp}{dy}\frac{dy}{dx} = p\frac{dp}{dy},$$

$$\frac{d^3y}{dx^3} = \frac{d}{dx}\left(p\frac{dp}{dy}\right) = p\frac{d}{dp}\left(p\frac{dp}{dy}\right) = p^2\frac{d^2p}{dy^2}+p\left(\frac{dp}{dy}\right)^2,$$

etc. Thus each derivative is replaced by derivatives of lower order and the order of the equation is reduced.

Example 11. *Solve the equation* $y\dfrac{d^2y}{dx^2}+\left(\dfrac{dy}{dx}\right)^2 = 3y$.

Writing $\dfrac{d^2y}{dx^2} = p\dfrac{dp}{dy}$, the equation can be written

$$yp\frac{dp}{dy}+p^2 = 3y,$$

and hence

$$\frac{dp^2}{dy}+\frac{2p^2}{y} = 6.$$

This is a first order linear equation for p^2 and the integrating factor is y^2. Multiplying by y^2,

$$\frac{d}{dy}(p^2y^2) = 6y^2,$$

giving

$$p^2y^2 = 2y^3+A,$$

and hence

$$p = \frac{dy}{dx} = \pm\left(2y+\frac{A}{y^2}\right)^{1/2}.$$

Thus

$$\int\frac{y\,dy}{(2y^3+A)^{1/2}} = \pm x+B,$$

where A and B are constants.

Second order differential equations in which the independent variable, the time t, is explicitly absent are of frequent occurrence in dynamics. Examples are $\ddot{x} = f(x)$, where the acceleration is a function of the distance, or $\ddot{\theta} = f(\theta)$, where the angular acceleration depends on the angle through which a body has turned. Such equations are easily solved by writing $\ddot{x} = p(dp/dx)$ where $p = \dot{x}$, or $\ddot{\theta} = p(dp/d\theta)$, where $p = \dot{\theta}$, and solving the resulting first order equation for p.

Example 12. *Solve the equation of the simple pendulum* $l\ddot{\theta}+g\sin\theta = 0$.

Writing $p = \dot{\theta}$, $\ddot{\theta} = p\dfrac{dp}{d\theta}$, and we have

$$p\frac{dp}{d\theta}+\frac{g}{l}\sin\theta = 0.$$

This is an equation with separable variables and on integration we have

$$\tfrac{1}{2}p^2-\frac{g}{l}\cos\theta = A.$$

If $\theta = \alpha$ when $\dot{\theta} = 0$, $A = -(g/l)\cos\alpha$, and hence

$$\frac{d\theta}{dt} = p = \left(\frac{2g}{l}\right)^{1/2}(\cos\theta-\cos\alpha)^{1/2},$$

so that

$$\int\frac{d\theta}{(\cos\theta-\cos\alpha)^{1/2}} = \left(\frac{2g}{l}\right)^{1/2}\int dt.$$

If T be the period, the quarter period is the time between $\theta = 0$ and $\theta = \alpha$, and hence

$$\left(\frac{g}{l}\right)^{1/2}\frac{T}{2} = \int_0^\alpha \frac{d\theta}{(\sin^2\frac{1}{2}\alpha - \sin^2\frac{1}{2}\theta)^{1/2}}.$$

Substituting $\sin\frac{1}{2}\theta = k\sin\phi$, where $k = \sin\frac{1}{2}\alpha$, we have

$$\left(\frac{g}{l}\right)^{1/2}\frac{T}{2} = 2\int_0^{\pi/2} \frac{d\phi}{(1-k^2\sin^2\phi)^{1/2}},$$

and hence (see §5.9)

$$T = 2\pi\left(\frac{l}{g}\right)^{1/2}\left\{1+\left(\frac{1}{2}\right)^2 k^2+\left(\frac{1.3}{2.4}\right)^2 k^4+\left(\frac{1.3.5}{2.4.6}\right)^2 k^6+\ldots\right\}.$$

Since $k = \sin\frac{1}{2}\alpha$, and α is small, a first approximation obtained by taking $k = 0$ is $T = 2\pi(l/g)^{1/2}$; a second approximation obtained by taking $k = \frac{1}{2}\alpha$ and ignoring α^4 and higher powers of α is

$$T = 2\pi(l/g)^{1/2}(1+\alpha^2/16).$$

(iii) *Equations of homogeneous form*

Euler's linear equation, considered in §3.9, is homogeneous in the sense that it is a sum of terms of the form $x^r(d^ry/dx^r)$ and if x^m is substituted for y and $m > r$, each term is a multiple of x^m. The substitution $x = e^t$, giving $xy' = \dot{y}$, $x^2y'' = \ddot{y}-\dot{y}$, etc., reduces Euler's equation to a linear equation with constant coefficients.

Certain non-linear equations can be solved or brought to a reducible form by the same substitution if they are homogeneous in the sense that each term in the equation yields the same power of x when x is substituted for y, for $x(dy/dx)$, for $x^2(d^2y/dx^2)$, etc. Thus the equation

$$xy\frac{d^2y}{dx^2}+x\left(\frac{dy}{dx}\right)^2+2y\frac{dy}{dx} = 0, \tag{1}$$

can be rewritten as

$$\frac{1}{x}\cdot y\left(x^2\frac{d^2y}{dx^2}\right)+\frac{1}{x}\cdot\left(x\frac{dy}{dx}\right)^2+\frac{2}{x}\cdot y\left(x\frac{dy}{dx}\right) = 0,$$

and on substitution each term yields x.

The first step in the solution of such equations is to make the substitution $x = e^t$. When this has been done it will be found that the independent variable t is explicitly absent, or, if not, it can be removed by a simple transformation of the dependent variable.

Example 13. *Solve the equation* $xyy''+x(y')^2+2yy' = 0$.

Putting $x = e^t$, $xy' = \dot{y}$, $x^2y'' = \ddot{y}-\dot{y}$, and the equation becomes

$$y(\ddot{y}-\dot{y})+\dot{y}^2+2y\dot{y} = 0.$$

Since t is explicitly absent we write $p = \dot{y}$ and $\ddot{y} = p\dfrac{dp}{dy}$, giving

$$yp\frac{dp}{dy}+p^2+py = 0.$$

This first order homogeneous equation is solved by writing $p = zy$, giving

$$z+y\frac{dz}{dy}+z+1 = 0.$$

Separating the variables,

$$\frac{dz}{2z+1}+\frac{dy}{y} = 0,$$

and hence

$$y^2(2z+1) = A,$$

that is

$$y(2p+y) = A.$$

Hence

$$p = \frac{dy}{dt} = \frac{1}{2}\left(\frac{A}{y}-y\right),$$

giving

$$\frac{2y\,dy}{A-y^2} = dt.$$

On integration this gives

$$\log_e(A-y^2)+t = B,$$

that is, since $x = e^t$,

$$x(A-y^2) = C,$$

where A and C are constants.

Example 14. *Solve the equation* $xyy''-x(y')^2+yy'-xy'+y = 0.$

The given equation can be written

$$y(x^2y'')-(xy')^2+y(xy')-x(xy')+xy = 0,$$

and in this form the equation is seen to be homogeneous in the sense used in this section. Putting $x = e^t$ we have

$$y(\ddot{y}-\dot{y})-\dot{y}^2+y\dot{y}-\dot{y}e^t+ye^t = 0.$$

Now let $y = ze^t$, then $\dot{y} = e^t(\dot{z}+z)$, $\ddot{y} = e^t(\ddot{z}+2\dot{z}+z)$ giving

$$z(\ddot{z}+\dot{z})-(\dot{z}+z)^2+z(\dot{z}+z)-(\dot{z}+z)+z = 0,$$

that is

$$z\ddot{z}-\dot{z}^2-\dot{z} = 0.$$

Putting $\dot{z} = p$, $\ddot{z} = p(dp/dz)$, this can be written

$$zp\frac{dp}{dz}-p^2-p = 0,$$

or

$$\frac{dp}{p+1}-\frac{dz}{z} = 0.$$

Integrating,

$$\log_e(p+1)-\log_e z = \log_e C,$$

that is

$$p+1 = Cz.$$

Then

$$\frac{dz}{dt} = p = Cz-1,$$

giving

$$\frac{dz}{Cz-1} = dt,$$

and integrating,

$$\log_e(Cz-1) = Ct+B.$$

Thus

$$Cz-1 = Ae^{Ct},$$

so that

$$z = \frac{Ax^C+1}{C},$$

and

$$y = xz = Ex^{C+1}+x/C,$$

where E and C are arbitrary constants.

Exercises 4(*e*)

Solve the differential equations:

1. $2xy''+y' = x$.
2. $xy'''+y'' = 1$.
3. $1+(y')^2 = (1+x^2)y''$. (L.U.)
4. $1+(y')^2 = 2xy'y''$, if $y = y' = 0$ when $x = 1$. (L.U.)
5. $y'' \sin x+y' \cos x = 0$.
6. $(x^2-1)y''+xy' = -1$.
7. $yy''+(y')^2 = 2y^2$, if $y = 1$ and $y' = \sqrt{2}$ when $x = 0$. (L.U.)
8. $2y''+(y')^2+y = 0$, if $y = 1$ and $y' = 0$ when $x = 0$. (L.U.)
9. $2y''+(y')^2-4y = 0$, if $y = 1$ and $y' = 0$ when $x = 0$. (L.U.)
10. $(1+y^2)y'' = y\{1+(y')^2\}$, if $y = 1$ and $y' = 0$ when $x = 0$. (L.U.)
11. $2yy''-3(y')^2-4y^2 = 0$.
12. $yy'' = 1+(y')^2$, if $y = 1$ and $y' = 0$ when $x = 0$.
13. $yy''+(y')^2-y' = 0$.
14. $2yy''+(y')^2 = 1$.
15. $y'' = 4 \sec^2 y \tan y$, if $y = 0$ and $y' = 2$ when $x = 0$.
16. $4yy''+2(y')^2 = 1$, if $y = 1$ and $y' = 0$ when $x = 0$.
17. $x^2y^2y''+x^2y+y^3 = x^3y'+xy^2y'$.
18. $y''+y = 2y^3$, if $y = 1$ and $y' = 0$ when $x = 0$.
19. $xyy''+x(y')^2 = nyy'$.
20. $2xy^2y''-x^2(y')^3-2xy(y')^2+3y^2y' = 0$.
21. The equation of motion of a particle falling freely under the earth's attraction is $\ddot{x} = -gR^2/x^2$, where x is the distance from the centre of the earth and R is the earth's radius. Solve this equation given that $x = a$ and $\dot{x} = 0$ when $t = 0$.
22. The equation of motion of a particle moving in a straight line is $\ddot{x} = -\mu^2(x+a^4/x^3)$. Solve this equation given that at time $t = 0$, $x = a$ and $\dot{x} = 0$ and prove that the motion is periodic with period π/μ.
23. A particle of unit mass moves in a straight line under the action of a force $\omega^2 x$ towards the origin in a medium in which the resistance is $k\dot{x}^2$. Show that if the particle is moving towards the origin and x is positive,

the equation of motion is $\ddot{x}-k\dot{x}^2+\omega^2x = 0$. If initially $x = a$ and $\dot{x} = 0$, show that the particle reaches the origin with velocity

$$(\omega\sqrt{2}/2k)\{1-(1+2ka)e^{-2ka}\}^{1/2}.$$

24. A uniform chain is coiled up on a table; one end passes over a smooth peg at a height a above the table and initially a length $2a$ hangs freely on the other side of the peg. Show that the equation of motion of the chain is $(x+a)\ddot{x}+\dot{x}^2 = (x-a)g$. Hence, show that the chain moves with constant acceleration $g/3$.

25. Prove that if (r, θ) are the polar coordinates with respect to the sun of a planet moving under an acceleration μ/r^2 towards the sun, its equation of motion is $\ddot{r}-h^2/r^3 = -\mu/r^2$, where $h = r^2\dot{\theta}$ is constant. If when $r = a$, $\dot{r}^2 = (\mu a-h^2)/a^2$, derive the equation

$$(dr/d\theta)^2 = -r^2+(2\mu/h^2)r^3-(\mu/h^2a)r^4,$$

and show that the equation of the orbit is $l/r = 1+e\cos(\theta+\alpha)$, where $l = h^2/\mu$, $e^2 = 1-h^2/\mu a$, and α is a constant.

CHAPTER 5

SERIES SOLUTIONS AND THE HYPERGEOMETRIC EQUATION

5.1 Introduction

It has usually been possible to express the solution of the differential equations so far considered in terms of elementary functions such as exponentials, sines, cosines, etc. However, in many problems of science and engineering, the differential equations are of such a form that this is not possible and the object of the present chapter is to indicate methods of obtaining solutions as convergent infinite series from which numerical values can, if required, be computed. Some differential equations occur so frequently in physical applications that it has been found convenient to consider their solutions as defining new functions and it is often worth while investigating the chief properties of the functions so defined. As an example of this procedure, a very general second order differential equation and the function defined by its solution form the subject matter of the second part of the chapter.

The basis of the method of obtaining a series solution can be seen by considering, for example, the second order linear differential equation

$$\frac{d^2y}{dx^2}+y = 0.$$

The solution is assumed to be given by $y = a_0+a_1x+a_2x^2+\ldots$, the series being assumed convergent and differentiable term by term for sufficiently small values of x. With these assumptions

$$\frac{dy}{dx} = a_1+2a_2x+3a_3x^2+\ldots,$$

and

$$\frac{d^2y}{dx^2} = 2a_2+2.3a_3x+3.4a_4x^2+\ldots.$$

Substituting the series for y and d^2y/dx^2 in the given differential equation and collecting together like powers of x yields the identity

$$(2a_2+a_0)+(2.3a_3+a_1)x+(3.4a_4+a_2)x^2+\ldots \equiv 0.$$

Since if a power series is identically zero all its coefficients are zero, equating to zero the term independent of x and the coefficients of $x, x^2, \ldots$, gives

$$2a_2+a_0 = 0, \quad 4.5a_5+a_3 = 0,$$
$$2.3a_3+a_1 = 0, \quad 5.6a_6+a_4 = 0,$$
$$3.4a_4+a_2 = 0, \quad \ldots\ldots\ldots\ldots,$$

and it follows that

$$a_2 = -\frac{a_0}{2}, \quad a_3 = -\frac{a_1}{2.3} = -\frac{a_1}{3!}, \quad a_4 = -\frac{a_2}{3.4} = \frac{a_0}{4!},$$

$$a_5 = -\frac{a_3}{4.5} = \frac{a_1}{5!}, \quad a_6 = -\frac{a_4}{5.6} = -\frac{a_0}{6!} \ldots$$

The required solution can therefore be written

$$y = a_0\left(1-\frac{x^2}{2!}+\frac{x^4}{4!}-\frac{x^6}{6!}+\ldots\right)+a_1\left(x-\frac{x^3}{3!}+\frac{x^5}{5!}-\ldots\right)$$

and the reader will recognize this as equivalent to the usual solution $y = a_0 \cos x+a_1 \sin x$, a_0 and a_1 being arbitrary constants.

5.2 Ordinary and singular points of a differential equation

A general form of the linear second order differential equation is

$$\frac{d^2y}{dx^2}+P\frac{dy}{dx}+Qy = 0, \tag{1}$$

where P, Q are functions of the real variable x, and there is no loss of generality in taking the coefficient of d^2y/dx^2 to be unity. This equation plays a very important part in physical problems and it is convenient here to introduce certain definitions and to state some important results applicable to equations of this type. With some small modifications, these are applicable also to linear equations of any order. Proofs of these results are outside the scope of the present work but they can be found, for instance, in Ince's valuable book.*

If both the functions P and Q can be expanded in Taylor series in the neighbourhood of $x = \alpha$, the differential equation (1) is said to possess an *ordinary point* at $x = \alpha$. Whenever the Taylor expansions of P and Q are valid for $|x-\alpha| < \beta$ (say), it can be shown that the Taylor expansion of the solution of the differential equation is valid for the same values of x. In this case, a solution of the equation can be expressed in the form

$$y = a_0+a_1(x-\alpha)+a_2(x-\alpha)^2+\ldots \tag{2}$$

* E. L. Ince, *Ordinary Differential Equations*, Longmans, Green and Co., London 1927, reprinted by Dover Publications Inc., New York, 1956.

and such a solution is said to be one relative to the ordinary point $x = \alpha$. The differential equation taken as an example in §5.1 is one for which the origin is an ordinary point and the general solution given there is a linear combination of two solutions of the type (2), each of the separate solutions being relative to the origin.

When either of the functions P or Q does not possess a Taylor series in the neighbourhood of $x = \alpha$, the differential equation is said to have a *singular point* at $x = \alpha$. In certain important practical cases

$$P = \frac{\lambda(x)}{x-\alpha}, \qquad Q = \frac{\mu(x)}{(x-\alpha)^2}, \tag{3}$$

where $\lambda(x)$ and $\mu(x)$ can be expanded in Taylor series near $x = \alpha$. In such cases, $x = \alpha$ is a singular point but the singularity is said to be *regular*. When P, Q are of the form given in (3), it can be shown that there is at least one solution of the differential equation of the form

$$y = a_0(x-\alpha)^\rho + a_1(x-\alpha)^{\rho+1} + a_2(x-\alpha)^{\rho+2} + \dots, \tag{4}$$

where $a_0 \neq 0$ and ρ is some constant, which is valid for $|x-\alpha| < \beta$ whenever the Taylor series for $\lambda(x)$ and $\mu(x)$ are valid for these values of x. The solution (4) is said to be one relative to the regular singular point $x = \alpha$.

The above condition for the validity of the series (4) means in effect that the range of convergence of (4) is not less than the smaller of the ranges of convergence of the expansions of the functions $\lambda(x)$ and $\mu(x)$. Thus the equation

$$x\frac{d^2y}{dx^2} + \frac{dy}{dx} + xy = 0$$

has everywhere ordinary points except at the origin, where there is a regular singularity, and the series solution relative to the origin in ascending powers of x will be convergent for all finite values of x.

It is usual to obtain series solutions of differential equations relative to the regular singular points of the equation for the following reasons:

(*a*) the range of convergence is sometimes greater than for expansions relative to ordinary points;

(*b*) the essential characteristics of the solutions are clearly indicated by the form of the expansions relative to regular singular points and the solutions are identified by their expansions;

(*c*) the solutions relative to two singular points will usually both be valid for some intermediate values of x and it is possible to obtain relations between the expansions relative to different singular points and so to obtain solutions for all values of x;

(*d*) attempts to obtain a series solution relative to an ordinary point lead, in general, to more complicated recurrence relations between the coefficients than arise in solutions relative to a regular singular point; thus for the hypergeometric equation considered later in this chapter, solutions relative to a regular singular point lead to a recurrence relation between two successive coefficients, while three coefficients will, in general, be involved in the recurrence relation for solutions relative to an ordinary point.

5.3 The indicial equation

The first step in finding a solution of a second order differential equation relative to a regular singular point $x = \alpha$ is to determine possible values for the index ρ in the solution (4). This is done by substituting the series (4) and its appropriate differential coefficients in the differential equation and equating to zero the resulting coefficient of the lowest power of $x-\alpha$. This leads to a quadratic equation, called the *indicial equation*, from which suitable values of ρ can be found. In the simplest cases, these values of ρ will give two different series solutions and the general solution of the differential equation is then given by a linear combination of the separate solutions. The complete procedure is shown in Example 1 below.

Example 1. *Find the general solution of the equation* $4x\dfrac{d^2y}{dx^2}+2\dfrac{dy}{dx}+y = 0.$

The origin is a regular singular point of this equation and, writing

$$y = a_0x^\rho+a_1x^{\rho+1}+a_2x^{\rho+2}+\ldots, \qquad (a_0 \neq 0),$$

we have

$$\frac{dy}{dx} = \rho a_0x^{\rho-1}+(\rho+1)a_1x^\rho+(\rho+2)a_2x^{\rho+1}+\ldots,$$

$$\frac{d^2y}{dx^2} = \rho(\rho-1)a_0x^{\rho-2}+(\rho+1)\rho a_1x^{\rho-1}+(\rho+2)(\rho+1)a_2x^\rho+\ldots.$$

Before substituting in the differential equation, it is convenient to rewrite it in the form

$$\left\{4x\frac{d^2y}{dx^2}+2\frac{dy}{dx}\right\}+\{y\} = 0.$$

If $a_nx^{\rho+n}$ is substituted for y, each term in the first bracket yields a multiple of $x^{\rho+n-1}$ while the second bracket gives a multiple of $x^{\rho+n}$ and, in this form, the differential equation is said to be arranged according to *weight*, the weights of the bracketed terms differing by unity. When the assumed series and its differential coefficients are substituted in the differential equation, the term containing the lowest power of x is obtained by writing $y = a_0x^\rho$ in the first bracket. Since the coefficient of the lowest power of x must be zero and, since by hypothesis a_0 is not zero, this gives

$$4\rho(\rho-1)+2\rho = 0, \quad \text{i.e.} \quad 2\rho(2\rho-1) = 0.$$

This is the indicial equation and its roots are $\rho = 0$, $\rho = \frac{1}{2}$. The term in $x^{\rho+n}$ obtained by writing $y = a_{n+1}x^{\rho+n+1}$ in the first bracket and $y = a_nx^{\rho+n}$ in the second. Equating to zero the coefficient of the term obtained in this way we have

$$\{4(\rho+n+1)(\rho+n)+2(\rho+n+1)\}a_{n+1}+a_n = 0$$

giving

$$a_{n+1} = -\frac{a_n}{2(\rho+n+1)(2\rho+2n+1)}.$$

This relation is true for $n = 0, 1, 2, \ldots$ and is called the *recurrence relation* for the coefficients. Using the first root $\rho = 0$ of the indicial equation, the recurrence relation gives

$$a_{n+1} = -\frac{a_n}{2(n+1)(2n+1)}$$

and hence

$$a_1 = -\frac{a_0}{2}, \quad a_2 = -\frac{a_1}{12} = \frac{a_0}{4!}, \quad a_3 = -\frac{a_2}{30} = -\frac{a_0}{6!},$$

and so on. Thus one solution of the differential equation is the series

$$a_0\left(1-\frac{x}{2!}+\frac{x^2}{4!}-\frac{x^3}{6!}+\ldots\right).$$

With the second root $\rho = \frac{1}{2}$ of the indicial equation, the recurrence relation becomes

$$a_{n+1} = -\frac{a_n}{(2n+3)(2n+2)}.$$

Replacing a_0 (which is quite arbitrary) by b_0, this gives

$$a_1 = -\frac{b_0}{3.2} = -\frac{b_0}{3!}, \quad a_2 = -\frac{a_1}{5.4} = \frac{b_0}{5!}, \quad a_3 = -\frac{a_2}{7.6} = -\frac{b_0}{7!},$$

etc. and a second solution is therefore

$$b_0 x^{1/2}\left(1-\frac{x}{3!}+\frac{x^2}{5!}-\frac{x^3}{7!}+\ldots\right).$$

The general solution of the equation is then found by adding these two solutions.

In the above example it was possible to obtain two distinct solutions of the differential equation. Difficulties arise when the roots of the indicial equation are equal and when they differ by an integer. With equal roots it is clear that the above method can only produce one solution. The nature of the difficulty when the roots differ by an integer is shown by taking, as an example, the equation

$$\frac{d^2y}{dx^2}+\frac{1}{x}\frac{dy}{dx}+\left(1-\frac{1}{x^2}\right)y = 0.$$

This equation has a regular singular point at $x = 0$ and, arranged according to weight, it can be written

$$\left\{x^2\frac{d^2y}{dx^2}+x\frac{dy}{dx}-y\right\}+\{x^2y\} = 0.$$

Substituting $y = a_0 x^\rho$ in the first bracket gives, as the indicial equation

$$\rho(\rho-1)+\rho-1 = 0, \quad \text{i.e.,} \ (\rho-1)(\rho+1) = 0,$$

so that possible values of ρ are ± 1. The recurrence relation is found by substituting $y = a_{n+2}x^{\rho+n+2}$ in the first bracket and $y = a_n x^{\rho+n}$ in the second to give, on equating to zero the resulting coefficient of $x^{\rho+n+2}$,

$$\{(\rho+n+2)(\rho+n+1)+(\rho+n+2)-1\}a_{n+2}+a_n = 0.$$

This can be written

$$a_{n+2} = -\frac{a_n}{(\rho+n+2)^2-1},$$

and the reader should note that, because the weights of the bracketed terms here differ by two, the recurrence relation now relates a_{n+2} to a_n instead of, as in the previous example, a_{n+1} to a_n. In this case the vanishing of the coefficient of x^ρ has been ensured by choosing ρ to satisfy the indicial equation and the vanishing of the coefficient of $x^{\rho+n+2}$, $n = 0, 1, 2, \ldots$, has been provided for by the recurrence relation. No provision has as yet been made for the vanishing of the coefficient of $x^{\rho+1}$ and this can be allowed for (by writing $y = a_1 x^{\rho+1}$ in the first bracket) through the extra condition

$$\{(\rho+1)\rho+(\rho+1)-1\}a_1 = 0, \quad \text{i.e.,} \ \rho(\rho+2)a_1 = 0.$$

Since $\rho = \pm 1$, this shows that $a_1 = 0$ and the recurrence relation then shows that $a_1 = a_3 = a_5 = \ldots = 0$ in both solutions. When $\rho = 1$, the recurrence relation becomes

$$a_{n+2} = -\frac{a_n}{(n+3)^2-1} = -\frac{a_n}{(n+2)(n+4)},$$

giving

$$a_2 = -\frac{a_0}{2.4} = -\frac{a_0}{2^2 1!\, 2!}, \quad a_4 = -\frac{a_2}{4.6} = \frac{a_0}{2^4 2!\, 3!},$$

and so on. Thus the solution corresponding to $\rho = 1$ is the series

$$a_0 x\left(1-\frac{x^2}{2^2 1!\, 2!}+\frac{x^4}{2^4 2!\, 3!}-\ldots\right).$$

When $\rho = -1$, the recurrence relation gives

$$a_{n+2} = -\frac{a_n}{(n+1)^2-1} = -\frac{a_n}{n(n+2)}$$

and this gives an infinite value for the coefficient a_2. A second solution of the equation cannot therefore be found by this method.

Methods exist for overcoming these difficulties but they are not pursued here in full detail. One method is given in §5.7 and specific examples in which difficulties arise in finding a second solution will be given later in this chapter and in the one which follows. A full treatment can be found in more advanced treatises such as the work by Ince to which reference has already been made. It is, however, worth stating the general form of the second solution when the roots of the indicial equation are either equal or when they differ by an integer. In such cases, if y_1 is a solution of a second order linear equation relative to a regular singular point $x = \alpha$, a distinct second solution y_2 is given by

$$y_2 = ay_1 \log_e(x-\alpha) + \sum_{n=0}^{\infty} b_n(x-\alpha)^{\nu+n},$$

and methods are available for the calculation of the constants a, ν, b_n. The occurrence of the logarithmic term in the general form of the second solution should be noted.

5.4 Solutions for large values of x

Many physical problems require solutions of linear differential equations which are valid for large values of the independent variable x. By using the transformation $x = 1/t$, the differential equation can be transformed into a linear equation in the new variable t and the solutions required will be those valid for small t.

If the transformed equation has an ordinary point at $t = 0$, the method already described will enable a solution of the form

$$y = a_0 + a_1 t + a_2 t^2 + \ldots$$

to be found. This can be written

$$y = a_0 + \frac{a_1}{x} + \frac{a_2}{x^2} + \ldots$$

and is of the required form in the original variable. When $t = 0$ is a regular singular point of the transformed equation, a solution

$$y = a_0 t^\rho + a_1 t^{\rho+1} + a_2 t^{\rho+2} + \ldots$$

leading to

$$y = \frac{a_0}{x^\rho} + \frac{a_1}{x^{\rho+1}} + \frac{a_2}{x^{\rho+2}} + \ldots$$

can be found by setting up the indicial equation and the recurrence relation of the transformed equation. In this case the original equation is said to have a regular singular point at infinity.

Example 2. *Find a series solution, valid for large values of x, of the equation*

$$(1-x^2)\frac{d^2y}{dx^2}-2x\frac{dy}{dx}+2y = 0.$$

Writing $x = 1/t$

$$\frac{dy}{dx} = \frac{dy}{dt}\frac{dt}{dx} = -\frac{1}{x^2}\frac{dy}{dt} = -t^2\frac{dy}{dt},$$

$$\frac{d^2y}{dx^2} = -t^2\frac{d}{dt}\left(-t^2\frac{dy}{dt}\right) = -t^2\left(-2t\frac{dy}{dt}-t^2\frac{d^2y}{dt^2}\right) = t^4\frac{d^2y}{dt^2}+2t^3\frac{dy}{dt};$$

substituting in the given equation, this gives

$$\left(1-\frac{1}{t^2}\right)\left(t^4\frac{d^2y}{dt^2}+2t^3\frac{dy}{dt}\right)-\frac{2}{t}\left(-t^2\frac{dy}{dt}\right)+2y = 0.$$

Arranged according to weight, this can be written

$$\left\{-t^2\frac{d^2y}{dt^2}+2y\right\}+\left\{t^4\frac{d^2y}{dt^2}+2t^3\frac{dy}{dt}\right\} = 0.$$

The indicial equation, given by writing $y = a_0t^\rho$ in the first bracket, is

$$-\rho(\rho-1)+2 = 0, \quad \text{i.e.,} \quad \rho^2-\rho-2 = 0,$$

and its roots are $\rho = 2$ and $\rho = -1$. The recurrence relation, obtained by writing $y = a_{n+2}t^{\rho+n+2}$ in the first bracket and $y = a_nt^{\rho+n}$ in the second, is

$$\{-(\rho+n+2)(\rho+n+1)+2\}a_{n+2}+\{(\rho+n)(\rho+n-1)+2(\rho+n)\}a_n = 0,$$

giving, after a little reduction,

$$a_{n+2} = \frac{\rho+n+1}{\rho+n+3}a_n.$$

So far we have only ensured that the coefficients of t^ρ and $t^{\rho+n+2}$ $(n = 0, 1, 2, \ldots)$ are to vanish and we still have to make that of $t^{\rho+1}$ vanish also. To do this we must have (by substituting $y = a_1t^{\rho+1}$ in the first bracket) $\{-(\rho+1)\rho+2\}a_1 = 0$ and hence $a_1 = 0$. It follows from the recurrence relation that, when $\rho = 2$,

$$a_{n+2} = \frac{n+3}{n+5}a_n,$$

giving $a_3 = a_5 = \ldots = 0$, and

$$a_2 = \frac{3}{5}a_0, \qquad a_4 = \frac{5}{7}a_2 = \frac{3}{7}a_0, \qquad \text{etc.}$$

Hence one solution is

$$y = a_0t^2\left(1+\frac{3}{5}t^2+\frac{3}{7}t^4+\ldots\right) = \frac{a_0}{x^2}\left(1+\frac{3}{5x^2}+\frac{3}{7x^4}+\ldots\right).$$

When we take the second root $\rho = -1$ of the indicial equation, the recurrence relation becomes

$$a_{n+2} = \frac{n}{n+2}a_n.$$

Since $a_1 = 0$, it follows that $a_3 = a_5 = \ldots = 0$ and, putting $n = 0$, $a_2 = 0$. Hence all the coefficients vanish except a_0, and the second solution is

$$y = a_0t^{-1} = a_0x.$$

5.5 The Gamma and Beta functions

The reader will have noticed that the factorial notation

$$n! = n(n-1)(n-2)\dots 3.2.1,$$

has proved useful in writing down the coefficients in some of the series solutions of the differential equations which have been taken as examples in the previous paragraphs. This notation is, however, meaningless when n is not a positive integer and a useful extension is provided by the *Gamma function.**

This function can be defined by

$$\Gamma(\alpha) = \int_0^\infty e^{-x}x^{\alpha-1}\,dx, \tag{1}$$

where $\alpha > 0$, and it follows immediately that

$$\Gamma(1) = \int_0^\infty e^{-x}\,dx = \Big[-e^{-x}\Big]_0^\infty = 1. \tag{2}$$

Integrating by parts,

$$\Gamma(\alpha+1) = \int_0^\infty e^{-x}x^{\alpha}\,dx = \Big[-e^{-x}x^{\alpha}\Big]_0^\infty + \alpha\int_0^\infty e^{-x}x^{\alpha-1}\,dx = \alpha\Gamma(\alpha), \tag{3}$$

the integrated term vanishing at both limits. When α is a positive integer n, repeated application of formula (3) and use of (2) show that

$$\begin{aligned}\Gamma(n+1) &= n\Gamma(n) = n(n-1)\Gamma(n-1) = \dots \\ &= n(n-1)(n-2)\dots 3.2.\Gamma(1) = n!.\end{aligned} \tag{4}$$

The function $\Gamma(\alpha)$ has been tabulated for values of α between 0 and 1. This table, together with equation (3), enables the value of the Gamma function for any positive value of α to be calculated: thus

$$\Gamma(\tfrac{7}{2}) = \tfrac{5}{2}\Gamma(\tfrac{5}{2}) = \tfrac{5}{2}.\tfrac{3}{2}\Gamma(\tfrac{3}{2}) = \tfrac{5}{2}.\tfrac{3}{2}.\tfrac{1}{2}\Gamma(\tfrac{1}{2}),$$

$$\Gamma(\tfrac{13}{4}) = \tfrac{9}{4}\Gamma(\tfrac{9}{4}) = \tfrac{9}{4}.\tfrac{5}{4}\Gamma(\tfrac{5}{4}) = \tfrac{9}{4}.\tfrac{5}{4}.\tfrac{1}{4}\Gamma(\tfrac{1}{4}).$$

Another useful formula, the proof of which is rather outside the scope of this book, is

$$\Gamma(\alpha)\Gamma(1-\alpha) = \pi\operatorname{cosec}(\alpha\pi). \tag{5}$$

With $\alpha = \frac{1}{2}$, this formula gives $\Gamma^2(\frac{1}{2}) = \pi\operatorname{cosec}(\frac{1}{2}\pi)$ so that

$$\Gamma(\tfrac{1}{2}) = \sqrt{\pi}, \tag{6}$$

* This function, unlike those discussed in the next chapter, does not satisfy a differential equation with rational coefficients.

since the integral defining $\Gamma(\frac{1}{2})$ is essentially positive. For negative values of α which are not negative integers, $\Gamma(\alpha)$ is defined by the relation (3). Thus, for example,

$$-\tfrac{3}{2}\Gamma(-\tfrac{3}{2}) = \Gamma(-\tfrac{1}{2}) \quad \text{and} \quad -\tfrac{1}{2}\Gamma(-\tfrac{1}{2}) = \Gamma(\tfrac{1}{2}) = \sqrt{\pi},$$

leading to

$$\Gamma(-\tfrac{3}{2}) = \tfrac{4}{3}\sqrt{\pi}.$$

Example 3. *Show that one solution of the differential equation*

$$2x(x-1)\frac{d^2y}{dx^2}+(4x-1)\frac{dy}{dx}-24y = 0$$

is the series

$$A\sum_{n=0}^{\infty}\frac{\Gamma(n+\frac{9}{2})\Gamma(n-\frac{5}{2})}{\Gamma(n+\frac{3}{2})\Gamma(n+1)}x^{n+1/2},$$

and find also a second solution.

Arranged according to weight, the differential equation can be written

$$-\left\{2x\frac{d^2y}{dx^2}+\frac{dy}{dx}\right\}+\left\{2x^2\frac{d^2y}{dx^2}+4x\frac{dy}{dx}-24y\right\} = 0.$$

The indicial equation is found by writing $y = a_0x^\rho$ in the first bracket and equating to zero the resulting coefficient of $x^{\rho-1}$; this gives $2\rho(\rho-1)+\rho = 0$ and the roots of this equation are $\rho = 0$ and $\rho = \frac{1}{2}$. The recurrence relation is obtained by writing $y = a_{n+1}x^{\rho+n+1}$ in the first bracket, $y = a_nx^{\rho+n}$ in the second bracket and equating to zero the resulting coefficient of $x^{\rho+n}$. This gives

$$\{-2(\rho+n+1)(\rho+n)-(\rho+n+1)\}a_{n+1} + \{2(\rho+n)(\rho+n-1)+4(\rho+n)-24\}a_n = 0,$$

leading to

$$a_{n+1} = \frac{(\rho+n+4)(\rho+n-3)}{(\rho+n+1)(\rho+n+\frac{1}{2})}a_n.$$

The solution in which we are first interested is derived from the root $\rho = \frac{1}{2}$ of the indicial equation. With this value of ρ, the recurrence relation becomes

$$a_{n+1} = \frac{(n+\frac{9}{2})(n-\frac{5}{2})}{(n+\frac{3}{2})(n+1)}a_n.$$

Hence,

$$a_1 = \frac{\frac{9}{2}(-\frac{5}{2})}{\frac{3}{2}.1}a_0, \quad a_2 = \frac{\frac{11}{2}(-\frac{3}{2})}{\frac{5}{2}.2}a_1 = \frac{\frac{11}{2}.\frac{9}{2}(-\frac{5}{2})(-\frac{3}{2})}{\frac{5}{2}.\frac{3}{2}.2.1}a_0,$$

$$a_3 = \frac{\frac{13}{2}(-\frac{1}{2})}{\frac{7}{2}.3}a_2 = \frac{\frac{13}{2}.\frac{11}{2}.\frac{9}{2}(-\frac{5}{2})(-\frac{3}{2})(-\frac{1}{2})}{\frac{7}{2}.\frac{5}{2}.\frac{3}{2}.3.2.1}a_0, \ldots$$

and it can be seen that

$$a_n = \frac{\Gamma(n+\frac{9}{2})}{\Gamma(\frac{9}{2})}\cdot\frac{\Gamma(n-\frac{5}{2})}{\Gamma(-\frac{5}{2})}\cdot\frac{\Gamma(\frac{3}{2})}{\Gamma(n+\frac{3}{2})}\cdot\frac{a_0}{\Gamma(n+1)}.$$

Writing

$$A = \frac{\Gamma(\frac{3}{2})}{\Gamma(\frac{9}{2})\Gamma(-\frac{5}{2})}a_0,$$

a solution of the differential equation is given by

$$y = A\sum_{n=0}^{\infty}\frac{\Gamma(n+\frac{9}{2})\Gamma(n-\frac{5}{2})}{\Gamma(n+\frac{3}{2})\Gamma(n+1)}x^{n+1/2}.$$

The second root, $\rho = 0$, of the indicial equation leads to the recurrence relation

$$a_{n+1} = \frac{(n+4)(n-3)}{(n+1)(n+\frac{1}{2})}a_n.$$

This gives $a_1 = -24a_0$, $a_2 = -\frac{10}{3}a_1 = 80a_0$, $a_3 = -\frac{4}{5}a_2 = -64a_0$, and $a_4 = a_5 = \ldots = 0$. Hence a second solution is the polynomial

$$a_0(1-24x+80x^2-64x^3).$$

Another function which will be useful later is the *Beta function* defined by

$$B(p,q) = \int_0^1 t^{p-1}(1-t)^{q-1}\,dt, \qquad (p>0, \quad q>0). \tag{7}$$

Writing $t = v/(1+v)$, this can be written in the alternative form

$$B(p,q) = \int_0^\infty v^{p-1}(1+v)^{-p-q}\,dv. \tag{8}$$

The Beta function can be expressed in terms of Gamma functions as follows. Writing $x = at$ $(a > 0)$ in the integral (1) defining $\Gamma(\alpha)$, it is easy to show that

$$\frac{\Gamma(\alpha)}{a^\alpha} = \int_0^\infty e^{-at}t^{\alpha-1}dt, \tag{9}$$

and, with $\alpha = p+q$, $a = 1+v$, this can be written

$$\Gamma(p+q)(1+v)^{-p-q} = \int_0^\infty e^{-(1+v)t}\,t^{p+q-1}\,dt.$$

Multiplying by v^{p-1} and integrating with respect to v between 0 and ∞,

$$\Gamma(p+q)\int_0^\infty v^{p-1}(1+v)^{-p-q}\,dv = \int_0^\infty v^{p-1}\,dv\int_0^\infty e^{-(1+v)t}\,t^{p+q-1}\,dt.$$

Interchanging the order of integration in the double integral on the right and using (8),

$$\Gamma(p+q)B(p,q) = \int_0^\infty e^{-t}\,t^{p+q-1}\,dt\int_0^\infty e^{-vt}\,v^{p-1}\,dv$$

$$= \int_0^\infty e^{-t}\,t^{p+q-1}\frac{\Gamma(p)}{t^p}dt, \quad \text{using (9)},$$

$$= \Gamma(p)\int_0^\infty e^{-t}\,t^{q-1}\,dt$$

$$= \Gamma(p)\Gamma(q). \tag{10}$$

Writing $t = \sin^2\theta$ in (7) and using (10), we have the important results

$$B(p,q) = \int_0^1 t^{p-1}(1-t)^{q-1}\,dt = 2\int_0^{\pi/2} \sin^{2p-1}\theta\cos^{2q-1}\theta\,d\theta$$

$$= \frac{\Gamma(p)\Gamma(q)}{\Gamma(p+q)}, \qquad (p>0, \quad q>0). \tag{11}$$

5.6 The convergence of series solutions

D'Alembert's ratio test, viz. the series $\sum_{n=1}^{\infty} u_n$ converges absolutely if $\lim_{n\to\infty}\left|\frac{u_{n+1}}{u_n}\right| < 1$, is a useful method of determining the range of values of the independent variable for which a given series solution is convergent. Applied to the series solution

$$y = a_0x^\rho + a_1x^{\rho+1} + a_2x^{\rho+2} + \ldots,$$

this test shows that the solution converges when

$$\lim_{n\to\infty}\left|\frac{a_{n+1}x}{a_n}\right| < 1.$$

As illustrations, consider the series solutions of Examples 1 and 2. In Example 1, the recurrence relation was

$$a_{n+1} = -\frac{a_n}{2(\rho+n+1)(2\rho+2n+1)},$$

so that

$$\lim_{n\to\infty}\left|\frac{a_{n+1}x}{a_n}\right| = \lim_{n\to\infty}\left\{\frac{|x|}{2(\rho+n+1)(2\rho+2n+1)}\right\} = 0, \quad \text{(all finite } x\text{)},$$

and clearly D'Alembert's test gives convergence for both roots ($\rho = 0$ and $\rho = \frac{1}{2}$) of the indicial equation. Hence both series solutions in this example are convergent for all finite values of x. This could also have been deduced from the result quoted in §5.2, since the only regular singular point of the differential equation of Example 1 occurs at the origin and hence the range of convergence of the solutions is given by $0 \leqq |x| < \infty$.

The series solution of Example 2 is not quite of the form assumed above. It is, in fact,

$$y = 3a_0\sum_{n=1}^{\infty}\frac{1}{(2n+1)x^{2n}},$$

and the series proceeds in descending powers of x^2 instead of in ascending powers of x. Direct application of the ratio test gives

$$\left|\frac{u_{n+1}}{u_n}\right| = \left(\frac{2n+1}{2n+3}\right)\frac{1}{x^2},$$

and the limit of this as $n \to \infty$ is $1/x^2$. The solution converges therefore when $x > 1$ and when $x < -1$. This again agrees with the result of §5.2, the regular singular points of the differential equation of Example 2 being at $x = \pm 1$ and at infinity.

5.7 The relation between the two solutions of a second order linear equation

The necessary condition that y_1 and y_2 shall be *independent* solutions of a second order linear differential equation, in the sense that one solution is not a constant multiple of the other, is that

$$\begin{vmatrix} y_1 & y_2 \\ \dfrac{dy_1}{dx} & \dfrac{dy_2}{dx} \end{vmatrix} \not\equiv 0.$$

The above determinant is called the *Wronskian* of the two functions y_1, y_2 and it is usually denoted by $W\{y_1, y_2\}$ or, more briefly, by W. As an example, the functions $\cos x$, $\sin x$ are independent solutions of the equation $y'' + y = 0$ for all values of x because

$$\begin{vmatrix} \cos x & \sin x \\ \dfrac{d}{dx}(\cos x) & \dfrac{d}{dx}(\sin x) \end{vmatrix} = \begin{vmatrix} \cos x & \sin x \\ -\sin x & \cos x \end{vmatrix}$$

$$= \cos^2 x + \sin^2 x = 1 \neq 0.$$

Further, if the Wronskian does vanish identically, either one of the solutions y_1, y_2 vanishes identically or else

$$y_1 \frac{dy_2}{dx} - y_2 \frac{dy_1}{dx} = 0.$$

This implies that $\dfrac{d}{dx}(y_1/y_2) = 0$ showing that the ratio of the two solutions is in this case a constant.

A useful relation between the two solutions y_1, y_2 can be found as follows. Since,

$$W = y_1 \frac{dy_2}{dx} - y_2 \frac{dy_1}{dx} = y_1{}^2 \frac{d}{dx}\left(\frac{y_2}{y_1}\right),$$

it follows that

$$y_2 = y_1 \int \frac{W}{y_1^2} dx. \tag{1}$$

Now suppose that the differential equation of which y_1, y_2 are solutions is

$$\frac{d^2y}{dx^2} + P\frac{dy}{dx} + Qy = 0$$

where P, Q are functions of x. Since both y_1 and y_2 satisfy this equation,

$$\frac{d^2y_1}{dx^2} + P\frac{dy_1}{dx} + Qy_1 = 0,$$

$$\frac{d^2y_2}{dx^2} + P\frac{dy_2}{dx} + Qy_2 = 0.$$

Multiplication by y_2, y_1 respectively and subtraction gives

$$y_2\frac{d^2y_1}{dx^2} - y_1\frac{d^2y_2}{dx^2} + P\left(y_2\frac{dy_1}{dx} - y_1\frac{dy_2}{dx}\right) = 0.$$

This can be written, using the definition of the Wronskian,

$$\frac{dW}{dx} + PW = 0,$$

and this first order equation for W has solution

$$W = c\,e^{-\int P\,dx}, \quad (c \text{ constant}).$$

Substituting in equation (1)

$$y_2 = cy_1 \int \frac{1}{y_1^2} e^{-\int P\,dx}\, dx. \tag{2}$$

Equation (2) is often a useful formula. It can, of course, only be used as a *practical* method of finding a second solution when the integrations involved can be carried out and an illustration of its use is given in Example 4 below. However, it will often throw light on the nature of the second solution even when the integrations cannot be completely effected. For example, in the important case in which $P = (x-\alpha)^{-1}$,

$$e^{-\int P\,dx} = e^{-\int (x-\alpha)^{-1}\,dx} = e^{-\log_e (x-\alpha)} = (x-\alpha)^{-1},$$

and equation (2) gives

$$\frac{y_2}{y_1} = c\int \frac{dx}{(x-\alpha)y_1^2}.$$

The solution y_1 is very often of the form $\sum_{n=0}^{\infty} a_n x^n$ and hence the ratio y_2/y_1 will in these cases contain a term in $\log_e (x-\alpha)$ (see also §5.3, page 117).

Example 4. *Given that* $y_1 = (x-1)^{-2}$ *is a solution of the equation*

$$x(x-1)^2 \frac{d^2y}{dx^2}+(x-1)\frac{dy}{dx}+(2-6x)y = 0,$$

find a second solution y_2.

Here

$$P = \frac{1}{x(x-1)} \equiv \frac{1}{x-1}-\frac{1}{x},$$

so that

$$\int P\,dx = \log_e \left(\frac{x-1}{x}\right) \quad \text{and} \quad e^{-\int P\,dx} = \frac{x}{x-1}.$$

Hence, using (2),

$$y_2 = \frac{c}{(x-1)^2}\int (x-1)^4\left(\frac{x}{x-1}\right) dx = \frac{c}{(x-1)^2}\int x(x-1)^3\,dx$$

$$= \frac{c}{(x-1)^2}\int \{(x-1)^4+(x-1)^3\}\,dx$$

$$= \frac{c}{(x-1)^2}\left\{\frac{(x-1)^5}{5}+\frac{(x-1)^4}{4}\right\}$$

$$= \frac{c}{20}(x-1)^2(4x+1).$$

Exercises 5(*a*)

1. Find the general solution, in series of ascending powers of x, of the equation

$$3x\frac{d^2y}{dx^2}+\frac{dy}{dx}-2y = 0.$$

2. Express, in series of ascending powers of x, the general solution of

$$x^2\frac{d^2y}{dx^2}+(x+x^2)\frac{dy}{dx}+(x-9)y = 0$$

and show that one series terminates. (L.U.)

3. Show that the equation $4xy''+6y'+y = 0$ has a solution in series of the form $\sum_{n=0}^{\infty} a_n x^n$ where $2n(2n+1)a_n = -a_{n-1}$ $(n \geqq 1)$. Obtain a second solution in series and state the general solution. (L.U.)

4. Find a solution, in ascending powers of x, of the equation

$$x\frac{d^2y}{dx^2}+(1+x)\frac{dy}{dx}+2y = 0.$$

5. Find one solution in series, relative to the regular singular point $x = 0$, of the equation

$$x(1-x)\frac{d^2y}{dx^2} = (1+3x)\frac{dy}{dx}+y.$$

6. Show that the equation

$$x^2\frac{d^2y}{dx^2}-(1-3x)\frac{dy}{dx}+y = 0$$

has a regular singularity at the point at infinity. Determine one solution of the equation as a series in descending powers of x.

7. Show that the equation

$$(1-x^2)\frac{d^2y}{dx^2}-2x\frac{dy}{dx}+a(a+1)y = 0$$

where a is a constant, has a regular singular point at $x = 1$. By the change of variable $x = 1-2\xi$, find a solution relative to this singularity as a series in ascending powers of $(1-x)$.

8. Show that the general solution of the equation

$$(1-x^2)\frac{d^2y}{dx^2}-7x\frac{dy}{dx}-9y = 0$$

can be written in the form

$$y = A\sum_{n=0}^{\infty}\frac{\Gamma^2(n+\frac{3}{2})}{\Gamma(2n+1)}(2x)^{2n}+B\sum_{n=0}^{\infty}\frac{\Gamma^2(n+2)}{\Gamma(2n+2)}(2x)^{2n+1},$$

where A and B are arbitrary constants.
By applying the ratio test, or otherwise, prove that the series converge for $-1 < x < 1$. (L.U.)

9. Deduce, from the appropriate general result quoted in §5.2, that the differential equation

$$(2x+x^3)\frac{d^2y}{dx^2}-\frac{dy}{dx}-6xy = 0$$

has at least one solution as a series of ascending powers of x which is valid when $|x| < \sqrt{2}$.

10. By setting up the indicial equation and the recurrence relation for the differential equation of Exercise 9 above, show that the equation has, in fact, two solutions as series in ascending powers of x which are valid when $|x| < \sqrt{2}$. Find these two solutions.

11. Verify that $y_1 = x-1$ is a solution of the differential equation

$$x(x-1)\frac{d^2y}{dx^2}+(x-1)\frac{dy}{dx} = y,$$

and find a second solution.

12. Show that a series solution of the differential equation

$$x(x-1)\frac{d^2y}{dx^2}+3x\frac{dy}{dx}+y = 0$$

is $x+2x^2+3x^3+\ldots+(n+1)x^{n+1}+\ldots$ and sum the series when $|x| < 1$. Derive a second solution. (L.U.)

13. Find the general series solution in ascending powers of x of the differential equation

$$(x^2-a^2)\frac{d^2y}{dx^2}+bx\frac{dy}{dx}+cy = 0,$$

and determine the region in which it is convergent.

14. Find a series solution, valid for large values of x, of the equation

$$(1-x^2)\frac{d^2y}{dx^2}-2x\frac{dy}{dx}+6y = 0.$$

Show that the second solution is a polynomial in x and find this polynomial.

15. Show that a series solution of *Airy's equation* $y''-xy = 0$ is

$$y = a_0\left(1+\frac{x^3}{2.3}+\frac{x^6}{2.3.5.6}+\ldots\right)+b_0\left(x+\frac{x^4}{3.4}+\frac{x^7}{3.4.6.7}+\ldots\right).$$

16. Show that *Weber's equation* $y''+(n+\frac{1}{2}-\frac{1}{4}x^2)y = 0$ is reduced by the substitution $y = e^{-x^2/4}v$ to the equation

$$\frac{d^2v}{dx^2}-x\frac{dv}{dx}+nv = 0.$$

Show also that two solutions of this latter equation are

$$v_1 = 1-\frac{n}{2!}x^2+\frac{n(n-2)}{4!}x^4-\frac{n(n-2)(n-4)}{6!}x^6+\ldots,$$

$$v_2 = x-\frac{n-1}{3!}x^3+\frac{(n-1)(n-3)}{5!}x^5-\frac{(n-1)(n-3)(n-5)}{7!}x^7+\ldots.$$

17. Show that the differential equation

$$4x^2\frac{d^2y}{dx^2}+(1-x^2)y = 0$$

has a solution

$$y_1 = x^{1/2}\left(1+\frac{x^2}{16}+\frac{x^4}{1024}+\ldots\right)$$

and that a second solution is of the form

$$y_2 = y_1\log_e x-\frac{x^{5/2}}{16}-\frac{3x^{9/2}}{2048}-\ldots.$$

5.8 The hypergeometric equation

The differential equation

$$x(1-x)\frac{d^2y}{dx^2}+\{\gamma-(\alpha+\beta+1)x\}\frac{dy}{dx}-\alpha\beta y = 0, \tag{1}$$

where α, β and γ are constants, occurs in problems on the flow of compressible fluids and has other applications in science and engineering. Other equations occurring in physical problems can, by a change of variable, be transformed into equations of this type. Some of these are discussed in the next chapter.

The differential equation (1) possesses regular singular points at $x = 0$, $x = 1$ and at the point at infinity, and we now consider solutions of the equation relative to these three points.

(*a*) *Solution relative to the singularity at* $x = 0$

Arranged according to weight, equation (1) can be written

$$\left\{x\frac{d^2y}{dx^2}+\gamma\frac{dy}{dx}\right\}-\left\{x^2\frac{d^2y}{dx^2}+(\alpha+\beta+1)x\frac{dy}{dx}+\alpha\beta y\right\} = 0.$$

The indicial equation, obtained by writing $y = a_0x^\rho$ in the first bracket is $\rho(\rho-1)+\gamma\rho = 0$, giving $\rho = 0$ and $\rho = 1-\gamma$. The recurrence relation is derived by writing $y = a_{n+1}x^{\rho+n+1}$ in the first bracket and $y = a_nx^{\rho+n}$ in the second, and is

$$\{(\rho+n+1)(\rho+n)+\gamma(\rho+n+1)\}a_{n+1}$$
$$-\{(\rho+n)(\rho+n-1)+(\alpha+\beta+1)(\rho+n)+\alpha\beta\}a_n = 0,$$

and this can be written

$$a_{n+1} = \frac{(\rho+n+\alpha)(\rho+n+\beta)}{(\rho+n+1)(\rho+n+\gamma)}a_n. \tag{2}$$

With the first root ($\rho = 0$) of the indicial equation, this becomes

$$a_{n+1} = \frac{(n+\alpha)(n+\beta)}{(n+1)(n+\gamma)}a_n \tag{3}$$

giving

$$a_1 = \frac{\alpha\,.\,\beta}{1\,.\,\gamma}a_0, \qquad a_2 = \frac{(\alpha+1)(\beta+1)}{2(\gamma+1)}a_1 = \frac{\alpha(\alpha+1)\beta(\beta+1)}{2!\gamma(\gamma+1)}a_0, \qquad \text{etc.,}$$

and one solution of equation (1) is therefore an arbitrary multiple of the series

$$1+\frac{\alpha\,.\,\beta}{1!\gamma}x+\frac{\alpha(\alpha+1)\beta(\beta+1)}{2!\gamma(\gamma+1)}x^2+\ldots. \tag{4}$$

This series is conveniently denoted by ${}_2F_1(\alpha, \beta; \gamma; x)$, the significance of the suffixes 2, 1, the comma and the semi-colons being that there are two constants α, β in the numerators and one constant γ in the denominators of the coefficients in the series. The series (4) is a generalization of the geometric series $1+x+x^2+\ldots$, to which it obviously reduces when $\alpha = 1$, $\beta = \gamma$, and for this reason it is termed the *hypergeometric series*. The function ${}_2F_1(\alpha, \beta; \gamma; x)$ and the differential equation (1) are known respectively as the *hypergeometric function* and the *hypergeometric equation.*

With the second root ($\rho = 1-\gamma$) of the indicial equation, the recurrence relation (2) becomes

$$a_{n+1} = \frac{(n+\alpha-\gamma+1)(n+\beta-\gamma+1)}{(n+2-\gamma)(n+1)} a_n$$

and this is the same as (3) when α, β, γ are replaced respectively by $\alpha-\gamma+1$, $\beta-\gamma+1$, $2-\gamma$. Hence the second solution is a multiple of $x^{1-\gamma}{}_2F_1(\alpha-\gamma+1, \beta-\gamma+1; 2-\gamma; x)$ and the general solution of the hypergeometric equation (1), valid in the neighbourhood of $x = 0$, is

$$y = A\,{}_2F_1(\alpha, \beta; \gamma; x) + Bx^{1-\gamma}{}_2F_1(\alpha-\gamma+1, \beta-\gamma+1; 2-\gamma; x), \quad (5)$$

A and B being arbitrary constants.

The above method of obtaining a general solution of the hypergeometric equation breaks down when $1-\gamma$ is zero or a positive integer since the roots of the indicial equation are then either equal or differ by an integer; these cases are not pursued here. The reader will find it useful to note that the equation

$$x(1-x)\frac{d^2y}{dx^2} + (q-px)\frac{dy}{dx} - ry = 0, \quad (6)$$

is of hypergeometric form with $q = \gamma$, $p = \alpha+\beta+1$, $r = \alpha\beta$; these relations are equivalent to taking α, β as the roots of the quadratic equation $z^2+(1-p)z+r = 0$ and $\gamma = q$. The general solution of the differential equation (6) can then be written down at once by using equation (5) with the appropriate values of α, β and γ.

(b) Solution relative to the singularity at $x = 1$

With the transformation $x = 1-t$, equation (1) becomes

$$t(1-t)\frac{d^2y}{dt^2} + \{\alpha+\beta-\gamma+1-(\alpha+\beta+1)t\}\frac{dy}{dt} - \alpha\beta y = 0.$$

This is identical with equation (1) if γ is replaced by $\alpha+\beta-\gamma+1$.

Remembering that $t = 1-x$, use of (5) then gives, as the required solution valid in the neighbourhood of $x = 1$,

$$y = A\,{}_2F_1(\alpha, \beta; \alpha+\beta-\gamma+1; 1-x)$$
$$+B(1-x)^{\gamma-\alpha-\beta}\,{}_2F_1(\gamma-\beta, \gamma-\alpha; \gamma-\alpha-\beta+1; 1-x). \quad (7)$$

(c) Solution relative to the singularity at the point at infinity

With the transformation $x = 1/t$, it follows (as in Example 2, page 118) that

$$\frac{dy}{dx} = -t^2\frac{dy}{dt}, \qquad \frac{d^2y}{dx^2} = t^4\frac{d^2y}{dt^2}+2t^3\frac{dy}{dt}.$$

Hence the hypergeometric equation (1) transforms into

$$\frac{1}{t}\left(1-\frac{1}{t}\right)\left(t^4\frac{d^2y}{dt^2}+2t^3\frac{dy}{dt}\right)+\left\{\gamma-(\alpha+\beta+1)\frac{1}{t}\right\}\left(-t^2\frac{dy}{dt}\right)-\alpha\beta y = 0,$$

and, after a little reduction, this can be arranged according to weight as

$$\left\{-t^2\frac{d^2y}{dt^2}+(\alpha+\beta-1)t\frac{dy}{dt}-\alpha\beta y\right\}+\left\{t^3\frac{d^2y}{dt^2}+(2-\gamma)t^2\frac{dy}{dt}\right\} = 0.$$

Assuming a solution of the form $y = \sum a_n t^{\rho+n}$, the indicial equation, obtained by writing $y = a_0 t^\rho$ in the first bracket, is

$$-\rho(\rho-1)+(\alpha+\beta-1)\rho-\alpha\beta = 0,$$

giving

$$\rho^2-(\alpha+\beta)\rho+\alpha\beta = 0,$$

that is,

$$(\rho-\alpha)(\rho-\beta) = 0,$$

and the values for ρ are α and β.

The recurrence relation, given by writing $y = a_{n+1}t^{\rho+n+1}$ in the first bracket and $y = a_n t^{\rho+n}$ in the second, is

$$\{-(\rho+n+1)(\rho+n)+(\alpha+\beta-1)(\rho+n+1)-\alpha\beta\}a_{n+1}$$
$$+\{(\rho+n)(\rho+n-1)+(2-\gamma)(\rho+n)\}a_n = 0,$$

leading to

$$a_{n+1} = \frac{(\rho+n)(\rho+n+1-\gamma)}{(\rho+n+1)(\rho+n-\alpha-\beta+1)+\alpha\beta}a_n.$$

With the first root ($\rho = \alpha$) of the indicial equation, the recurrence relation becomes

$$a_{n+1} = \frac{(\alpha+n)(\alpha-\gamma+1+n)}{(1+n)(\alpha-\beta+1+n)}a_n,$$

giving

$$a_1 = \frac{\alpha(\alpha-\gamma+1)}{1!(\alpha-\beta+1)}a_0,$$

$$a_2 = \frac{(\alpha+1)(\alpha-\gamma+2)}{2(\alpha-\beta+2)}a_1$$

$$= \frac{\alpha(\alpha+1)(\alpha-\gamma+1)(\alpha-\gamma+2)}{2!(\alpha-\beta+1)(\alpha-\beta+2)}a_0,$$

and so on. Remembering that $x = 1/t$, one solution is therefore $x^{-\alpha}{}_2F_1(\alpha,\ \alpha-\gamma+1;\ \alpha-\beta+1;\ x^{-1})$ and the second solution can be found in a similar way. The general solution, valid for large x, is

$$y = Ax^{-\alpha}{}_2F_1(\alpha, \alpha-\gamma+1; \alpha-\beta+1; x^{-1}) + Bx^{-\beta}{}_2F_1(\beta, \beta-\gamma+1; \beta-\alpha+1; x^{-1}), \tag{8}$$

A and B being arbitrary constants.

5.9 Some properties of the hypergeometric function

Introducing the notation

$$(\alpha)_n = \alpha(\alpha+1)(\alpha+2)\ldots(\alpha+n-1), \qquad (\alpha)_0 = 1, \tag{1}$$

the hypergeometric function defined in §5.8 can be expressed in the form

$${}_2F_1(\alpha, \beta; \gamma; x) = \sum_{n=0}^{\infty} \frac{(\alpha)_n(\beta)_n}{n!\,(\gamma)_n}x^n, \tag{2}$$

and, in this section, some of its properties are investigated. It should first be noted that, provided γ is not zero or a negative integer, the series in (2) converges absolutely when $|x| < 1$ while, for absolute convergence when $|x| = 1$, the additional restriction $\gamma > \alpha+\beta$ is necessary; these restrictions will be assumed to apply in what follows. The reader should also note that the hypergeometric function reduces to a polynomial in x when either α or β is a negative integer. These polynomials are called *Jacobi polynomials* and are discussed later (see §6.14).

(a) Elementary properties

Simple properties of the hypergeometric function which follow immediately from its definition are:

$${}_2F_1(\alpha, \beta; \gamma; 0) = 1 \tag{3}$$

and

$$ {}_2F_1(\beta, \alpha; \gamma; x) = {}_2F_1(\alpha, \beta; \gamma; x). \tag{4}$$

Further,

$$\begin{aligned}
\frac{d}{dx}\{{}_2F_1(\alpha, \beta; \gamma; x)\} &= \sum_{n=1}^{\infty} \frac{(\alpha)_n(\beta)_n}{(n-1)!\,(\gamma)_n} x^{n-1} \\
&= \sum_{n=0}^{\infty} \frac{(\alpha)_{n+1}(\beta)_{n+1}}{n!\,(\gamma)_{n+1}} x^n \\
&= \frac{\alpha\beta}{\gamma} \sum_{n=0}^{\infty} \frac{(\alpha+1)_n(\beta+1)_n}{n!\,(\gamma+1)_n} x^n \\
&= \frac{\alpha\beta}{\gamma}\, {}_2F_1(\alpha+1, \beta+1; \gamma+1; x), \qquad (5)
\end{aligned}$$

since

$$\begin{aligned}
(\alpha)_{n+1} &= \alpha(\alpha+1)\ldots(\alpha+n-1)(\alpha+n) \\
&= \alpha\{(\alpha+1)(\alpha+2)\ldots(\alpha+1+n-1)\} = \alpha(\alpha+1)_n, \quad \text{etc.}
\end{aligned}$$

Example 5. *The solution of the differential equation* $y''+n^2y = 0$ *such that* $y = 1$, $y' = 0$ *when* $x = 0$ *is* $y = \cos nx$. *By writing* $t = \sin^2 x$ *in the differential equation, deduce that*

$$\cos nx = {}_2F_1(\tfrac{1}{2}n, -\tfrac{1}{2}n; \tfrac{1}{2}; \sin^2 x), \qquad (-\tfrac{1}{2}\pi < x < \tfrac{1}{2}\pi).$$

Writing $t = \sin^2 x$,

$$\begin{aligned}
\frac{dy}{dx} &= \frac{dy}{dt}\frac{dt}{dx} = 2 \sin x \cos x \frac{dy}{dt}, \\
\frac{d^2y}{dx^2} &= 2(\cos^2 x - \sin^2 x)\frac{dy}{dt} + 4 \sin^2 x \cos^2 x \frac{d^2y}{dt^2} \\
&= 2(1-2\sin^2 x)\frac{dy}{dt} + 4 \sin^2 x(1-\sin^2 x)\frac{d^2y}{dt^2} \\
&= 2(1-2t)\frac{dy}{dt} + 4t(1-t)\frac{d^2y}{dt^2}.
\end{aligned}$$

The differential equation $y''+n^2y = 0$ therefore transforms into

$$t(1-t)\frac{d^2y}{dt^2} + (\tfrac{1}{2}-t)\frac{dy}{dt} + \frac{n^2}{4}y = 0.$$

This is the hypergeometric equation with $\alpha = \frac{1}{2}n$, $\beta = -\frac{1}{2}n$, $\gamma = \frac{1}{2}$ and its solution, valid when $-1 < t < 1$ is given by equation (5) of §5.8. Substituting for α, β, γ and remembering that $t = \sin^2 x$,

$$y = A\, {}_2F_1(\tfrac{1}{2}n, -\tfrac{1}{2}n; \tfrac{1}{2}; \sin^2 x) + B \sin x\, {}_2F_1(\tfrac{1}{2}+\tfrac{1}{2}n, \tfrac{1}{2}-\tfrac{1}{2}n; \tfrac{3}{2}; \sin^2 x).$$

Since $y = 1$ and $y' = 0$ when $x = 0$, use of equations (3) and (5) above shows that $A = 1$, $B = 0$ and the required result follows.

Example 5 above is an instance of the expression of an elementary function in terms of the hypergeometric function. Some other examples, which the reader will be able to verify by writing down the expansions of the functions on the left, are

$$(1-x)^n = {}_2F_1(-n, \beta; \beta; x),$$
$$x^{-1}\log_e(1+x) = {}_2F_1(1, 1; 2; -x),$$
$$x^{-1}\sin^{-1}x = {}_2F_1(\tfrac{1}{2}, \tfrac{1}{2}; \tfrac{3}{2}; x^2),$$
$$x^{-1}\tan^{-1}x = {}_2F_1(\tfrac{1}{2}, 1; \tfrac{3}{2}; -x^2).$$

By expanding the integrand and integrating term by term it is also easy to verify that, for the complete elliptic integrals $K(k)$, $E(k)$ defined as below,

$$K(k) = \int_0^{\pi/2} \frac{d\phi}{(1-k^2\sin^2\phi)^{\frac{1}{2}}} = \frac{\pi}{2}\,{}_2F_1(\tfrac{1}{2}, \tfrac{1}{2}; 1; k^2),$$

$$E(k) = \int_0^{\pi/2} (1-k^2\sin^2\phi)^{\frac{1}{2}}\,d\phi = \frac{\pi}{2}\,{}_2F_1(-\tfrac{1}{2}, \tfrac{1}{2}; 1; k^2),$$

with $k^2 < 1$ in each case. The reader will find further examples in Exercises 5(*b*).

(*b*) *Recursion formulae*

The six functions ${}_2F_1(\alpha \pm 1, \beta; \gamma; x)$, ${}_2F_1(\alpha, \beta \pm 1; \gamma; x)$ and ${}_2F_1(\alpha, \beta; \gamma \pm 1; x)$ are said to be *contiguous* to the function ${}_2F_1(\alpha, \beta; \gamma; x)$. There are fifteen relations, with coefficients linear functions of x, between ${}_2F_1(\alpha, \beta; \gamma; x)$ and any two functions contiguous to it; these relations are often known as the *Gauss recursion formulae*. A typical relation is

$$(\beta-\alpha)\,{}_2F_1(\alpha, \beta; \gamma; x) + \alpha\,{}_2F_1(\alpha+1, \beta; \gamma; x) = \beta\,{}_2F_1(\alpha, \beta+1; \gamma; x), \qquad (6)$$

and this is easily verified by expanding the hypergeometric functions as power series and comparing the coefficients of x^n on the two sides. Thus, the coefficient of x^n on the left-hand side of (6) is

$$(\beta-\alpha)\frac{(\alpha)_n(\beta)_n}{n!\,(\gamma)_n} + \alpha\frac{(\alpha+1)_n(\beta)_n}{n!\,(\gamma)_n}$$
$$= \frac{(\alpha)_n(\beta)_n}{n!\,(\gamma)_n}\left\{\beta-\alpha+\alpha\left(\frac{\alpha+n}{\alpha}\right)\right\}$$
$$= \frac{(\alpha)_n(\beta)_n}{n!\,(\gamma)_n}(\beta+n) = \beta\frac{(\alpha)_n(\beta+1)_n}{n!\,(\gamma)_n},$$

and this is the coefficient of x^n in $\beta\,{}_2F_1(\alpha, \beta+1; \gamma; x)$. Further relations of this type, which can be verified in a similar way, will be found in Exercises 5(*b*), 4.

(c) The integral representation of the hypergeometric function

Using the notation $(\alpha)_n = \alpha(\alpha+1)(\alpha+2)\ldots(\alpha+n-1)$, we have, for suitable values of t and x,

$$(1-tx)^{-\alpha} = \sum_{n=0}^{\infty} \frac{(\alpha)_n t^n x^n}{n!}.$$

Hence, interchanging the order of integration and summation,

$$\begin{aligned}\frac{\Gamma(\gamma)}{\Gamma(\beta)\Gamma(\gamma-\beta)}&\int_0^1 t^{\beta-1}(1-t)^{\gamma-\beta-1}(1-tx)^{-\alpha}\,dt \\ &= \frac{\Gamma(\gamma)}{\Gamma(\beta)\Gamma(\gamma-\beta)}\sum_{n=0}^{\infty}\frac{(\alpha)_n x^n}{n!}\int_0^1 t^{\beta+n-1}(1-t)^{\gamma-\beta-1}\,dt \\ &= \frac{\Gamma(\gamma)}{\Gamma(\beta)\Gamma(\gamma-\beta)}\sum_{n=0}^{\infty}\frac{(\alpha)_n x^n}{n!}\cdot\frac{\Gamma(\beta+n)\Gamma(\gamma-\beta)}{\Gamma(\gamma+n)},\end{aligned}$$

using equation (11) of §5.5 and assuming that $\gamma > \beta > 0$. Since

$$\frac{\Gamma(\beta+n)}{\Gamma(\beta)} = (\beta)_n, \qquad \frac{\Gamma(\gamma+n)}{\Gamma(\gamma)} = (\gamma)_n,$$

the coefficient of x^n on the right-hand side is easily seen to reduce to

$$\frac{(\alpha)_n(\beta)_n}{n!\,(\gamma)_n}$$

and the right-hand side is therefore ${}_2F_1(\alpha, \beta; \gamma; x)$. Hence we have the important *integral representation* of the hypergeometric function

$${}_2F_1(\alpha,\beta;\gamma;x) = \frac{\Gamma(\gamma)}{\Gamma(\beta)\Gamma(\gamma-\beta)}\int_0^1 t^{\beta-1}(1-t)^{\gamma-\beta-1}(1-tx)^{-\alpha}\,dt \qquad (7)$$

provided $\gamma > \beta > 0$.

Writing $x = 1$ in (7)

$$\begin{aligned}{}_2F_1(\alpha,\beta;\gamma;1) &= \frac{\Gamma(\gamma)}{\Gamma(\beta)\Gamma(\gamma-\beta)}\int_0^1 t^{\beta-1}(1-t)^{\gamma-\alpha-\beta-1}\,dt \\ &= \frac{\Gamma(\gamma)}{\Gamma(\beta)\Gamma(\gamma-\beta)}\cdot\frac{\Gamma(\beta)\Gamma(\gamma-\alpha-\beta)}{\Gamma(\gamma-\alpha)},\end{aligned}$$

again using equation (11) of §5.5 and assuming further that $\gamma > \alpha+\beta$. Hence, when $\gamma > \beta > 0$ and $\gamma > \alpha+\beta$,

$${}_2F_1(\alpha,\beta;\gamma;1) = \frac{\Gamma(\gamma)\Gamma(\gamma-\alpha-\beta)}{\Gamma(\gamma-\alpha)\Gamma(\gamma-\beta)}; \qquad (8)$$

this relation will be required later.

Example 6. *Show that when* $\gamma > \beta > 0$

$${}_2F_1\left(\alpha, \gamma-\beta; \gamma; \frac{x}{x-1}\right) = (1-x)^{\alpha}\, {}_2F_1(\alpha, \beta; \gamma; x).$$

Since

$$1-x(1-u) \equiv (1-x)\left\{1-\left(\frac{x}{x-1}\right)u\right\},$$

the integral representation (7) gives, with $t = 1-u$

$$\begin{aligned} {}_2F_1(\alpha, \beta; \gamma; x) &= \frac{\Gamma(\gamma)}{\Gamma(\beta)\Gamma(\gamma-\beta)}\int_0^1 (1-u)^{\beta-1}u^{\gamma-\beta-1}(1-x)^{-\alpha}\left\{1-\left(\frac{x}{x-1}\right)u\right\}^{-\alpha} du \\ &= \frac{(1-x)^{-\alpha}\Gamma(\gamma)}{\Gamma(\beta)\Gamma(\gamma-\beta)}\int_0^1 u^{\gamma-\beta-1}(1-u)^{\beta-1}\left\{1-\left(\frac{x}{x-1}\right)u\right\}^{-\alpha} du. \end{aligned}$$

Using the integral representation (7) to express the integral on the right as a hypergeometric function of argument $x/(x-1)$, this can be written

$${}_2F_1(\alpha, \beta; \gamma; x) = \frac{(1-x)^{-\alpha}\Gamma(\gamma)}{\Gamma(\beta)\Gamma(\gamma-\beta)}\cdot\frac{\Gamma(\gamma-\beta)\Gamma(\beta)}{\Gamma(\gamma)}\, {}_2F_1\left(\alpha, \gamma-\beta; \gamma; \frac{x}{x-1}\right) \tag{9}$$

and the required result follows immediately.

The result of this example is one of the twenty-four homographic transformations of the hypergeometric function which can be made by transforming x into $(ct+d)/(et+f)$, leaving the singularities unchanged.

(d) Linear relations between hypergeometric functions

From the general theory of series solutions relative to regular singular points (§5.2), the intervals of convergence of the solutions (5) and (7) of §5.8 are respectively $-1 < x < 1$ and $0 < x < 2$. There is therefore a common interval $0 < x < 1$ for which both solutions are valid. Since not more than two solutions of a second order linear differential equation are linearly independent, a linear relation must exist between the hypergeometric functions occurring in the solutions (5) and (7). As an example,

$$\begin{aligned} {}_2F_1(\alpha, \beta; \gamma; x) = A\, &{}_2F_1(\alpha, \beta; \alpha+\beta-\gamma+1; 1-x) \\ &+ B(1-x)^{\gamma-\alpha-\beta}\, {}_2F_1(\gamma-\beta, \gamma-\alpha; \gamma-\alpha-\beta+1; 1-x), \end{aligned} \tag{10}$$

where A and B are constants.

The values of the constants can be found by setting $x = 1$ and $x = 0$ in turn. With $x = 1$, we have

$${}_2F_1(\alpha, \beta; \gamma; 1) = A\, {}_2F_1(\alpha, \beta; \alpha+\beta-\gamma+1; 0).$$

Using equations (3) and (8), this gives for $\gamma > \beta > 0$ and $\gamma > \alpha + \beta$,

$$A = \frac{\Gamma(\gamma)\Gamma(\gamma-\alpha-\beta)}{\Gamma(\gamma-\alpha)\Gamma(\gamma-\beta)}.$$

Putting $x = 0$ in (10) and again using (3) and (8),

$$1 = A\frac{\Gamma(\alpha+\beta-\gamma+1)\Gamma(1-\gamma)}{\Gamma(\beta-\gamma+1)\Gamma(\alpha-\gamma+1)} + B\frac{\Gamma(\gamma-\alpha-\beta+1)\Gamma(1-\gamma)}{\Gamma(1-\alpha)\Gamma(1-\beta)}.$$

Substituting the value of A already found and solving for B,

$$B = \frac{\Gamma(1-\alpha)\Gamma(1-\beta)}{\Gamma(1-\gamma)\Gamma(\gamma-\alpha-\beta+1)}\left\{1 - \frac{\Gamma(\gamma)\Gamma(1-\gamma)\Gamma(\gamma-\alpha-\beta)\Gamma(1-\gamma+\alpha+\beta)}{\Gamma(\gamma-\alpha)\Gamma(1-\gamma+\alpha)\Gamma(\gamma-\beta)\Gamma(1-\gamma+\beta)}\right\}$$

$$= \frac{\Gamma(\gamma)\Gamma(\alpha+\beta-\gamma)}{\Gamma(\alpha)\Gamma(\beta)} \cdot \frac{\sin\gamma\pi\sin(\alpha+\beta-\gamma)\pi}{\sin\alpha\pi\sin\beta\pi}\left\{1 - \frac{\sin(\gamma-\alpha)\pi\sin(\gamma-\beta)\pi}{\sin\gamma\pi\sin(\gamma-\alpha-\beta)\pi}\right\},$$

when use is made of equation (5) of §5.5. It is easy to show that this reduces to

$$B = \frac{\Gamma(\gamma)\Gamma(\alpha+\beta-\gamma)}{\Gamma(\alpha)\Gamma(\beta)},$$

and relation (10) can be written

$${}_2F_1(\alpha,\beta;\gamma;x) = \frac{\Gamma(\gamma)\Gamma(\gamma-\alpha-\beta)}{\Gamma(\gamma-\alpha)\Gamma(\gamma-\beta)}\,{}_2F_1(\alpha,\beta;\alpha+\beta-\gamma+1;1-x)$$

$$+\frac{\Gamma(\gamma)\Gamma(\alpha+\beta-\gamma)}{\Gamma(\alpha)\Gamma(\beta)}(1-x)^{\gamma-\alpha-\beta}\,{}_2F_1(\gamma-\beta,\gamma-\alpha;\gamma-\alpha-\beta+1;1-x). \quad (11)$$

Exercises 5(*b*)

1. Show that

   (i) $\log_e\left(\frac{1+x}{1-x}\right) = 2x\,{}_2F_1(\tfrac{1}{2}, 1; \tfrac{3}{2}; x^2)$,

   (ii) $(1-x)^{-2a-1}(1+x) = {}_2F_1(2a, a+1; a; x)$.

2. If n is a positive integer, prove that

$$\frac{d^n}{dx^n}\{{}_2F_1(a, \beta; \gamma; x)\} = \frac{(a)_n(\beta)_n}{(\gamma)_n}\,{}_2F_1(a+n, \beta+n; \gamma+n; x).$$

   Show also that

$$\frac{d}{dx}\{x^{\gamma-1}\,{}_2F_1(a, \beta; \gamma; x)\} = (\gamma-1)x^{\gamma-2}\,{}_2F_1(a, \beta; \gamma-1; x).$$

3. Use the method of Example 5, page 132, to show that

   $\sin nx = n \sin x\,{}_2F_1(\tfrac{1}{2}+\tfrac{1}{2}n, \tfrac{1}{2}-\tfrac{1}{2}n; \tfrac{3}{2}; \sin^2 x)$, $(-\tfrac{1}{2}\pi < x < \tfrac{1}{2}\pi)$.

4. Verify the following three examples of the Gauss recursion formulae:

   (i) $(\gamma-a-1)\,{}_2F_1(a, \beta; \gamma; x)+a\,{}_2F_1(a+1, \beta; \gamma; x)$
   $-(\gamma-1)\,{}_2F_1(a, \beta; \gamma-1; x) = 0;$

(ii) $\gamma(1-x)\,{}_2F_1(\alpha, \beta; \gamma; x)+(\gamma-\beta)x\,{}_2F_1(\alpha, \beta; \gamma+1; x)$
$$-\gamma\,{}_2F_1(\alpha-1, \beta; \gamma; x) = 0;$$
(iii) $(\beta-\alpha)(1-x)\,{}_2F_1(\alpha, \beta; \gamma; x)+(\gamma-\beta)\,{}_2F_1(\alpha, \beta-1; \gamma; x)$
$$-(\gamma-\alpha)\,{}_2F_1(\alpha-1, \beta; \gamma; x) = 0.$$

5. Verify the formulae:

(i) $\gamma\,{}_2F_1(\alpha, \beta-1; \gamma; x)-\gamma\,{}_2F_1(\alpha-1, \beta; \gamma; x)$
$$+(\alpha-\beta)x\,{}_2F_1(\alpha, \beta; \gamma+1; x) = 0;$$
(ii) $\gamma\,{}_2F_1(\alpha, \beta; \gamma; x)-\gamma\,{}_2F_1(\alpha, \beta+1; \gamma; x)$
$$+\alpha x\,{}_2F_1(\alpha+1, \beta+1; \gamma+1; x) = 0;$$
(iii) $\gamma\,{}_2F_1(\alpha, \beta; \gamma; x)-(\gamma-\beta)\,{}_2F_1(\alpha, \beta; \gamma+1; x)$
$$-\beta\,{}_2F_1(\alpha, \beta+1; \gamma+1; x) = 0.$$

6. If n is a positive integer, show that
$$\int_0^{\pi/2} (1-x\sin^2\theta)^n\,d\theta = \tfrac{1}{2}\pi\,{}_2F_1(-n, \tfrac{1}{2}; 1; x).$$

7. By using the transformation $t = (1-\mu)/(1-x\mu)$ in the integral representation of the hypergeometric function [§5.9 (*c*)], show that
$${}_2F_1(\alpha, \beta; \gamma; x) = (1-x)^{\gamma-\alpha-\beta}\,{}_2F_1(\gamma-\alpha, \gamma-\beta; \gamma; x).$$

8. If $\gamma > \lambda > 0$ and $x \neq 1$, show that
$${}_2F_1(\alpha, \beta; \gamma; x) = \frac{\Gamma(\gamma)}{\Gamma(\lambda)\Gamma(\gamma-\lambda)}\int_0^1 t^{\lambda-1}(1-t)^{\gamma-\lambda-1}\,{}_2F_1(\alpha, \beta; \lambda; xt)\,dt.$$

9. Show that the substitution $y = (1-x)^n z$ transforms the hypergeometric equation into another hypergeometric equation if $n = \gamma-\alpha-\beta$. Hence, show that
$$(1-x)^{\gamma-\alpha-\beta}\,{}_2F_1(\gamma-\alpha, \gamma-\beta; \gamma; x)$$
and
$$x^{1-\gamma}(1-x)^{\gamma-\alpha-\beta}\,{}_2F_1(1-\alpha, 1-\beta; 2-\gamma; x)$$
are solutions of the hypergeometric equation.

10. Show that
$${}_2F_1(\alpha, \beta; \gamma+1; 1) = \frac{\gamma(\gamma-\alpha-\beta)}{(\gamma-\alpha)(\gamma-\beta)}\,{}_2F_1(\alpha, \beta; \gamma; 1).$$

11. Deduce from Example 5, page 132, that
$$\frac{\sin(n\sin^{-1}\theta)}{n(1-\theta^2)^{1/2}} = {}_2F_1(1+\tfrac{1}{2}n, 1-\tfrac{1}{2}n; \tfrac{3}{2}; \theta^2).$$

12. The *incomplete Beta function* $B_x(p, q)$ is defined by
$$B_x(p, q) = \int_0^x t^{p-1}(1-t)^{q-1}\,dt.$$
Show that
$$B_x(p, q) = p^{-1}x^p\,{}_2F_1(p, 1-q; p+1; x)$$
and deduce that
$$B_1(p, q) = \frac{\Gamma(p)\Gamma(q)}{\Gamma(p+q)}.$$

13. Show that, when $\alpha > 0$, $\beta > 0$,

$$\int_0^1 \{(1-t)^\alpha(1+t)^\beta+(1-t)^\beta(1+t)^\alpha\}\,dt = \int_{-1}^1 (1-t)^\alpha(1+t)^\beta\,dt = 2^{\alpha+\beta+1}B(\alpha+1, \beta+1).$$

Deduce that

$$(\alpha+1)\,{}_2F_1(-\alpha, 1; \beta+2; -1)+(\beta+1)\,{}_2F_1(-\beta, 1; \alpha+2; -1) = 2^{\alpha+\beta+1}\frac{\Gamma(\alpha+2)\Gamma(\beta+2)}{\Gamma(\alpha+\beta+2)}.$$

14. If $0 < x < 1$, show that the general solution of the equation

$$x(1-x)\frac{d^2y}{dx^2}+\tfrac{1}{2}(\alpha+\beta+1)(1-2x)\frac{dy}{dx}-\alpha\beta y = 0$$

is

$$A\,{}_2F_1\{\tfrac{1}{2}\alpha, \tfrac{1}{2}\beta; \tfrac{1}{2}; (1-2x)^2\}+B(1-2x)\,{}_2F_1\{\tfrac{1}{2}+\tfrac{1}{2}\alpha, \tfrac{1}{2}+\tfrac{1}{2}\beta; \tfrac{3}{2}; (1-2x)^2\}$$

where A and B are arbitrary constants.

15. If $a > b > c$, $p > 0$, $q > 0$ and

$$I = \int_a^b (t-a)^{p-1}(t-b)^{q-1}(t-c)^{r-1}\,dt,$$

show that

$$I = (-1)^p(a-b)^{p+q-1}(a-c)^{r-1}\frac{\Gamma(p)\Gamma(q)}{\Gamma(p+q)}\,{}_2F_1\left(1-r, p; p+q; \frac{a-b}{a-c}\right).$$

16. Find the necessary value of n so that x^n shall be one solution of the equation

$$2x(1-x)\frac{d^2y}{dx^2}+(3-4x)\frac{dy}{dx}-\tfrac{1}{2}y = 0,$$

and use the formula (2) of §5.7 to find the second solution. Deduce an expression for $\sin^{-1} x$ in terms of a hypergeometric function.

17. Show that *Elliott's equation*

$$\frac{d^2y}{dx^2}-\left\{\frac{n(n+1)}{\sin^2 x}+h\right\}y = 0$$

has a solution $(\sin x)^{-n}\,{}_2F_1(-\tfrac{1}{2}n-\tfrac{1}{2}m, -\tfrac{1}{2}n+\tfrac{1}{2}m; \tfrac{1}{2}-n; \sin^2 x)$, m being an integer and $h = -m^2$.

CHAPTER 6

SOME SPECIAL FUNCTIONS

6.1 Introduction

The functions discussed in this chapter are usually known as the special functions of mathematical physics, the adjective special being appropriate because the functions appear in special, rather than in general, physical problems. These functions arise as solutions of second order differential equations and most of the equations concerned can be derived from the hypergeometric equation by a change of variable or by a limiting process. The functions considered here are, however, of such fundamental importance that they warrant more detailed study than merely a statement that they are special cases of the hypergeometric function and it is the purpose of this chapter to discuss the properties of the more important special functions.

6.2 Legendre's equation

Legendre's differential equation

$$(1-x^2)\frac{d^2y}{dx^2}-2x\frac{dy}{dx}+\nu(\nu+1)y = 0, \qquad (1)$$

in which ν is a positive constant, is of great importance in potential theory and in many other fields. A particularly important special case arises when ν is an integer n and we first show that the equation then has a solution which is a polynomial of degree n in x. To do this, let $y_1 = (x^2-1)^n$ and use the notation D^r to denote the operation d^r/dx^r. Then $Dy_1 = 2nx(x^2-1)^{n-1}$ and we have

$$(x^2-1)\,Dy_1-2nxy_1 = 0.$$

Differentiating $(n+1)$ times by Leibnitz' theorem,

$$(x^2-1)D^{n+2}y_1+2(n+1)xD^{n+1}y_1+n(n+1)D^ny_1$$
$$-2nxD^{n+1}y_1-2n(n+1)D^ny_1 = 0,$$

and this reduces to

$$(x^2-1)D^{n+2}y_1+2xD^{n+1}y_1-n(n+1)D^ny_1 = 0.$$

This shows that D^ny_1 is a solution of equation (1) when $\nu = n$ and we conclude that any multiple of $\dfrac{d^n}{dx^n}\{(x^2-1)^n\}$ satisfies Legendre's equation in this case.

The *Legendre polynomial* $P_n(x)$ of degree n is defined by the equation

$$P_n(x) = \frac{1}{2^n n!}\frac{d^n}{dx^n}\{(x^2-1)^n\}, \tag{2}$$

and we have shown that $P_n(x)$ satisfies Legendre's equation when $\nu = n$. Equation (2) is known as *Rodrigues' formula* and the multiplying factor $1/(2^n n!)$ is chosen so that $P_n(1) = 1$ for all integers n, this being convenient in subsequent work. The Legendre polynomials for $n = 0, 1, 2, 3, 4, \ldots$ are easily found from equation (2) to be

$$P_0(x) = 1, \qquad P_3(x) = \tfrac{1}{2}(5x^3-3x),$$
$$P_1(x) = x, \qquad P_4(x) = \tfrac{1}{8}(35x^4-30x^2+3),$$
$$P_2(x) = \tfrac{1}{2}(3x^2-1), \qquad \ldots\ldots\ldots\ldots\ldots\ldots,$$

and some of the more important properties of these polynomials are discussed in §6.3.

When ν is not an integer, the above method of finding a solution of Legendre's equation is not, of course, available. However, by writing $t = \frac{1}{2}-\frac{1}{2}x$, the equation transforms into

$$t(1-t)\frac{d^2y}{dt^2}+(1-2t)\frac{dy}{dt}+\nu(\nu+1)y = 0, \tag{3}$$

and this is the hypergeometric equation with $\alpha = \nu+1$, $\beta = -\nu$, $\gamma = 1$. With these values of α, β and γ, equation (5) of §5.8 shows that a solution, relative to the regular singular point $t = 0$ (or $x = 1$), is ${}_2F_1(\nu+1, -\nu; 1; t)$. Hence, remembering that $t = \frac{1}{2}-\frac{1}{2}x$, one solution of Legendre's equation, usually denoted by $P_\nu(x)$, is given by

$$P_\nu(x) = {}_2F_1(\nu+1, -\nu; 1; \tfrac{1}{2}-\tfrac{1}{2}x), \tag{4}$$

this solution being valid when $-1 < x < 3$. The function $P_\nu(x)$ is called the *Legendre function of the first kind of order* ν and it will be shown in equation (4) of §6.3 that it reduces to the Legendre polynomial $P_n(x)$ when ν is an integer n. There is, of course, a second solution of Legendre's equation: however, this solution is usually only required in practical applications when $|x| > 1$ and we shall only discuss it for such values of x.

Solutions of Legendre's equation relative to the regular singular point at infinity can be investigated by writing $x^2 = t$. With this substitution,

$$\frac{dy}{dx} = \frac{dy}{dt}\frac{dt}{dx} = 2x\frac{dy}{dt} = 2t^{\frac{1}{2}}\frac{dy}{dt},$$

$$\frac{d^2y}{dx^2} = 2\frac{dy}{dt}+4t\frac{d^2y}{dt^2},$$

and the differential equation (1) becomes

$$(1-t)\left(4t\frac{d^2y}{dt^2}+2\frac{dy}{dt}\right)-2t^{\frac{1}{2}}\left(2t^{\frac{1}{2}}\frac{dy}{dt}\right)+\nu(\nu+1)y=0,$$

that is,

$$t(1-t)\frac{d^2y}{dt^2}+(\tfrac{1}{2}-\tfrac{3}{2}t)\frac{dy}{dt}+\frac{\nu(\nu+1)}{4}y=0.$$

This is the hypergeometric equation with $\alpha=-\frac{1}{2}\nu$, $\beta=\frac{1}{2}+\frac{1}{2}\nu$, $\gamma=\frac{1}{2}$ and, by equation (8) of §5.8, its solutions, valid when $|t|>1$, are multiples of

$$t^{\frac{1}{2}\nu}{}_2F_1(-\tfrac{1}{2}\nu,\tfrac{1}{2}-\tfrac{1}{2}\nu;\tfrac{1}{2}-\nu;t^{-1})$$

and

$$t^{-\frac{1}{2}-\frac{1}{2}\nu}{}_2F_1(\tfrac{1}{2}+\tfrac{1}{2}\nu,1+\tfrac{1}{2}\nu;\tfrac{3}{2}+\nu;t^{-1}). \tag{5}$$

Remembering that $t=x^2$, and writing

$$P_\nu(x)=\frac{\Gamma(2\nu+1)}{2^\nu\Gamma^2(\nu+1)}x^\nu{}_2F_1\left(-\tfrac{1}{2}\nu,\tfrac{1}{2}-\tfrac{1}{2}\nu;\tfrac{1}{2}-\nu;\frac{1}{x^2}\right), \tag{6}$$

it will be shown in equation (3) §6.3 that this reduces to the Legendre polynomial when ν is an integer. Formula (6), being valid when $|x|>1$, provides a convenient extension of formula (4). A second solution of Legendre's equation, valid when $|x|>1$, can be taken to be the function $Q_\nu(x)$ defined by

$$Q_\nu(x)=\frac{\Gamma(\frac{1}{2})\Gamma(\nu+1)}{2^{\nu+1}\Gamma(\nu+\frac{3}{2})}x^{-\nu-1}{}_2F_1\left(\tfrac{1}{2}+\tfrac{1}{2}\nu,1+\tfrac{1}{2}\nu;\tfrac{3}{2}+\nu;\frac{1}{x^2}\right), \tag{7}$$

since it is a constant multiple of the second solution given in (5) when we write $t=x^2$. The function $Q_\nu(x)$ is called the *Legendre function of the second kind of order* ν, and the general solution of Legendre's equation (1) can be written

$$y=AP_\nu(x)+BQ_\nu(x), \tag{8}$$

A and B being arbitrary constants.

6.3 Some properties of Legendre polynomials

The Legendre polynomials $P_n(x)$ are often used in applied mathematics and this section is devoted to investigating some of their properties. We take, as their definition,

$$P_n(x)=\frac{1}{2^n n!}\frac{d^n}{dx^n}\{(x^2-1)^n\}, \tag{1}$$

and use the notation $D\equiv d/dx$ as necessary.

(a) The expression of $P_n(x)$ as a hypergeometric function

Expanding $(x^2-1)^n$ by the binomial theorem, equation (1) gives

$$P_n(x) = \frac{1}{2^n n!} D^n \left\{ x^{2n} - nx^{2n-2} + \frac{n(n-1)}{2!} x^{2n-4} - \dots \right\}. \tag{2}$$

Now, if $r < n$, $D^n x^r = 0$ and, if $r \geqq n$,

$$D^n x^r = r(r-1)(r-2)\dots(r-n+1)x^{r-n} = \frac{r!}{(r-n)!} x^{r-n}.$$

Hence,

$$D^n x^{2n} = \frac{(2n)!}{n!} x^n, \qquad D^n x^{2n-4} = \frac{(2n-4)!}{(n-4)!} x^{n-4},$$

$$D^n x^{2n-2} = \frac{(2n-2)!}{(n-2)!} x^{n-2}, \qquad \dots\dots\dots\dots ,$$

so that (2) gives

$$P_n(x) = \frac{1}{2^n n!} \left\{ \frac{(2n)!}{n!} x^n - \frac{n(2n-2)!}{1!(n-2)!} x^{n-2} + \frac{n(n-1)(2n-4)!}{2!(n-4)!} x^{n-4} - \dots \right\}$$

$$= \frac{(2n)!\, x^n}{2^n n!\, n!} \left\{ 1 + \frac{(-\frac{1}{2}n)(\frac{1}{2}-\frac{1}{2}n)}{1!(\frac{1}{2}-n)} \cdot \frac{1}{x^2} + \frac{(-\frac{1}{2}n)(1-\frac{1}{2}n)(\frac{1}{2}-\frac{1}{2}n)(\frac{3}{2}-\frac{1}{2}n)}{2!(\frac{1}{2}-n)(\frac{3}{2}-n)} \cdot \frac{1}{x^4} + \dots \right\}$$

$$= \frac{\Gamma(2n+1)}{2^n \Gamma^2(n+1)} x^n\, {}_2F_1\left(-\tfrac{1}{2}n, \tfrac{1}{2}-\tfrac{1}{2}n; \tfrac{1}{2}-n; \frac{1}{x^2} \right), \tag{3}$$

see §5.8 (4).

The polynomial $P_n(x)$ can also be expressed in terms of a hypergeometric function of argument $\frac{1}{2}-\frac{1}{2}x$. This can be done by writing $x = 1-2t$ and noting that

$$(x^2-1)^n = 2^{2n} t^n (t-1)^n \quad \text{and} \quad d/dx = -\tfrac{1}{2}\, d/dt.$$

Hence,

$$P_n(x) = \frac{1}{2^n n!} \frac{d^n}{dx^n} (x^2-1)^n = \frac{1}{n!} \frac{d^n}{dt^n} \{ t^n (1-t)^n \}$$

$$= \frac{1}{n!} \frac{d^n}{dt^n} \left\{ t^n - \frac{n}{1!} t^{n+1} + \frac{n(n-1)}{2!} t^{n+2} - \frac{n(n-1)(n-2)}{3!} t^{n+3} + \dots \right\}$$

$$= \frac{1}{n!}\left\{n! - \frac{n}{1!}\cdot\frac{(n+1)!}{1!}t + \frac{n(n-1)}{2!}\cdot\frac{(n+2)!}{2!}t^2 - \frac{n(n-1)(n-2)}{3!}\cdot\frac{(n+3)!}{3!}t^3 + \dots\right\}$$

$$= 1 + \frac{(n+1)(-n)}{1!\ 1!}t + \frac{(n+1)(n+2)(-n)(1-n)}{2!\quad 2!}t^2 + \frac{(n+1)(n+2)(n+3)(-n)(1-n)(2-n)}{3!\qquad 3!}t^3 + \dots$$

$$= {}_2F_1(n+1, -n; 1; t) = {}_2F_1(n+1, -n; 1; \tfrac{1}{2}-\tfrac{1}{2}x). \tag{4}$$

Equation (4) is known as *Murphy's formula.* With $x = \cos\theta$, this gives the useful result

$$P_n(\cos\theta) = {}_2F_1(n+1, -n; 1; \sin^2\tfrac{1}{2}\theta). \tag{5}$$

Numerical values of $P_n(x)$ for $n = 2, 3, \dots, 12$ and x ranging from 0 to 6 at intervals of 0.01 can be found in the British Association Mathematical Tables, Part-Volume A, 1946. These tables also give the values for $n = 2, 3, \dots, 6$ when x ranges from 6 to 11 at intervals of 0.1.

(b) Recurrence formulae

A group of recurrence formulae relating Legendre polynomials of degrees $n+1$, n, $n-1$ and their derivatives can be established by manipulating the fundamental formula (1). Thus

$$P_{n+1}(x) = \frac{1}{2^{n+1}(n+1)!}D^{n+1}\{(x^2-1)^{n+1}\}$$

$$= \frac{1}{2^{n+1}(n+1)!}D^n\{D(x^2-1)^{n+1}\}$$

$$= \frac{1}{2^n n!}D^n\{x(x^2-1)^n\}$$

$$= \frac{1}{2^n n!}[xD^n\{(x^2-1)^n\} + nD^{n-1}\{(x^2-1)^n\}],$$

using Leibnitz' theorem,

$$= xP_n(x) + \frac{1}{2^n(n-1)!}D^{n-1}\{(x^2-1)^n\}. \tag{6}$$

Differentiating with respect to x and using primes to denote derivatives,

$$P'_{n+1}(x) = P_n(x) + xP_n'(x) + \frac{1}{2^n(n-1)!}D^n\{(x^2-1)^n\}$$
$$= P_n(x) + xP'_n(x) + nP_n(x),$$

giving, as the first of the required formulae,

$$P'_{n+1}(x) - xP'_n(x) - (n+1)P_n(x) = 0. \tag{7}$$

Starting with the identity

$$(x^2-1) + 2nx^2 - n(x^2-1) - 2n \equiv (n+1)(x^2-1),$$

we have, on multiplying by $(x^2-1)^{n-1}$ and differentiating $(n-1)$ times with respect to x,

$$D^{n-1}\{(x^2-1)^n\} + 2nD^{n-1}\{x^2(x^2-1)^{n-1}\} - nD^{n-1}\{(x^2-1)^n\}$$
$$-2nD^{n-1}\{(x^2-1)^{n-1}\} = (n+1)D^{n-1}\{(x^2-1)^n\}.$$

Since $D^{n-1}\{(x^2-1)^n + 2nx^2(x^2-1)^{n-1}\} = D^{n-1}[D\{x(x^2-1)^n\}]$, this can be written

$$D^n\{x(x^2-1)^n\} - nD^{n-1}\{(x^2-1)^n\} - 2nD^{n-1}\{(x^2-1)^{n-1}\}$$
$$= (n+1)D^{n-1}\{(x^2-1)^n\},$$

that is,

$$xD^n\{(x^2-1)^n\} - 2nD^{n-1}\{(x^2-1)^{n-1}\} = (n+1)D^{n-1}\{(x^2-1)^n\}.$$

Dividing by $2^n n!$, this becomes, on using (1),

$$xP_n(x) - P_{n-1}(x) = \frac{n+1}{2^n n!}D^{n-1}\{(x^2-1)^n\},$$

and substituting for $D^{n-1}\{(x^2-1)^n\}$ from equation (6), we have

$$xP_n(x) - P_{n-1}(x) = \frac{n+1}{n}\{P_{n+1}(x) - xP_n(x)\}.$$

A slight rearrangement gives the second recurrence formula in the form

$$(n+1)P_{n+1}(x) - (2n+1)xP_n(x) + nP_{n-1}(x) = 0. \tag{8}$$

Three other useful recurrence formulae are

$$xP'_n(x) - P'_{n-1}(x) - nP_n(x) = 0, \tag{9}$$

$$P'_{n+1}(x) - (2n+1)P_n(x) - P'_{n-1}(x) = 0, \tag{10}$$

$$(x^2-1)P'_n(x) - nxP_n(x) + nP_{n-1}(x) = 0. \tag{11}$$

Omitting the full details, which can be filled in by the reader, these can be obtained as follows:

(i) differentiation of equation (8) with respect to x and the use of (7) to eliminate $P'_{n+1}(x)$ leads to the relation (9);

(ii) the addition of equations (7) and (9) immediately yields relation (10);

(iii) replacement of n by $n-1$ in equation (7) and the elimination of $P'_{n-1}(x)$ between the equation so obtained and equation (9) gives relation (11).

It is worth pointing out that the Legendre functions of the first and second kinds $P_\nu(x)$ and $Q_\nu(x)$ satisfy the same five recurrence formulae but a proof will not be attempted here.

Example 1. *Prove that* $P_5'(x) = 9P_4(x)+5P_2(x)+P_0(x)$. (L.U.)

Setting $n = 4$ and 2 in the recurrence relation (10),

$$P_5'(x)-P_3'(x) = 9P_4(x),$$
$$P_3'(x)-P_1'(x) = 5P_2(x).$$

Remembering that $P_1(x) = x$ and hence $P_1'(x) = 1 = P_0(x)$, the required result follows by addition.

It is worth noticing that this example is a special case of the more general result (which can be proved in the same way or by the method used in Example 3, page 149),

$$P'_{n+1}(x) = (2n+1)P_n(x)+(2n-3)P_{n-2}(x)+(2n-7)P_{n-4}(x)+\ldots.$$

(c) *The generating function for $P_n(x)$*

Assuming that the quantities involved are such that the expansion by the binomial theorem and the subsequent rearrangement of terms are valid, we have

$$\begin{aligned}(1-2xh+h^2)^{-\frac{1}{2}} &= \{1-h(2x-h)\}^{-\frac{1}{2}}\\ &= 1+\tfrac{1}{2}h(2x-h)+\tfrac{3}{8}h^2(2x-h)^2+\ldots\\ &= 1+hx+\tfrac{1}{2}h^2(3x^2-1)+\ldots\\ &= P_0(x)+hP_1(x)+h^2P_2(x)+\ldots.\end{aligned}$$

This suggests that the coefficient of h^n in the expansion of $(1-2xh+h^2)^{-\frac{1}{2}}$ in ascending powers of h might well be $P_n(x)$. That this is so can be shown as follows.

Let $y = (1-2xh+h^2)^{-\frac{1}{2}}$ so that

$$(1-2xh+h^2)\frac{dy}{dh}+(h-x)y = 0,$$

that is,

$$\frac{dy}{dh}-x\left(2h\frac{dy}{dh}+y\right)+\left(h^2\frac{dy}{dh}+hy\right) = 0.$$

If $y = a_0+a_1h+a_2h^2+\ldots+a_nh^n+\ldots$, substituting in this differential equation for y and equating to zero the coefficient of h^n gives

$$(n+1)a_{n+1}-(2n+1)xa_n+na_{n-1} = 0.$$

Now we have shown that $a_0 = P_0(x)$, $a_1 = P_1(x)$, $a_2 = P_2(x)$ and, the above relation between a_{n+1}, a_n and a_{n-1} being identical with the recurrence relation (8) for the Legendre polynomials, it follows that $a_n = P_n(x)$. Hence, with suitable restrictions on the values of x and h, we have established the important result

$$(1-2xh+h^2)^{-\frac{1}{2}} = \sum_{n=0}^{\infty} h^nP_n(x), \tag{12}$$

and $(1-2xh+h^2)^{-\frac{1}{2}}$ is referred to as the *generating function* of the Legendre polynomials. In most practical applications, $|x| \leqq 1$ and it can be shown then that $|P_n(x)| \leqq 1$ (see Exercises 6(*a*), 6). For such values of x, the relation (12) is valid when $|h| < 1$.

The above result has an application in potential theory. If a unit mass is situated at a point A with polar coordinates $(a, 0)$ (Fig. 9), the gravitational potential ψ at the point $P(r, \theta)$ is given by

$$\psi = \frac{1}{PA} = \frac{1}{(r^2-2ra\cos\theta+a^2)^{\frac{1}{2}}}.$$

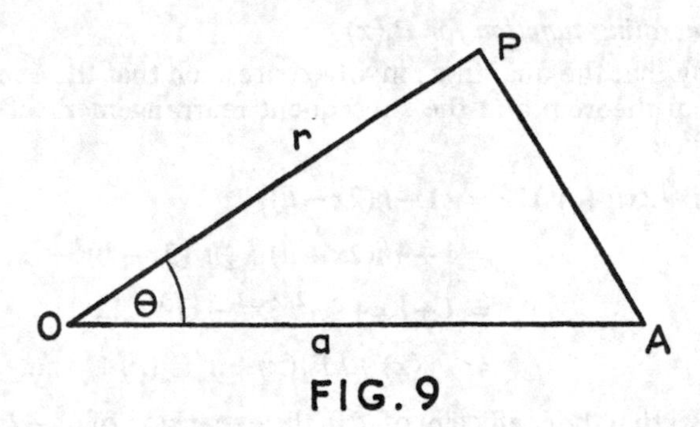

FIG. 9

It is often desirable to expand ψ in ascending or descending powers of r and this is easily done by using (12). Thus, taking $x = \cos\theta$ and $h = r/a$, a/r in turn, we have

$$\left.\begin{aligned} \psi &= \frac{1}{a}\sum_{n=0}^{\infty}\left(\frac{r}{a}\right)^n P_n(\cos\theta), \qquad r < a, \\ \psi &= \frac{1}{r}\sum_{n=0}^{\infty}\left(\frac{a}{r}\right)^n P_n(\cos\theta), \qquad r > a. \end{aligned}\right\} \tag{13}$$

Example 2. *Find the values of $P_n(1)$, $P_n(-1)$, $P_n(0)$.*

By setting $x = 1$ in equation (12),

$$\sum_{n=0}^{\infty} h^n P_n(1) = (1-2h+h^2)^{-1/2} = (1-h)^{-1} = \sum_{n=0}^{\infty} h^n,$$

so that, by equating the coefficients of h^n, $P_n(1) = 1$. Similarly,

$$\sum_{n=0}^{\infty} h^n P_n(-1) = (1+2h+h^2)^{-1/2} = (1+h)^{-1} = \sum_{n=0}^{\infty} (-1)^n h^n,$$

leading to

$$P_n(-1) = (-1)^n.$$

Finally,

$$\sum_{n=0}^{\infty} h^n P_n(0) = (1+h^2)^{-1/2} = \sum_{n=0}^{\infty} (-1)^n \frac{(2n)!}{2^{2n} n!\, n!} h^{2n},$$

giving

$$P_{2n+1}(0) = 0, \qquad P_{2n}(0) = (-1)^n \frac{(2n)!}{2^{2n} n!\, n!}.$$

(*d*) *Definite integrals involving Legendre polynomials*

Integrals of the form

$$I(n) = \int_{-1}^{1} f(x) P_n(x)\, dx \tag{14}$$

are often required and these can be evaluated by making use of Rodrigues' formula. Thus

$$I(n) = \frac{1}{2^n n!} \int_{-1}^{1} f(x) \frac{d^n}{dx^n} \{(x^2-1)^n\}\, dx,$$

and an integration by parts gives

$$I(n) = \frac{1}{2^n n!} \left[f(x) \frac{d^{n-1}}{dx^{n-1}} \{(x^2-1)^n\} \right]_{-1}^{1} - \frac{1}{2^n n!} \int_{-1}^{1} f'(x) \frac{d^{n-1}}{dx^{n-1}} \{(x^2-1)^n\}\, dx.$$

Provided $f(x)$ is finite when $x = \pm 1$, the first term on the right vanishes since

$$\frac{d^{n-1}}{dx^{n-1}} \{(x^2-1)^n\} = \frac{d^{n-1}}{dx^{n-1}} \{(x-1)^n (x+1)^n\},$$

and, using Leibnitz' theorem, each term will contain at least the first power of $(x-1)$ and $(x+1)$. Hence

$$I(n) = -\frac{1}{2^n n!} \int_{-1}^{1} f'(x) \frac{d^{n-1}}{dx^{n-1}} \{(x^2-1)^n\}\, dx.$$

By continuing this process we ultimately obtain

$$I(n) = \frac{(-1)^n}{2^n n!}\int_{-1}^{1} (x^2-1)^n f^{(n)}(x)\,dx. \tag{15}$$

As an important example of this result, take $f(x) = P_m(x)$ where $m < n$. $P_m(x)$ being a polynomial in x of degree m less than n, it follows that $f^{(n)}(x) \equiv 0$ and therefore

$$\int_{-1}^{1} P_m(x)P_n(x)\,dx = 0, \qquad m \neq n. \tag{16}$$

If $f(x) = P_n(x)$,

$$f^{(n)}(x) = \frac{d^n}{dx^n}\left[\frac{1}{2^n n!}\frac{d^n}{dx^n}\{(x^2-1)^n\}\right] = \frac{1}{2^n n!}\frac{d^{2n}}{dx^{2n}}\{(x^2-1)^n\} = \frac{(2n)!}{2^n n!}.$$

Hence, using (14) and (15),

$$\begin{aligned}\int_{-1}^{1}\{P_n(x)\}^2\,dx &= \frac{(2n)!}{2^{2n}n!n!}\int_{-1}^{1}(1-x^2)^n\,dx\\ &= \frac{(2n)!}{2^{2n}n!n!}\int_{0}^{1}(2\mu)^n(2-2\mu)^n 2\,d\mu, \quad x = 2\mu-1,\\ &= \frac{2(2n)!}{n!n!}\int_0^1 \mu^n(1-\mu)^n\,d\mu = \frac{2(2n)!}{n!n!}\cdot\frac{\Gamma(n+1)\Gamma(n+1)}{\Gamma(2n+2)}\\ &= \frac{2}{2n+1},\end{aligned} \tag{17}$$

when use is made of equation (11) §5.5.

(e) Expansions in terms of Legendre polynomials

Properties similar to those discussed in (*d*) above for the Legendre polynomials can be proved for other functions encountered in mathematical physics, and we digress for a moment to give some definitions.

If a sequence of functions $F_0(x), F_1(x), \ldots, F_n(x), \ldots$ possesses the property

$$\int_a^b F_m(x)F_n(x)\,dx = 0, \qquad m \neq n,$$

the functions are said to be *orthogonal* in the interval (a, b). If, in addition,

$$\int_a^b \{F_n(x)\}^2\,dx = 1,$$

for all values of n, the functions are said to be *normalized* and to form an *orthonormal set*. In view of the results (16) and (17), we can therefore say that the Legendre polynomials $P_n(x)$ are orthogonal in the interval $(-1, 1)$ and that the functions $(n+\frac{1}{2})^{\frac{1}{2}}P_n(x)$ form an orthonormal set in this interval.

We now consider the possibility of expanding a function $f(x)$, defined in the range $-1 \leqq x \leqq 1$, in a series of Legendre polynomials and assume that

$$f(x) = a_0P_0(x)+a_1P_1(x)+a_2P_2(x)+ \ldots +a_nP_n(x)+ \ldots, \tag{18}$$

the series being taken to converge uniformly in the range stated. The coefficient a_n can be found by multiplying by $P_n(x)$ and integrating with respect to x term by term over the given range. Carrying out this process,

$$\int_{-1}^{1} f(x)P_n(x)\,dx = a_n\int_{-1}^{1} \{P_n(x)\}^2\,dx,$$

all the other terms on the right vanishing in view of the result (16). Using the result (17), this gives

$$a_n = (n+\tfrac{1}{2})\int_{-1}^{1} f(x)P_n(x)\,dx \tag{19}$$

and the coefficients are therefore known when the integrals in (19) have been evaluated. The expansion given by equations (18) and (19) is analogous to the Fourier series expansion of a periodic function and the coefficients are calculated in a similar manner. It is emphasized that the analysis given above is purely formal. For a rigorous discussion and a precise statement of the conditions under which equations (18) and (19) are valid, reference should be made to a standard text.*

Example 3. *If p is a positive integer or zero, show that*

$$P'_{2p+1}(x) = P_0(x)+5P_2(x)+9P_4(x)+ \ldots +(4p+1)P_{2p}(x).$$

$P_{2p+1}(x)$ is a polynomial in x of degree $2p+1$, the indices of the various powers of x all being odd. Hence $P'_{2p+1}(x)$ is a polynomial of degree $2p$, the indices of the various powers of x all being even and we can write

$$P'_{2p+1}(x) = a_0P_0+a_2P_2(x)+ \ldots +a_{2p}P_{2p}(x). \tag{20}$$

Multiplying by $P_{2n}(x)$, $(n = 0, 1, 2, \ldots, p)$ and integrating with respect to x between $-1, 1$, leads to

$$\int_{-1}^{1} P'_{2p+1}(x)P_{2n}(x)\,dx = a_{2n}\int_{-1}^{1} \{P_{2n}(x)\}^2\,dx = \frac{2a_{2n}}{4n+1}, \quad (n = 0, 1, 2, \ldots, p),$$

when use is made of equation (17) and the orthogonal property (16). Hence, integrating by parts

$$\frac{2a_{2n}}{4n+1} = \Big[P_{2p+1}(x)P_{2n}(x)\Big]_{-1}^{1} - \int_{-1} P_{2p+1}(x)P'_{2n}(x)\,dx. \tag{21}$$

* See, for example, E. W. Hobson, *Spherical and Ellipsoidal Harmonics*, Cambridge, 1931.

Now, reasoning as in the derivation of equation (20),

$$P'_{2n}(x) = b_1P_1(x)+b_3P_3(x)+\ldots+b_{2n-1}P_{2n-1}(x)$$

where the b's are numerical coefficients. Since $n \leqq p$, when this is substituted in the integral on the right of (21), the orthogonal property of the polynomials shows that the integral vanishes and equation (21) gives

$$\frac{2a_{2n}}{4n+1} = P_{2p+1}(1)P_{2n}(1)-P_{2p+1}(-1)P_{2n}(-1) = 1-(-1)^{2p+2n+1} = 2,$$

when the results of Example 2 are used. Hence $a_{2n} = 4n+1$ and the required result follows.

6.4 Legendre's associated equation

The second order differential equation

$$(1-x^2)\frac{d^2y}{dx^2}-2x\frac{dy}{dx}+\left\{n(n+1)-\frac{m^2}{1-x^2}\right\}y = 0, \tag{1}$$

in which m and n are constants is known as *Legendre's associated equation* and its solutions are called *associated Legendre functions.* It reduces to Legendre's equation when $m = 0$. The equation arises in physical problems in which spherical polar coordinates are appropriate and, in practice, m, n are positive integers and $m \leqq n$. We consider below the solution for such values of m and n.

Writing

$$y = (x^2-1)^{\frac{1}{2}m}u \tag{2}$$

and using D to denote the operation d/dx, we have

$$Dy = (x^2-1)^{\frac{1}{2}m}Du+mx(x^2-1)^{\frac{1}{2}m-1}u,$$

and

$$D^2y = (x^2-1)^{\frac{1}{2}m}D^2u+2mx(x^2-1)^{\frac{1}{2}m-1}Du$$
$$+\{m(x^2-1)^{\frac{1}{2}m-1}+m(m-2)x^2(x^2-1)^{\frac{1}{2}m-2}\}u.$$

Substitution in equation (1), division by $(x^2-1)^{\frac{1}{2}m}$ and a slight reduction leads to

$$(1-x^2)\frac{d^2u}{dx^2}-2(1+m)x\frac{du}{dx}+(n-m)(n+m+1)u = 0. \tag{3}$$

If v satisfies the ordinary Legendre equation

$$(1-x^2)\frac{d^2v}{dx^2}-2x\frac{dv}{dx}+n(n+1)v = 0,$$

and we use Leibnitz' theorem to differentiate n times with respect to x, we have

$$(1-x^2)D^{m+2}v-2mxD^{m+1}v-m(m-1)D^mv$$
$$-2xD^{m+1}v-2mD^mv+n(n+1)D^mv = 0,$$

giving

$$(1-x^2)D^{m+2}v-2(1+m)xD^{m+1}v+(n-m)(n+m+1)D^m v = 0. \quad (4)$$

A comparison of equations (3) and (4) shows that we can take $u = D^m v$ and, combining this with (2), we have shown that a solution of Legendre's associated equation (1) is given by

$$y = (x^2-1)^{\frac{1}{2}m}D^m v \quad (5)$$

where v is a solution of the ordinary Legendre equation of order n.

Using the symbols $P_n^m(x)$, $Q_n^m(x)$ to denote independent solutions of Legendre's associated equation, equation (5) shows that we can write

$$P_n^m(x) = (x^2-1)^{\frac{1}{2}m}\frac{d^m}{dx^m}\{P_n(x)\}, \quad (6)$$

$$Q_n^m(x) = (x^2-1)^{\frac{1}{2}m}\frac{d^m}{dx^m}\{Q_n(x)\}. \quad (7)$$

These solutions have properties analogous to those of the Legendre polynomials and some of these will be found in Exercises 6(*a*).

Exercises 6(a)

1. Use equation (7) of §6.2 to show that, for $x > 1$,

$$Q_0(x) = \frac{1}{2}\log_e\left(\frac{x+1}{x-1}\right),$$

$$Q_1(x) = \frac{1}{2}x\log_e\left(\frac{x+1}{x-1}\right)-1.$$

2. Use the results of Exercise 1 above and the recurrence formulae to show that, for $x > 1$,

$$Q_2(x) = \frac{1}{4}(3x^2-1)\log_e\left(\frac{x+1}{x-1}\right)-\frac{3}{2}x,$$

$$Q_3(x) = \frac{1}{4}(5x^3-3x)\log_e\left(\frac{x+1}{x-1}\right)-\frac{5}{2}x^2+\frac{2}{3}.$$

3. Show that the results of Exercises 1 and 2 above are particular cases of the general formula

$$Q_n(x) = \frac{1}{2}P_n(x)\log_e\left(\frac{x+1}{x-1}\right)-W_{n-1}(x),$$

where $W_{-1}(x) = 0$ and

$$W_{n-1}(x) = \frac{2n-1}{1.n}P_{n-1}(x)+\frac{2n-5}{3(n-1)}P_{n-3}(x)+\frac{2n-9}{5(n-2)}P_{n-5}(x)+\ldots,$$

for $n = 1, 2, 3, \ldots$, the series terminating at $P_1(x)$ or $P_0(x)$.

4. If $|h| > 1$, show that

$$h(1-2xh+h^2)^{-3/2} = \sum_{n=1}^{\infty} h^n P_n'(x).$$

Deduce that

$$P_n'(1) = \tfrac{1}{2}n(n+1), \quad P_n'(-1) = (-1)^{n-1}\tfrac{1}{2}n(n+1).$$

5. Show that $y = P_n(\cos\theta)$ satisfies the differential equation

$$\frac{d^2y}{d\theta^2} + \cot\theta\,\frac{dy}{d\theta} + n(n+1)y = 0.$$

6. By using the generating function for the Legendre polynomials and writing $(1-2h\cos\theta+h^2) \equiv (1-he^{i\theta})(1-he^{-i\theta})$, show that

$$P_n(\cos\theta) = \frac{1.3\ldots(2n-1)}{2.4\ldots(2n)}\left\{2\cos n\theta + \frac{1.(2n)}{2.(2n-1)}2\cos(n-2)\theta + \frac{1.3.(2n)(2n-2)}{2.4.(2n-1)(2n-3)}2\cos(n-4)\theta + \ldots\right\}.$$

Deduce that $|P_n(\cos\theta)| \leqq 1$.

7. If m and n are integers and $I_{m,n} = \int_{-1}^{1} x^m P_n(x)\,dx$, show that $I_{m,n}$ has the value zero when $m < n$ and when $m-n > 0$ is odd. Show also that, when $m-n \geqq 0$ is even

$$I_{m,n} = \frac{\Gamma(m+1)\Gamma(\frac{1}{2}m-\frac{1}{2}n+\frac{1}{2})}{2^n\Gamma(m-n+1)\Gamma(\frac{1}{2}m+\frac{1}{2}n+\frac{3}{2})}.$$

8. Show that

(i) $\displaystyle\int_{-1}^{1} xP_n(x)P_{n+1}(x)\,dx = \frac{2(n+1)}{(2n+1)(2n+3)}$;

(ii) $\displaystyle\int_{-1}^{1} (x^2-1)P_{n+1}(x)P_n'(x)\,dx = \frac{2n(n+1)}{(2n+1)(2n+3)}$.

9. Verify the following expansions as far as the last terms given:

(i) $\displaystyle(2p+1)x^{2p} = P_0(x)+5\left(\frac{2p}{2p+3}\right)P_2(x)+9\left\{\frac{2p(2p-2)}{(2p+3)(2p+5)}\right\}P_4(x)+\ldots,$

p being a positive integer;

(ii) $\displaystyle\frac{2}{\pi}(1-x^2)^{-1/2} = P_0(x)+5\left(\frac{1}{2}\right)^2 P_2(x)+9\left(\frac{1.3}{2.4}\right)^2 P_4(x)+\ldots.$

10. If n is a non-negative integer and $I(n) = \int_0^1 P_n(x)\,dx$, show that

$$I(0) = 1, \quad I(2n) = 0,$$

$$I(2n+1) = \frac{(-1)^n(2n)!}{2^{2n+1}n!\,(n+1)!}.$$

If $f(x) = 0$ when $-1 \leqq x < 0$ and $f(x) = 1$ when $0 < x \leqq 1$, deduce that

$$f(x) = \frac{1}{2}+\frac{3}{2^2}P_1(x)-\frac{7.2!}{2^4.2!1!}P_3(x)+\frac{11.4!}{2^6.3!2!}P_5(x)- \ldots.$$

11. Using the known result

$$\int_0^\pi \frac{d\phi}{1+a\cos\phi} = \frac{\pi}{(1-a^2)^{1/2}}, \qquad (|a| < 1),$$

and writing $a = \mp h(x^2-1)^{1/2}/(1-hx)$, show formally that an integral representation of the Legendre polynomial is

$$P_n(x) = \frac{1}{\pi}\int_0^\pi \{x\pm(x^2-1)^{1/2}\cos\phi\}^n\,d\phi.$$

(This representation is known as *Laplace's first integral*).

12. Show, by means of the recurrence formulae, that

$$n\{P_n(x)Q_{n-1}(x)-Q_n(x)P_{n-1}(x)\} = (n-1)\{P_{n-1}(x)Q_{n-2}(x)-Q_{n-1}(x)P_{n-2}(x)\}.$$

Deduce that

$$P_n(x)Q_{n-1}(x)-Q_n(x)P_{n-1}(x) = 1/n.$$

13. A *nearly* spherical surface is formed by the rotation of the polar curve $r = a\{1+\varepsilon P_n(\cos\theta)\}$, ε small and $n \geqq 1$, about the initial line. Neglecting powers of ε above the second, show that the area of the surface so generated is

$$4\pi a^2\left\{1+\varepsilon^2\left(\frac{n^2+n+2}{4n+2}\right)\right\}.$$

14. Show that when n is a positive integer and $x > 1$, a second solution of Legendre's equation is given by

$$y = P_n(x)\int_x^\infty \frac{dt}{(t^2-1)\{P_n(t)\}^2}.$$

Verify the formula

$$Q_n(x) = P_n(x)\int_x^\infty \frac{dt}{(t^2-1)\{P_n(t)\}^2}$$

for $n = 1$.

15. By expanding $(x-\mu)^{-1}$ by the binomial theorem and using the results of Exercise 7 above, show that the integral

$$\int_{-1}^1 \frac{P_n(\mu)}{x-\mu}\,d\mu, \qquad (|x| > 1, \qquad n \text{ a positive integer}),$$

is equal to

$$\frac{1}{2^n x^{n+1}}\sum_{m=0}^\infty \frac{\Gamma(n+2m+1)\Gamma(m+\frac{1}{2})}{\Gamma(2m+1)\Gamma(n+m+\frac{3}{2})}x^{-2m}.$$

Deduce *Neumann's formula*,

$$Q_n(x) = \frac{1}{2}\int_{-1}^1 \frac{P_n(\mu)}{x-\mu}\,d\mu.$$

16. Use the recurrence formulae to prove *Christoffel's two summation formulae*:

(i) $\sum_{r=0}^{n}(2r+1)P_r(x)P_r(y) = (n+1)\left\{\dfrac{P_n(x)P_{n+1}(y)-P_n(y)P_{n+1}(x)}{y-x}\right\};$

(ii) $\sum_{r=0}^{n}(2r+1)P_r(x)Q_r(y) = \dfrac{1+(n+1)\{P_n(x)Q_{n+1}(y)-Q_n(y)P_{n+1}(x)\}}{y-x}.$

17. Show that the associated Legendre function $P_n{}^m(x)$ can be expressed in terms of the hypergeometric function by the relation

$$P_n{}^m(x) = \frac{\Gamma(n+m+1)}{2^m m!\,\Gamma(n-m+1)}(x^2-1)^{\frac{1}{2}m}\,{}_2F_1(m-n, n+m+1; m+1; \tfrac{1}{2}-\tfrac{1}{2}x).$$

18. If $P_n{}^m(x)$ is the associated Legendre function, prove that

$$P_n{}^{m+2}(x)+\frac{2(m+1)x}{(x^2-1)^{1/2}}P_n{}^{m+1}(x)+(m-n)(m+n+1)P_n{}^m(x) = 0.$$

19. Prove the following recurrence formulae:

(i) $P_{n+1}{}^{m+1}(x)-P_{n-1}{}^{m+1}(x) = (2n+1)(x^2-1)^{1/2}P_n{}^m(x);$

(ii) $(n-m+1)P_{n+1}{}^m(x)-(2n+1)xP_n{}^m(x)+(n+m)P_{n-1}{}^m(x) = 0.$

20. Show that

(i) $\displaystyle\int_{-1}^{1} P_n{}^m(x)P_{n'}{}^m(x)\,dx = 0, \quad n \neq n';$

(ii) $\displaystyle\int_{-1}^{1} \{P_n{}^m(x)\}^2\,dx = (-1)^m\frac{2}{2n+1}\cdot\frac{\Gamma(n+m+1)}{\Gamma(n-m+1)}.$

6.5 Bessel's equation

The differential equation

$$\frac{d^2y}{dx^2}+\frac{1}{x}\frac{dy}{dx}+\left(1-\frac{\nu^2}{x^2}\right)y = 0, \tag{1}$$

in which ν is a constant, is known as *Bessel's equation* and its solutions are called *Bessel functions*. These functions were used by Bessel in the nineteenth century in a problem of dynamical astronomy, but they had, in fact, been used by other workers much earlier. The present importance of the functions lies almost entirely in the fact that the differential equation (1) occurs so frequently in the boundary-value problems of mathematical physics. We first consider the series solution of this equation.

The differential equation (1) has a regular singular point at the origin and we seek a solution of the form $y = a_0x^\rho+a_1x^{\rho+1}+\ldots$. Arranged according to weight, the equation can be written

$$\left\{x^2\frac{d^2y}{dx^2}+x\frac{dy}{dx}-\nu^2y\right\}+\{x^2y\} = 0,$$

and the indicial equation is obtained by writing $y = a_0 x^\rho$ in the first bracket. This gives $\rho(\rho-1)+\rho-\nu^2 = 0$, leading to $\rho^2-\nu^2 = 0$ and hence $\rho = \pm\nu$. The recurrence relation, given by writing $y = a_{r+2}x^{\rho+r+2}$ in the first bracket and $y = a_r x^{\rho+r}$ in the second is

$$\{(\rho+r+2)(\rho+r+1)+(\rho+r+2)-\nu^2\}a_{r+2}+a_r = 0,$$

giving

$$a_{r+2} = \frac{-a_r}{(\rho+r+2-\nu)(\rho+r+2+\nu)}. \tag{2}$$

Equating to zero the coefficient of $x^{\rho+1}$ leads to

$$\{(\rho+1)\rho+\rho+1-\nu^2\}a_1 = 0,$$

showing that $a_1 = 0$ and it follows from (2) that

$$a_3 = a_5 = a_7 = \ldots = 0.$$

With $\rho = \nu$, the recurrence relation gives

$$a_{r+2} = \frac{-a_r}{(r+2)(2\nu+r+2)},$$

so that

$$a_2 = -\frac{a_0}{2^2(\nu+1)}, \quad a_4 = \frac{-a_2}{2^2 2(\nu+2)} = \frac{a_0}{2^4 2(\nu+1)(\nu+2)},$$

$$a_6 = \frac{-a_4}{2^2 3(\nu+3)} = \frac{-a_0}{2^6 3!(\nu+1)(\nu+2)(\nu+3)}, \quad \ldots\ldots\ldots\ldots\ldots\ldots\ldots$$

Hence one solution of Bessel's equation is the series

$$a_0 x^\nu\left\{1-\frac{x^2}{2^2 1!(\nu+1)}+\frac{x^4}{2^4 2!(\nu+1)(\nu+2)}-\ldots\right\}.$$

Taking $a_0 = 1/\{2^\nu\Gamma(\nu+1)\}$, we have the standard solution

$$J_\nu(x) = \frac{(x/2)^\nu}{\Gamma(\nu+1)}\left\{1-\frac{(x/2)^2}{1!(\nu+1)}+\frac{(x/2)^4}{2!(\nu+1)(\nu+2)}-\ldots\right\}$$

$$= \sum_{r=0}^{\infty}\frac{(-1)^r(x/2)^{\nu+2r}}{r!\,\Gamma(\nu+r+1)}. \tag{3}$$

$J_\nu(x)$ is called the *Bessel function of the first kind of order* ν and is given by equation (3) for all values of ν.

With the second root $\rho = -\nu$ of the indicial equation, the recurrence relation (2) becomes

$$a_{r+2} = \frac{-a_r}{(r+2)(-2\nu+r+2)}. \tag{4}$$

If ν is not an integer, this leads to an independent second solution (see Exercises 6(*b*), 2) which can be written

$$J_{-\nu}(x) = \sum_{r=0}^{\infty} \frac{(-1)^r (x/2)^{-\nu+2r}}{r!\,\Gamma(-\nu+r+1)}, \tag{5}$$

and the complete solution of Bessel's equation is then

$$y = AJ_{\nu}(x) + BJ_{-\nu}(x), \tag{6}$$

A and B being arbitrary constants. If ν is a positive integer, the recurrence relation (4) breaks down when $r = 2(\nu-1)$ and a second solution then has to be found by other methods (see §6.7). There is a difficulty also when $\nu = 0$, in which case the two roots of the indicial equation are equal; the second solution in this case is discussed in §6.6.

Example 4. *If n is a positive integer, show that the formal expression for $J_{-n}(x)$ gives* $J_{-n}(x) = (-1)^n J_n(x)$.

Writing $\nu = -n$ in equation (3)

$$J_{-n}(x) = \sum_{r=0}^{\infty} \frac{(-1)^r (x/2)^{-n+2r}}{r!\,\Gamma(-n+r+1)}.$$

Now $\Gamma(n-r)\Gamma(1-n+r) = \pi \operatorname{cosec}(n-r)\pi$ (equation (5) §5.5), and hence $\Gamma(1-n+r)$ is infinite when $r = 0, 1, 2, \ldots, n-1$. Therefore

$$J_{-n}(x) = \sum_{r=n}^{\infty} \frac{(-1)^r (x/2)^{-n+2r}}{r!\,\Gamma(-n+r+1)}.$$

Writing $r = s+n$, this becomes

$$J_{-n}(x) = \sum_{s=0}^{\infty} \frac{(-1)^{s+n} (x/2)^{n+2s}}{(s+n)!\,\Gamma(s+1)} = (-1)^n \sum_{s=0}^{\infty} \frac{(-1)^s (x/2)^{n+2s}}{\Gamma(1+s+n)\,s!}$$

$$= (-1)^n J_n(x).$$

Example 5. *Show that*

$$J_{1/2}(x) = \left(\frac{2}{\pi x}\right)^{1/2} \sin x, \qquad J_{-1/2}(x) = \left(\frac{2}{\pi x}\right)^{1/2} \cos x.$$

Putting $\nu = \frac{1}{2}$ in (3),

$$J_{1/2}(x) = \frac{(x/2)^{1/2}}{\Gamma(\frac{3}{2})}\left\{1 - \frac{(x/2)^2}{1.\frac{3}{2}} + \frac{(x/2)^4}{1.2.\frac{3}{2}.\frac{5}{2}} - \frac{(x/2)^6}{1.2.3.\frac{3}{2}.\frac{5}{2}.\frac{7}{2}} + \ldots\right\}$$

$$= \frac{(x/2)^{1/2}}{(\pi^{1/2}/2)}\left\{1 - \frac{x^2}{1.2.3} + \frac{x^4}{1.2.3.4.5} - \frac{x^6}{1.2.3.4.5.6.7} + \ldots\right\}$$

$$= \left(\frac{2}{\pi x}\right)^{1/2}\left\{x - \frac{x^3}{3!} + \frac{x^5}{5!} - \frac{x^7}{7!} + \ldots\right\}$$

$$= \left(\frac{2}{\pi x}\right)^{1/2} \sin x.$$

The second result can be obtained similarly by writing $\nu = \frac{1}{2}$ in equation (5).

The results obtained in Example 5 above are special cases of an important general theorem which states that $J_\nu(x)$ is expressible in finite terms by means of algebraic and trigonometrical functions of x whenever ν is half of an odd integer. Further examples are

$$J_{\frac{3}{2}}(x) = \left(\frac{2}{\pi x}\right)^{\frac{1}{2}}\left(\frac{\sin x}{x} - \cos x\right),$$

$$J_{-\frac{5}{2}}(x) = \left(\frac{2}{\pi x}\right)^{\frac{1}{2}}\left\{\frac{3\sin x}{x} + \left(\frac{3}{x^2} - 1\right)\cos x\right\}.$$

The functions $J_{(n+\frac{1}{2})}(x)$ and $J_{-(n+\frac{1}{2})}(x)$, where n is a positive integer or zero, are called *spherical Bessel functions*; they have important applications in problems of wave motion in which spherical polar coordinates are appropriate.

6.6 Bessel functions of order zero

When $\nu = 0$, the first solution of Bessel's equation is given by equation (3) of §6.5 as

$$J_0(x) = \sum_{r=0}^{\infty} \frac{(-1)^r (x/2)^{2r}}{r!\,\Gamma(r+1)}$$

$$= 1 - \frac{x^2}{2^2} + \frac{x^4}{2^2 . 4^2} - \frac{x^6}{2^2 . 4^2 . 6^2} + \ldots. \tag{1}$$

The method of §6.5 being not now available as a means of obtaining an independent second solution, we take this solution to be $Y_0(x)$ where

$$Y_0(x) = \frac{2}{\pi}\left(\log_e \frac{x}{2} + \gamma\right) J_0(x) - V_0(x), \tag{2}$$

γ being Euler's constant* and $V_0(x)$ a series in ascending powers of x whose coefficients are to be found. Using (2), we have

$$Y_0'(x) = \frac{2}{\pi}\left(\log_e \frac{x}{2} + \gamma\right) J_0'(x) + \frac{2}{\pi x} J_0(x) - V_0'(x),$$

$$Y_0''(x) = \frac{2}{\pi}\left(\log_e \frac{x}{2} + \gamma\right) J_0''(x) + \frac{4}{\pi x} J_0'(x) - \frac{2}{\pi x^2} J_0(x) - V_0''(x).$$

* $\gamma = \lim_{n\to\infty}\left(1 + \frac{1}{2} + \frac{1}{3} + \ldots + \frac{1}{n} - \log_e n\right) = 0.5772\ldots.$

Substituting in Bessel's equation with $\nu = 0$, we find that $Y_0(x)$ is a solution if

$$\frac{2}{\pi}\left(\log_e \frac{x}{2}+\gamma\right)\left\{J_0''(x)+\frac{1}{x}J_0'(x)+J_0(x)\right\}+\frac{4}{\pi x}J_0'(x)$$

$$-\left\{V_0''(x)+\frac{1}{x}V_0'(x)+V_0(x)\right\} = 0.$$

Since $J_0(x)$ satisfies Bessel's equation, the first term vanishes and $V_0(x)$ must therefore satisfy the differential equation

$$V_0''(x)+\frac{1}{x}V_0'(x)+V_0(x) = \frac{4}{\pi x}J_0'(x)$$

$$= \frac{2}{\pi}\sum_{r=1}^{\infty}\frac{(-1)^r(x/2)^{2r-2}}{r!(r-1)!}, \tag{3}$$

when use is made of equation (1).

To solve equation (3) let

$$V_0(x) = \frac{2}{\pi}\sum_{r=0}^{\infty} b_r\left(\frac{x}{2}\right)^{2r},$$

so that

$$\frac{1}{x}V_0'(x) = \frac{2}{\pi}\sum_{r=1}^{\infty}\tfrac{1}{2}rb_r\left(\frac{x}{2}\right)^{2r-2},$$

and

$$V_0''(x) = \frac{2}{\pi}\sum_{r=1}^{\infty} r(r-\tfrac{1}{2})b_r\left(\frac{x}{2}\right)^{2r-2}.$$

Hence

$$V_0''(x)+\frac{1}{x}V_0'(x)+V_0(x) = \frac{2}{\pi}\sum_{r=1}^{\infty}(r^2b_r+b_{r-1})\left(\frac{x}{2}\right)^{2r-2}.$$

Substituting in (3) and equating coefficients of x^{2r-2}, we have

$$r^2b_r+b_{r-1} = \frac{(-1)^r}{r!(r-1)!}, \qquad r = 1, 2, 3, \ldots.$$

Taking $b_0 = 0$, the above recurrence relation gives $b_1 = -1$ and

$$4b_2+b_1 = \frac{1}{2!}, \qquad 9b_3+b_2 = -\frac{1}{2!3!}, \qquad 16b_4+b_3 = \frac{1}{3!4!}, \ldots.$$

These give successively

$$b_2 = \frac{1}{(2!)^2}(1+\tfrac{1}{2}), \qquad b_3 = -\frac{1}{(3!)^2}(1+\tfrac{1}{2}+\tfrac{1}{3}), \qquad b_4 = \frac{1}{(4!)^2}(1+\tfrac{1}{2}+\tfrac{1}{3}+\tfrac{1}{4}),$$

and, in general,

$$b_r = \frac{(-1)^r}{(r!)^2}\left(1+\frac{1}{2}+\frac{1}{3}+\ldots+\frac{1}{r}\right).$$

Substituting in (2), we find that $Y_0(x)$ is given by

$$Y_0(x) = \frac{2}{\pi}\left(\log_e \frac{x}{2}+\gamma\right)J_0(x)-\frac{2}{\pi}\sum_{r=1}^{\infty}\frac{(-1)^r(x/2)^{2r}}{r!r!}\left(1+\frac{1}{2}+\frac{1}{3}+\ldots+\frac{1}{r}\right). \quad (4)$$

The function $Y_0(x)$ as given by equation (4) is called the *Bessel function of the second kind of order zero* and the complete solution of Bessel's equation of order zero

$$\frac{d^2y}{dx^2}+\frac{1}{x}\frac{dy}{dx}+y = 0, \quad (5)$$

is

$$y = AJ_0(x)+BY_0(x), \quad (6)$$

$J_0(x)$, $Y_0(x)$ being given respectively by equations (1) and (4) and A, B being arbitrary constants. Tabulated values of $J_0(x)$ and $Y_0(x)$ can be found, for example, in the British Association Mathematical Tables,

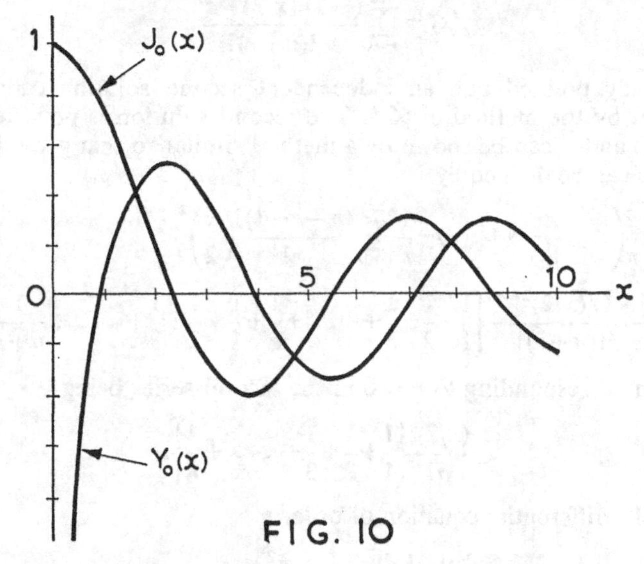

FIG. 10

Vol. 6, 1937. Both $J_0(x)$ and $Y_0(x)$ are oscillatory functions and their graphs are shown in Fig. 10. It should be noted that both $J_0(x)$ and $Y_0(x)$ vanish at an infinite sequence of values of x. In this respect they behave similarly to the trigonometrical functions cos x and sin x which vanish respectively when $x = (r+\frac{1}{2})\pi$ and $x = r\pi$. In practical applications, the positive values of x for which the Bessel functions vanish are of great importance and these values are known as the positive *zeros* of the functions. If we denote the rth positive zero of $J_0(x)$, $Y_0(x)$ by α_r, β_r respectively, it can be shown that these zeros interlace, that is, $\beta_r < \alpha_r < \beta_{r+1}$ for all r. Unlike those of the trigonometrical functions, the positive zeros of the Bessel functions are not equally spaced along the x-axis, but it can be shown that α_r, β_r approximate respectively to $(r-\frac{1}{4})\pi$ and $(r-\frac{3}{4})\pi$ for large values of r. Numerical values of the zeros can be found in the volume of tables already mentioned. It is easily seen that

$$J_0(x) \simeq 1, \qquad Y_0(x) \simeq \frac{2}{\pi}\left(\log_e \frac{x}{2}+\gamma\right)$$

for small values of x. Thus $J_0(x) \to 1$, $Y_0(x) \to -\infty$ as $x \to 0$, and it can be shown that both functions tend to zero as x tends to infinity.

6.7 Bessel functions of integral order

When the constant ν in Bessel's equation is a positive integer n, the first solution of the equation, which is given by equation (3) of §6.5, can be written in the form

$$J_n(x) = \sum_{r=0}^{\infty} \frac{(-1)^r (x/2)^{n+2r}}{r!(n+r)!}. \tag{1}$$

As already pointed out, an independent second solution cannot be obtained by the method of §6.5. The second solution is now denoted by $Y_n(x)$ and it can be shown by a method similar to that given in §6.6 that this can be defined by

$$Y_n(x) = \frac{2}{\pi}\left(\log_e \frac{x}{2}+\gamma\right)J_n(x) - \frac{1}{\pi}\sum_{r=0}^{n-1} \frac{(n-r-1)!}{r!}\left(\frac{x}{2}\right)^{2r-n}$$
$$-\frac{1}{\pi}\sum_{r=0}^{\infty} \frac{(-1)^r (x/2)^{2r+n}}{r!(n+r)!}\left\{\frac{1}{1}+\frac{1}{2}+\frac{1}{3}+\ldots+\frac{1}{r}+\frac{1}{1}+\frac{1}{2}+\frac{1}{3}+\ldots+\frac{1}{n+r}\right\}, \tag{2}$$

the term corresponding to $r = 0$ in the second series being

$$\frac{(x/2)^n}{n!}\left\{\frac{1}{1}+\frac{1}{2}+\frac{1}{3}+\ldots+\frac{1}{n}\right\}.$$

Bessel's differential equation of order n,

$$\frac{d^2y}{dx^2}+\frac{1}{x}\frac{dy}{dx}+\left(1-\frac{n^2}{x^2}\right)y = 0, \tag{3}$$

has the general solution

$$y = AJ_n(x)+BY_n(x) \tag{4}$$

where A, B are arbitrary constants. The functions $J_n(x)$, $Y_n(x)$, defined by equations (1), (2), are known respectively as Bessel functions of *the first and second kinds of order n.* In physical applications, the most important Bessel functions are those of order zero already discussed and those of order unity, and it is worth setting out equations (1) and (2) in rather more detail when $n = 1$. We have, in this case,

$$J_1(x) = \sum_{r=0}^{\infty} \frac{(-1)^r (x/2)^{2r+1}}{r!(r+1)!}$$

$$= \frac{x}{2} - \frac{x^3}{2^3 1!\,2!} + \frac{x^5}{2^5 2!\,3!} - \ldots, \tag{5}$$

$$Y_1(x) = \frac{2}{\pi}\left(\log_e \frac{x}{2} + \gamma\right) J_1(x)$$

$$- \frac{2}{\pi x} - \frac{1}{\pi} \sum_{r=0}^{\infty} \frac{(-1)^r (x/2)^{2r+1}}{r!(r+1)!} \left\{ 2\left(1 + \frac{1}{2} + \ldots + \frac{1}{r}\right) + \frac{1}{r+1} \right\}, \tag{6}$$

the first term in the series on the right being $x/2$.

Tabulated values of $J_n(x)$, $Y_n(x)$ for $n = 1$ are given in Vol. 6 of the British Association Mathematical Tables, and values of the functions for integral values of n up to $n = 20$ have been given in a later volume (Vol. 10, 1952).

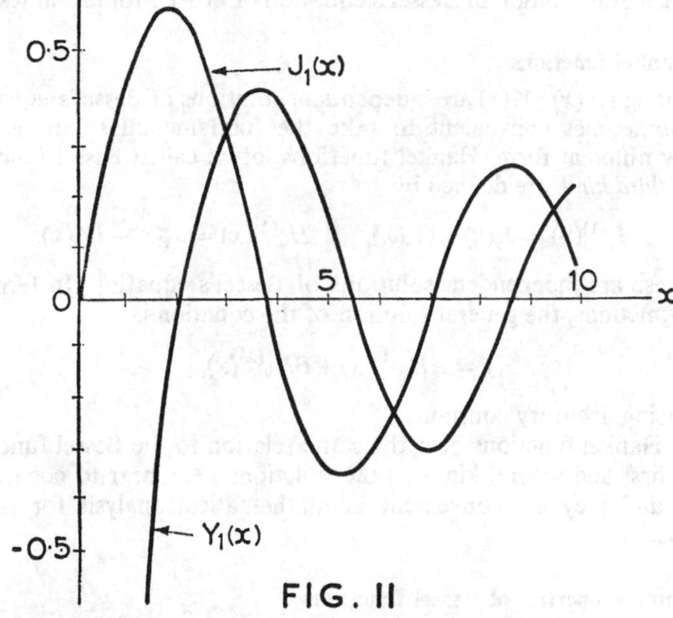

FIG. 11

Rough graphs showing the behaviour of the functions of order unity are shown in Fig. 11. It can be seen from equations (5) and (6) that

$J_1(x)$ behaves like $x/2$ and $Y_1(x)$ like $-2/(\pi x)$ for small values of x. Both the functions tend to zero as x tends to infinity and both functions have an infinite sequence of zeros which have been extensively tabulated.

To avoid the necessity of having to write the general solution of Bessel's equation in the two different forms

$$y = AJ_\nu(x)+BJ_{-\nu}(x), \quad \nu \text{ not zero or a positive integer,}$$

$$y = AJ_n(x)+BY_n(x), \quad n \text{ zero or a positive integer,}$$

it is possible to define the second solution when ν is not zero or a positive integer by the formula

$$Y_\nu(x) = \frac{J_\nu(x)\cos\nu\pi - J_{-\nu}(x)}{\sin\nu\pi}. \tag{7}$$

It can be shown that, when ν tends to zero or a positive integer n, the expression on the right of (7) gives the functions $Y_0(x)$ and $Y_n(x)$ respectively. It can also be shown that $J_\nu(x)$ and $Y_\nu(x)$ (as defined by (7)) are independent solutions of Bessel's equation and hence we may take

$$y = AJ_\nu(x)+BY_\nu(x) \tag{8}$$

as the general solution of Bessel's equation of order ν for *all* values of ν.

6.8 Hankel functions

Although $J_\nu(x)$, $Y_\nu(x)$ are independent solutions of Bessel's equation, it is sometimes convenient to take the fundamental solutions in a slightly different form. Hankel functions, often called Bessel functions of the *third kind*, are defined by

$$H_\nu^{(1)}(x) = J_\nu(x)+iY_\nu(x), \qquad H_\nu^{(2)}(x) = J_\nu(x)-iY_\nu(x), \tag{1}$$

and these are independent solutions of Bessel's equation. In terms of these functions, the general solution of the equation is

$$y = AH_\nu^{(1)}(x)+BH_\nu^{(2)}(x), \tag{2}$$

A, B being arbitrary constants.

The Hankel functions bear the same relation to the Bessel functions of the first and second kinds as the functions $e^{\pm i\nu x}$ bear to $\cos\nu x$ and $\sin\nu x$ and they are convenient in mathematical analysis for similar reasons.

6.9 Some properties of Bessel functions

Bessel functions possess properties of which great use can be made in the discussion of physical problems. A few of the more important of these are given in this section, but for an extensive treatment the student is referred to the standard text by Watson.*

(*a*) *Recurrence formulae*

We start with equation (3) of §6.5, which can be written in the form

$$x^{-\nu}J_\nu(x) = 2^{-\nu}\sum_{r=0}^{\infty}\frac{(-1)^r(x/2)^{2r}}{r!\Gamma(\nu+r+1)}.$$

Hence

$$\frac{d}{dx}\{x^{-\nu}J_\nu(x)\} = 2^{-\nu}\sum_{r=1}^{\infty}\frac{(-1)^r r(x/2)^{2r-1}}{r!\Gamma(\nu+r+1)}$$

$$= -2^{-\nu}\sum_{r=0}^{\infty}\frac{(-1)^r(x/2)^{2r+1}}{r!\Gamma(\nu+r+2)} = -x^{-\nu}J_{\nu+1}(x). \tag{1}$$

This can be written in the form

$$x^{-\nu}J_\nu'(x) - \nu x^{-\nu-1}J_\nu(x) = -x^{-\nu}J_{\nu+1}(x),$$

and it follows that

$$\frac{\nu}{x}J_\nu(x) - J_\nu'(x) = J_{\nu+1}(x); \tag{2}$$

this is the first of the required recurrence formulae. A useful particular case of this is $J_0'(x) = -J_1(x)$, this result following immediately when $\nu = 0$.

Again, starting with equation (3) of §6.5 in the form

$$x^\nu J_\nu(x) = 2^\nu\sum_{r=0}^{\infty}\frac{(-1)^r(x/2)^{2\nu+2r}}{r!\Gamma(\nu+r+1)},$$

we have

$$\frac{d}{dx}\{x^\nu J_\nu(x)\} = 2^\nu\sum_{r=0}^{\infty}\frac{(-1)^r(\nu+r)(x/2)^{2\nu+2r-1}}{r!\Gamma(\nu+r+1)}$$

$$= x^\nu\sum_{r=0}^{\infty}\frac{(-1)^r(x/2)^{\nu-1+2r}}{r!\Gamma(\nu+r)} = x^\nu J_{\nu-1}(x). \tag{3}$$

This can be written

$$x^\nu J_\nu'(x) + \nu x^{\nu-1}J_\nu(x) = x^\nu J_{\nu-1}(x),$$

giving, as the second recurrence formula

$$\frac{\nu}{x}J_\nu(x) + J_\nu'(x) = J_{\nu-1}(x). \tag{4}$$

* Watson, G. N., *Theory of Bessel Functions*, Cambridge University Press, 1944.

Two other useful recurrence formulae follow by addition and subtraction of (2) and (4). Thus

$$\frac{2\nu}{x}J_\nu(x) = J_{\nu-1}(x)+J_{\nu+1}(x), \tag{5}$$

and

$$2J_\nu'(x) = J_{\nu-1}(x)-J_{\nu+1}(x). \tag{6}$$

The functions of the second kind $Y_\nu(x)$ and the Hankel functions $H_\nu^{(1)}(x)$, $H_\nu^{(2)}(x)$ satisfy the same recurrence formulae. For example, equation (7) of §6.7 gives

$$Y_{\nu-1}(x) = \frac{J_{\nu-1}(x)\cos(\nu-1)\pi - J_{-\nu+1}(x)}{\sin(\nu-1)\pi} = \frac{J_{\nu-1}(x)\cos\nu\pi + J_{-\nu+1}(x)}{\sin\nu\pi},$$

$$Y_{\nu+1}(x) = \frac{J_{\nu+1}(x)\cos(\nu+1)\pi - J_{-\nu-1}(x)}{\sin(\nu+1)\pi} = \frac{J_{\nu+1}(x)\cos\nu\pi + J_{-\nu-1}(x)}{\sin\nu\pi}.$$

Hence, by addition and use of (5), we have

$$Y_{\nu-1}(x)+Y_{\nu+1}(x) = \frac{(2\nu/x)J_\nu(x)\cos\nu\pi+(-2\nu/x)J_{-\nu}(x)}{\sin\nu\pi}$$

$$= \frac{2\nu}{x}Y_\nu(x).$$

The other recurrence formulae can be proved in the same way.

Example 6. *Show that*

$$J_{-5/2}(x) = \left(\frac{2}{\pi x}\right)^{1/2}\left\{\frac{3\sin x}{x}+\left(\frac{3}{x^2}-1\right)\cos x\right\}.$$

Using the results (Example 5)

$$J_{1/2}(x) = (2/\pi x)^{1/2}\sin x, \qquad J_{-1/2}(x) = (2/\pi x)^{1/2}\cos x$$

and, taking $\nu = -\frac{1}{2}$ in equation (5),

$$J_{-3/2}(x) = -\frac{1}{x}J_{-1/2}(x)-J_{1/2}(x) = -\left(\frac{2}{\pi x}\right)^{1/2}\left\{\frac{\cos x}{x}+\sin x\right\}.$$

Using these results in the equation obtained by taking $\nu = -\frac{3}{2}$ in (5),

$$J_{-5/2}(x) = -\frac{3}{x}J_{-3/2}(x)-J_{-1/2}(x) = \left(\frac{2}{\pi x}\right)^{1/2}\left\{\frac{3\cos x}{x^2}+\frac{3\sin x}{x}-\cos x\right\}$$

and the required result follows immediately.

(b) The generating function for the Bessel coefficients

Provided t is not zero, the functions $e^{\frac{1}{2}xt}$ and $e^{-\frac{1}{2}(x/t)}$ can be expanded in powers of t, and the product of these expansions gives

$$\exp\{\tfrac{1}{2}x(t-t^{-1})\} = \sum_{r=0}^{\infty}\frac{1}{r!}\left(\frac{xt}{2}\right)^r \sum_{s=0}^{\infty}\frac{1}{s!}\left(-\frac{x}{2t}\right)^s$$

$$= \sum_{r=0}^{\infty}\sum_{s=0}^{\infty}\frac{(-1)^s}{r!\,s!}\left(\frac{x}{2}\right)^{r+s} t^{r-s},$$

term by term multiplication of the series being permissible because of the absolute convergence of the separate series. The coefficient of t^n (n a positive integer or zero) is found by taking $r = n+s$ and making s vary from 0 to infinity; thus the coefficient of t^n

$$= \sum_{s=0}^{\infty}\frac{(-1)^s}{(n+s)!\,s!}\left(\frac{x}{2}\right)^{n+2s} = J_n(x),$$

by equation (3) of §6.5. The coefficient of t^{-n} (n a positive integer) is found by taking $r = -n+s$ and making s vary from n to infinity. Thus the coefficient of t^{-n}

$$= \sum_{s=n}^{\infty}\frac{(-1)^s}{(-n+s)!\,s!}\left(\frac{x}{2}\right)^{-n+2s} = J_{-n}(x),$$

see Example 4.

Hence

$$\exp\{\tfrac{1}{2}x(t-t^{-1})\} = \sum_{-\infty}^{\infty} t^n J_n(x), \tag{7}$$

and the exponential function on the left can be regarded as the *generating function* of $J_n(x)$. Because of the form of (7), the functions $J_n(x)$ where $n = 0, 1, 2, 3, \ldots$, are often called the *Bessel coefficients.*

Example 7. *Show that*

$$\exp(-ix\sin\theta) = J_0(x)+2\sum_{n=1}^{\infty} J_{2n}(x)\cos 2n\theta - 2i\sum_{n=0}^{\infty} J_{2n+1}(x)\sin(2n+1)\theta.$$

Writing $t = -e^{i\theta}$ in (7),

$$\exp\{-\tfrac{1}{2}x(e^{i\theta}-e^{-i\theta})\} = \sum_{-\infty}^{\infty}(-1)^n e^{ni\theta}J_n(x)$$

$$= J_0(x)+\sum_{n=1}^{\infty}(-1)^n\{e^{ni\theta}+(-1)^n e^{-ni\theta}\}J_n(x), \tag{8}$$

since $J_{-n}(x) = (-1)^n J_n(x)$. The required result follows when we note that $e^{i\theta} - e^{-i\theta} = 2i \sin\theta$ and

$$(-1)^n e^{ni\theta} + e^{-ni\theta} = 2\cos n\theta, \quad n \text{ even},$$
$$= -2i \sin n\theta, \quad n \text{ odd}.$$

(c) Bessel's integral

Equation (8) can be written

$$\exp(-ix\sin\theta) = J_0(x) + \sum_{n=0}^{\infty}\{(-1)^n e^{ni\theta} + e^{-ni\theta}\}J_n(x). \qquad (9)$$

Since $\int_0^{2\pi} e^{ir\theta}\,d\theta = 2\pi$ when $r = 0$ and since this integral vanishes when r is an integer, multiplication of (9) by $e^{in\theta}$ and integration with respect to θ between 0 and 2π gives formally

$$\int_0^{2\pi} \exp\{i(n\theta - x\sin\theta)\}\,d\theta = 2\pi J_n(x), \qquad (10)$$

where $n = 0, 1, 2, 3, \ldots$ and this can, in fact, be proved rigorously. Since $J_n(x)$ is real, we can show by equating real and imaginary parts that

$$J_n(x) = \frac{1}{2\pi}\int_0^{2\pi} \cos(n\theta - x\sin\theta)\,d\theta$$
$$= \frac{1}{\pi}\int_0^{\pi} \cos(n\theta - x\sin\theta)\,d\theta, \qquad (11)$$

and the expression (11) gives *Bessel's integral* for $J_n(x)$.

(d) Some integrals involving Bessel functions

Using the result (3),

$$\int_0^a x^\nu J_{\nu-1}(x)\,dx = \Big[x^\nu J_\nu(x)\Big]_0^a,$$

and if ν is real and positive, equation (3) of §6.5 shows that $x^\nu J_\nu(x) \to 0$ as $x \to 0$. Hence

$$\int_0^a x^\nu J_{\nu-1}(x)\,dx = a^\nu J_\nu(a), \qquad (\nu > 0), \qquad (12)$$

and a similar integral can be obtained from the result (1). Other integrals of this type can be evaluated by such devices as integration by parts and use of the recurrence formulae. A typical example is given below.

Example 8. *Show that*

$$\int_0^a x^3 J_0(x)\,dx = a^3 J_1(a) - 2a^2 J_2(a) = 2a^2 J_0(a) + a(a^2-4)J_1(a). \qquad \text{(L.U.)}$$

From equation (3) with $\nu = 1$, $xJ_0(x) = (d/dx)\{xJ_1(x)\}$ and hence

$$\int_0^a x^3 J_0(x)\,dx = \int_0^a x^2 \frac{d}{dx}\{xJ_1(x)\}\,dx$$

$$= \Big[x^3 J_1(x)\Big]_0^a - \int_0^a 2x^2 J_1(x)\,dx,$$

on integrating by parts. Using equation (12) with $\nu = 2$ to evaluate the integral on the right,

$$\int_0^a x^3 J_0(x)\,dx = a^3 J_1(a) - 2a^2 J_2(a). \tag{13}$$

The second part of the required result can be found by using the recurrence formula (5) with $\nu = 1$ and $x = a$. This gives $2a^{-1}J_1(a) = J_0(a) + J_2(a)$ so that

$$2a^2 J_2(a) = 4aJ_1(a) - 2a^2 J_0(a).$$

Combining this with (13),

$$\int_0^a x^3 J_0(x)\,dx = 2a^2 J_0(a) + a(a^2-4)J_1(a).$$

Another important integral is $\displaystyle\int_0^a xJ_\nu(\lambda x)J_\nu(\mu x)\,dx$ and this can be evaluated as follows. If $u = J_\nu(\lambda x)$, then u satisfies Bessel's equation when u is written in place of y and λx in place of x. This can be written in the form

$$x^2 \frac{d^2u}{dx^2} + x\frac{du}{dx} + (\lambda^2 x^2 - \nu^2)u = 0.$$

Similarly $v = J_\nu(\mu x)$ satisfies the equation

$$x^2 \frac{d^2v}{dx^2} + x\frac{dv}{dx} + (\mu^2 x^2 - \nu^2)v = 0.$$

Multiplying by v/x, u/x respectively and subtracting

$$(\lambda^2 - \mu^2)xuv = x\left(u\frac{d^2v}{dx^2} - v\frac{d^2u}{dx^2}\right) + \left(u\frac{dv}{dx} - v\frac{du}{dx}\right)$$

$$= \frac{d}{dx}\left\{x\left(u\frac{dv}{dx} - v\frac{du}{dx}\right)\right\}.$$

Substituting for u, v and integrating with respect to x between 0 and a,

$$(\lambda^2-\mu^2)\int_0^a xJ_\nu(\lambda x)J_\nu(\mu x)\,dx = \left[x\left\{J_\nu(\lambda x)\frac{d}{dx}J_\nu(\mu x) - J_\nu(\mu x)\frac{d}{dx}J_\nu(\lambda x)\right\}\right]_0^a$$

$$= a\{\mu J_\nu(\lambda a)J_\nu{}'(\mu a) - \lambda J_\nu(\mu a)J_\nu{}'(\lambda a)\}, \tag{14}$$

and this provides a means of evaluating the integral except when $\lambda^2 = \mu^2$.

To evaluate the integral $\int_0^a xJ_\nu^2(\lambda x)\,dx$, equation (14) gives

$$\int_0^a xJ_\nu^2(\lambda x)\,dx = \lim_{\mu\to\lambda}\left\{\frac{a\mu J_\nu(\lambda a)J_\nu'(\mu a) - a\lambda J_\nu(\mu a)J_\nu'(\lambda a)}{\lambda^2-\mu^2}\right\}$$

$$= \lim_{\mu\to\lambda}\left[-\frac{a}{2\mu}\frac{\partial}{\partial\mu}\{\mu J_\nu(\lambda a)J_\nu'(\mu a) - \lambda J_\nu(\mu a)J_\nu'(\lambda a)\}\right]$$

$$= -\frac{a}{2\lambda}\lim_{\mu\to\lambda}\{J_\nu(\lambda a)J_\nu'(\mu a) + a\mu J_\nu(\lambda a)J_\nu''(\mu a) - \lambda aJ_\nu'(\mu a)J_\nu'(\lambda a)\}$$

$$= \frac{a^2}{2}\left[\{J_\nu'(\lambda a)\}^2 - J_\nu(\lambda a)J_\nu''(\lambda a) - \frac{1}{\lambda a}J_\nu(\lambda a)J_\nu'(\lambda a)\right].$$

Since $J_\nu(\lambda a)$ satisfies Bessel's equation,

$$J_\nu''(\lambda a) + \frac{1}{\lambda a}J_\nu'(\lambda a) + \left(1 - \frac{\nu^2}{\lambda^2 a^2}\right)J_\nu(\lambda a) = 0$$

and, using this, we find

$$\int_0^a xJ_\nu^2(\lambda x)\,dx = \frac{a^2}{2}\left[\{J_\nu'(\lambda a)\}^2 + \left(1 - \frac{\nu^2}{\lambda^2 a^2}\right)J_\nu^2(\lambda a)\right]. \tag{15}$$

(e) *Fourier-Bessel series*

Taking $\lambda = \alpha_r$, $\mu = \alpha_s$ where α_r and α_s are two positive roots of the equation $J_\nu(\alpha a) = 0$ and substituting in (14), (15), we have

$$\int_0^a xJ_\nu(\alpha_r x)J_\nu(\alpha_s x)\,dx = 0, \qquad (r \neq s), \tag{16}$$

$$\int_0^a xJ_\nu^2(\alpha_r x)\,dx = \tfrac{1}{2}a^2\{J_\nu'(\alpha_r a)\}^2. \tag{17}$$

Hence the functions $x^{\frac{1}{2}}J_\nu(\alpha_r x)$, $r = 1, 2, 3, \ldots$, are orthogonal in the interval $(0, a)$, see §6.3 (e), and the possibility exists of expanding an arbitrary function $f(x)$ in the form

$$f(x) = b_1 J_\nu(\alpha_1 x) + b_2 J_\nu(\alpha_2 x) + \ldots + b_r J_\nu(\alpha_r x) + \ldots, \quad (0 < x < a). \tag{18}$$

Assuming this to be possible, multiplication of (18) by $xJ_\nu(\alpha_r x)$ and the use of (16), (17) leads to

$$\int_0^a xf(x)J_\nu(\alpha_r x)\,dx = b_r \,.\, \tfrac{1}{2}a^2\{J_\nu'(\alpha_r a)\}^2, \tag{19}$$

the other terms on the right-hand side vanishing by virtue of (16). The recurrence formula (2) gives $(\nu/\alpha_r a)J_\nu(\alpha_r a) - J_\nu'(\alpha_r a) = J_{\nu+1}(\alpha_r a)$ and, since $J_\nu(\alpha_r a) = 0$, it follows that $J_\nu'(\alpha_r a) = -J_{\nu+1}(\alpha_r a)$. Thus (19) gives

$$b_r = \frac{2}{a^2 J_{\nu+1}^2(\alpha_r a)}\int_0^a xf(x)J_\nu(\alpha_r x)\,dx. \tag{20}$$

Equation (20) gives a formula from which the coefficients in the series on the right of (18) can be calculated and, from the analogy with Fourier series, this series is called a *Fourier-Bessel series*.

The analysis given above is, of course, purely formal and no attempt has been made to discuss the conditions under which such an expansion of an arbitrary function is legitimate. A full discussion of these will be found in Watson's standard treatise. By choosing α_r to be the roots of equations other than $J_\nu(\alpha_r a) = 0$, other expansions can be obtained; an example will be found in Exercises 6(*b*).

Example 9. *If a_r satisfies the equation $J_1(a_r) = 0$, $r = 1, 2, 3, \ldots$, show that*

$$\tfrac{1}{2}x = \sum_{r=1}^{\infty} \frac{J_1(a_r x)}{a_r J_2(a_r)}, \qquad (0 < x < 1).$$

Here $\nu = 1$, $a = 1$ and writing $\tfrac{1}{2}x = \sum_{r=1}^{\infty} b_r J_1(a_r x)$ equation (20) shows that the coefficients b_r are calculated from

$$b_r = \frac{2}{J_2^2(a_r)}\int_0^1 \frac{x^2}{2} J_1(a_r x)\,dx.$$

Writing $a_r x = t$, this gives

$$b_r = \frac{1}{a_r^3 J_2^2(a_r)}\int_0^{a_r} t^2 J_1(t)\,dt = \frac{1}{a_r^3 J_2^2(a_r)}\{a_r^2 J_2(a_r)\},$$

using equation (12) with $\nu = 2$, $a = a_r$. Hence

$$b_r = \frac{1}{a_r J_2(a_r)}$$

and the required result is established.

(*f*) *Asymptotic values*

In physical problems it is often desirable to be able to approximate to the Bessel functions $J_\nu(x)$ and $Y_\nu(x)$ when x is large. For such values of x, the series defining these functions converge slowly but useful approximations can be obtained as follows.

Writing $y = ue^{ix}$ in Bessel's equation we obtain

$$\left(x^2\frac{d^2u}{dx^2}+x\frac{du}{dx}-\nu^2u\right)+i\left(2x^2\frac{du}{dx}+xu\right)=0, \tag{21}$$

and we seek a solution of the form $u = a_0x^\rho+a_1x^{\rho-1}+a_2x^{\rho-2}+\ldots$. Substituting in (21), the term with the highest power of x is $ia_0(2\rho+1)x^{\rho+1}$ and, equating its coefficient to zero, $\rho = -\frac{1}{2}$. With this value of ρ, the recurrence relation between the coefficients a_{r+1} and a_r is found in the usual way to be

$$a_{r+1} = i\left\{\frac{4\nu^2-(2r+1)^2}{2^3(r+1)}\right\}a_r. \tag{22}$$

The resulting series is not convergent but it can, nevertheless, be used to find numerical values of the solution.

For large values of x, we can take a first approximation to the solution of Bessel's equation to be $y = ue^{ix} = a_0x^{-\frac{1}{2}}e^{ix}$ and another approximate solution can similarly be found to be $b_0x^{-\frac{1}{2}}e^{-ix}$. The standard solutions $J_\nu(x)$, $Y_\nu(x)$ can be identified with combinations of these approximate solutions by writing

$$J_\nu(x) \sim \left(\frac{2}{\pi x}\right)^{\frac{1}{2}}\cos\left(x-\frac{\pi}{4}-\nu\frac{\pi}{2}\right), \quad Y_\nu(x) \sim \left(\frac{2}{\pi x}\right)^{\frac{1}{2}}\sin\left(x-\frac{\pi}{4}-\nu\frac{\pi}{2}\right). \tag{23}$$

It is easy to verify that when $\nu = \frac{1}{2}$, these approximations yield the exact solutions $J_{\frac{1}{2}}(x) = (2/\pi x)^{\frac{1}{2}}\sin x$, $Y_{\frac{1}{2}}(x) = -(2/\pi x)^{\frac{1}{2}}\cos x$. The results (23) can, of course, be improved by using the recurrence relation (22) to calculate the coefficients of further terms in the series giving u in descending powers of x. Such series are called *asymptotic* series and the reader is referred to more advanced treatises for a full discussion.

6.10 Modified Bessel functions

Writing ix in place of x, Bessel's equation becomes

$$\frac{d^2y}{dx^2}+\frac{1}{x}\frac{dy}{dx}-\left(1+\frac{\nu^2}{x^2}\right)y=0, \tag{1}$$

and this "modified" equation plays a significant role in science and engineering. Working in exactly the same way as in §6.5, one solution of this equation is given by the function*

* The introduction of $J_\nu(ix)$ in the second part of equation (2) is useful but it is, of course, only formal as we have not defined the Bessel function for imaginary arguments. However, this procedure can be justified by appealing to works discussing Bessel functions of the complex variable.

$$I_\nu(x) = \sum_{r=0}^{\infty} \frac{(x/2)^{\nu+2r}}{r!\Gamma(\nu+r+1)} = (i)^{-\nu} J_\nu(ix). \tag{2}$$

The function $I_\nu(x)$ is called the *modified Bessel function of the first kind of order* ν and it is a solution of Bessel's modified equation for all values of ν.

As was the case with Bessel's equation, certain difficulties arise in finding a second solution of the modified equation when $\nu = n$ (n zero or a positive integer). It is conventional to denote the second solution by $K_\nu(x)$ and to define it by the relation

$$\frac{2}{\pi} K_\nu(x) = \operatorname{cosec} \nu\pi \{I_{-\nu}(x) - I_\nu(x)\}, \tag{3}$$

or by the limit of the expression on the right when $\nu = n$. The function $K_\nu(x)$ so defined is known as the *modified Bessel function of the second kind of order* ν and it can be shown that

$$K_0(x) = -\left(\log_e \frac{x}{2} + \gamma\right) I_0(x) + \sum_{r=1}^{\infty} \frac{(x/2)^{2r}}{r!r!}\left(1 + \frac{1}{2} + \frac{1}{3} + \ldots + \frac{1}{r}\right), \tag{4}$$

while, for $n = 1, 2, 3, \ldots$,

$$K_n(x) = (-1)^{n+1}\left(\log_e \frac{x}{2} + \gamma\right) I_n(x) + \frac{1}{2}\sum_{r=0}^{n-1} \frac{(-1)^r (n-r-1)!}{r!}\left(\frac{x}{2}\right)^{2r-n}$$
$$+(-1)^n \frac{1}{2}\sum_{r=0}^{\infty} \frac{(x/2)^{2r+n}}{r!(n+r)!}\left\{\frac{1}{1} + \frac{1}{2} + \frac{1}{3} + \ldots + \frac{1}{r} + \frac{1}{1} + \frac{1}{2} + \frac{1}{3} + \ldots + \frac{1}{n+r}\right\}, \tag{5}$$

the term corresponding to $r = 0$ in the second series being

$$\frac{(x/2)^n}{n!}\left\{\frac{1}{1} + \frac{1}{2} + \frac{1}{3} + \ldots + \frac{1}{n}\right\}.$$

In terms of these functions, the general solution of the modified equation (1) is, for all values of ν,

$$y = AI_\nu(x) + BK_\nu(x), \tag{6}$$

where A, B are arbitrary constants.

The modified functions which occur most frequently in practical applications are those of orders zero and unity: rough graphs are shown on the next page.

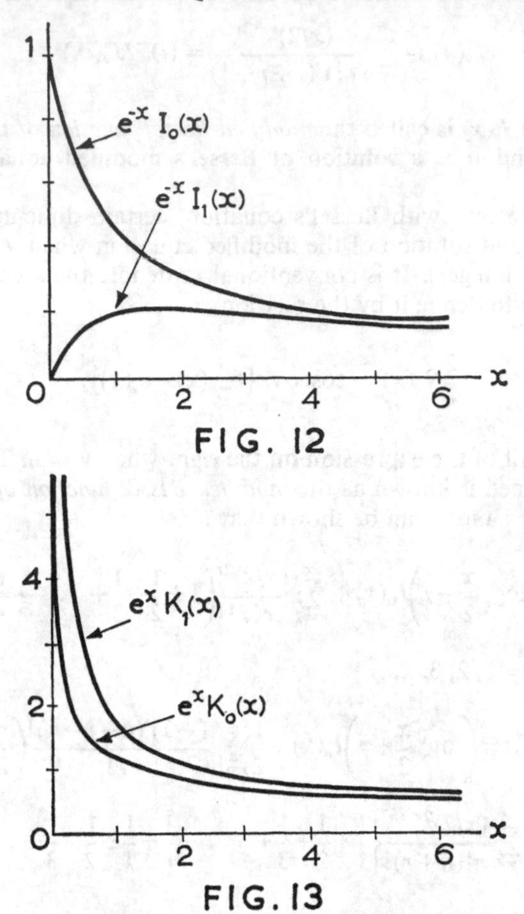

FIG. 12

FIG. 13

It is more convenient to plot $e^{-x}I_n(x)$ and $e^x K_n(x)$ than the functions themselves because of the behaviour of the functions for large values of x. This can be shown to be

$$I_\nu(x) \sim \frac{1}{(2\pi x)^{\frac{1}{2}}} e^x, \qquad K_\nu(x) \sim \left(\frac{\pi}{2x}\right)^{\frac{1}{2}} e^{-x}. \tag{7}$$

The modified functions $I_\nu(x)$, $K_\nu(x)$ bear to the exponential functions $e^{\pm x}$ similar relations to those which the functions $J_\nu(x)$, $Y_\nu(x)$ bear to the trigonometrical functions, and the modified functions have no zeros for real values of x. The proofs of some useful properties of $I_\nu(x)$, $K_\nu(x)$ are left to the reader (see Exercises 6(b), 13, 14, 15). Numerical values can be found in the volumes of tables to which reference has already been made.

6.11 The ber and bei functions

The differential equation

$$\frac{d^2y}{dx^2}+\frac{1}{x}\frac{dy}{dx}-iy=0 \qquad (1)$$

arises in problems in electrical engineering. It can be derived from Bessel's modified equation by writing $i^{\frac{1}{2}}x$ in place of x. A formal solution of this equation (remarks similar to those given in the footnote on p. 170 applying) is

$$y=AI_0(i^{\frac{1}{2}}x)+BK_0(i^{\frac{1}{2}}x). \qquad (2)$$

The arguments of the modified Bessel functions being complex, Kelvin found it convenient in practice to introduce four functions defined by

$$\text{ber}(x)+i\text{bei}(x)=I_0(i^{\frac{1}{2}}x), \qquad (3)$$

$$\text{ker}(x)+i\text{kei}(x)=K_0(i^{\frac{1}{2}}x), \qquad (4)$$

and all these four new functions are real when x is real.

From the series defining $I_0(x)$, it can be shown formally that

$$\text{ber}(x)=1-\frac{(x/2)^4}{(2!)^2}+\frac{(x/2)^8}{(4!)^2}-\ldots,$$

$$\text{bei}(x)=\frac{(x/2)^2}{(1!)^2}-\frac{(x/2)^6}{(3!)^2}+\frac{(x/2)^{10}}{(5!)^2}-\ldots.$$

Similar, but rather more complicated, formulae can be deduced for the functions ker(x) and kei(x). Short tables giving numerical values of these functions can be found in, for instance, *Bessel Functions for Engineers*, N. W. McLachlan, Oxford University Press, 1934.

6.12 Some transformations of Bessel's equation

Many practical problems can be reduced to the solution of a differential equation which, by suitable transformations, can itself be reduced to Bessel's equation. We give below an example of a useful transformation of this type.

Writing $x=\beta t^\gamma$, where β and γ are constants, we have

$$\frac{dy}{dx}=\frac{dy}{dt}\Big/\frac{dx}{dt}=\frac{1}{\beta\gamma t^{\gamma-1}}\frac{dy}{dt},$$

giving

$$x\frac{dy}{dx}=\frac{t}{\gamma}\frac{dy}{dt}.$$

Hence Bessel's equation of order ν, which can be written in the form

$$x\frac{d}{dx}\left(x\frac{dy}{dx}\right)+(x^2-\nu^2)y=0, \tag{1}$$

becomes

$$t\frac{d}{dt}\left(t\frac{dy}{dt}\right)+(\beta^2\gamma^2t^{2\gamma}-\nu^2\gamma^2)y=0. \tag{2}$$

Now write $y = t^\alpha u$, where α is a constant. Then

$$t\frac{dy}{dt}=t^{\alpha+1}\frac{du}{dt}+\alpha t^\alpha u,$$

and

$$t\frac{d}{dt}\left(t\frac{dy}{dt}\right)=t^{\alpha+2}\frac{d^2u}{dt^2}+(2\alpha+1)t^{\alpha+1}\frac{du}{dt}+\alpha^2t^\alpha u.$$

Substitution in (2) gives, after division by t^α,

$$t^2\frac{d^2u}{dt^2}+(2\alpha+1)t\frac{du}{dt}+\{\beta^2\gamma^2t^{2\gamma}+\alpha^2-\nu^2\gamma^2\}u=0. \tag{3}$$

The general solution of equation (1) being $y = AJ_\nu(x)+BY_\nu(x)$, since $u = t^{-\alpha}y$, $x = \beta t^\gamma$, we deduce that the general solution of equation (3) is

$$u=t^{-\alpha}\{AJ_\nu(\beta t^\gamma)+BY_\nu(\beta t^\gamma)\}. \tag{4}$$

Particular cases of practical interest are:

(i) $\alpha = -\frac{1}{2}$, $\gamma = \frac{1}{2}$, $\nu = 1$.

Equation (3) becomes

$$t\frac{d^2u}{dt^2}+\tfrac{1}{4}\beta^2u=0, \tag{5}$$

with solution $u = t^{\frac{1}{2}}\{AJ_1(\beta t^{\frac{1}{2}})+BY_1(\beta t^{\frac{1}{2}})\}$. This equation arises in considering the stability of a tapered strut.

(ii) $\alpha = -\frac{1}{2}$, $\beta = \frac{2}{3}i$, $\gamma = \frac{3}{2}$, $\nu = \frac{1}{3}$.

In this case equation (3) simplifies to

$$\frac{d^2u}{dt^2}=tu \tag{6}$$

and, since here ν is not an integer and β is imaginary, the solution of (6) can be written

$$u=t^{\frac{1}{2}}\{AI_{\frac{1}{3}}(\tfrac{2}{3}t^{\frac{3}{2}})+BI_{-\frac{1}{3}}(\tfrac{2}{3}t^{\frac{3}{2}})\}.$$

Equation (6) is known as *Airy's equation* and it was first encountered in a study of the intensity of light in the neighbourhood of a caustic.

Two functions $Ai(t)$, $Bi(t)$, known as *Airy functions*, have been defined by taking linear combinations of the functions $t^{\frac{1}{2}}I_{\frac{1}{3}}(\frac{2}{3}t^{\frac{3}{2}})$, $t^{\frac{1}{2}}I_{-\frac{1}{3}}(\frac{2}{3}t^{\frac{3}{2}})$ and these have been tabulated in the British Association Mathematical Tables, Part-Volume B, 1946.

(iii) $\alpha = \frac{1}{2}(m-1)$, $\gamma = \frac{1}{2}$, $\nu = m-1$, *m an integer.*

Equation (3) now becomes

$$t\frac{d^2u}{dt^2}+m\frac{du}{dt}+\tfrac{1}{4}\beta^2u = 0, \tag{7}$$

and its general solution is, from (4),

$$u = t^{-\frac{1}{2}(m-1)}\{AJ_{m-1}(\beta t^{\frac{1}{2}})+BY_{m-1}(\beta t^{\frac{1}{2}})\}. \tag{8}$$

Equations of the form (7) appear in problems concerned with the vibration of tapering beams.

Example 10. *The deflection y at time t of a point distant x from one end of a vibrating tapering beam is given by $y = u\cos\omega t$ where ω is a constant and u satisfies the fourth order differential equation*

$$\frac{d^2}{dx^2}\left(x^4\frac{d^2u}{dx^2}\right) = k^4x^2u, \qquad (k\ constant).$$

Find the general solution of this equation.

The given equation is, after division by x^2,

$$x^2\frac{d^4u}{dx^4}+8x\frac{d^3u}{dx^3}+12\frac{d^2u}{dx^2}-k^4u = 0,$$

and this can be written in the form

$$\left(x\frac{d^2}{dx^2}+3\frac{d}{dx}-k^2\right)\left(x\frac{d^2}{dx^2}+3\frac{d}{dx}+k^2\right)u = 0.$$

This may be seen by differentiating out this expression and it can be verified that the order of the brackets can be interchanged. Hence solutions of the fourth order differential equation are given by any solutions of either

$$x\frac{d^2u}{dx^2}+3\frac{du}{dx}+k^2u = 0, \tag{9}$$

or,

$$x\frac{d^2u}{dx^2}+3\frac{du}{dx}-k^2u = 0. \tag{10}$$

By putting $m = 3$, $\beta = 2k$, $t = x$ in equations (7) and (8), the solution of equation (9) is

$$u = x^{-1}\{AJ_2(2kx^{1/2})+BY_2(2kx^{1/2})\}.$$

The solution of equation (10) can be obtained in a similar way but, as we now have to write $\beta = 2ik$, modified Bessel functions will be involved and

$$u = x^{-1}\{CI_2(2kx^{1/2})+DK_2(2kx^{1/2})\}.$$

Thus the complete solution, involving four arbitrary constants, is

$$u = x^{-1}\{AJ_2(2kx^{1/2})+BY_2(2kx^{1/2})+CI_2(2kx^{1/2})+DK_2(2kx^{1/2})\}.$$

Exercises 6(*b*)

1. Assuming a series solution of the form $y = \sum_{n=0}^{\infty} a_n x^{\rho+n}$, show that the solution of the differential equation

$$x\frac{d^2y}{dx^2}+\frac{dy}{dx}-y = 0$$

is $y = Au+Bv$ where

$$u = \sum_{n=0}^{\infty} \frac{x^n}{(n!)^2} \quad \text{and} \quad v = u \log_e x - 2\sum_{n=1}^{\infty} \frac{x^n}{(n!)^2}\left(1+\frac{1}{2}+\frac{1}{3}+\ldots+\frac{1}{n}\right).$$

(L.U.)

2. By writing Bessel's equation in the form $\frac{1}{x}\frac{d}{dx}\left(x\frac{dy}{dx}\right)+\left(1-\frac{\nu^2}{x^2}\right)y = 0,$ show that

$$\frac{d}{dx}[x\{J_\nu(x)J_{-\nu}'(x)-J_\nu'(x)J_{-\nu}(x)\}] = 0.$$

By using the series for $J_\nu(x)$ deduce that

$$J_\nu(x)J_{-\nu}'(x)-J_\nu'(x)J_{-\nu}(x) = -\frac{2\sin\nu\pi}{\pi x}.$$

Hence show that $J_\nu(x)$, $J_{-\nu}(x)$ are independent solutions of Bessel's equation whenever ν is not an integer or zero.

3. Use the definition of $Y_\nu(x)$ and the result of Exercise 2 above to show that

$$J_\nu(x)Y_\nu'(x)-Y_\nu(x)J_\nu'(x) = \frac{2}{\pi x}.$$

Hence show that the solution y of Bessel's equation of order ν which is such that $y = \lambda$ and $y' = \mu$ when $x = a$ is given by

$$\frac{2y}{\pi a} = \{\lambda Y_\nu'(a)-\mu Y_\nu(a)\}J_\nu(x)-\{\lambda J_\nu'(a)-\mu J_\nu(a)\}Y_\nu(x).$$

4. If m is a positive integer show that

$$\left(\frac{1}{x}\frac{d}{dx}\right)^m \{x^\nu J_\nu(x)\} = x^{\nu-m}J_{\nu-m}(x),$$

$$\left(\frac{1}{x}\frac{d}{dx}\right)^m \{x^{-\nu} J_\nu(x)\} = (-1)^m x^{-\nu-m}J_{\nu+m}(x).$$

5. Use the generating function for the Bessel coefficients to show that

$$\exp(ix\cos\theta) = J_0(x)-2\{J_2(x)\cos 2\theta-J_4(x)\cos 4\theta+\ldots\}$$
$$+2i\{J_1(x)\cos\theta-J_3(x)\cos 3\theta+\ldots\}.$$

The current in a rectifying valve is $a\exp(b\cos\omega t)$ where a and b are constants: prove that the mean value of the current over a period is $aJ_0(ib)$. (L.U.)

6. Show that:

(i) $J_{1/2}(x)J_{-1/2}(x) = \dfrac{\sin 2x}{\pi x}$; (ii) $J_{1/2}^2(x)+J_{-1/2}^2(x) = \dfrac{2}{\pi x}$;

(iii) $Y_{1/2}(x) = -\left(\dfrac{2}{\pi x}\right)^{1/2} \cos x$; (iv) $Y_{n+1/2}(x) = (-1)^{n+1}J_{-n-1/2}(x)$,

n being a positive integer.

7. Prove that:

(i) $\displaystyle\int_0^a x^{-1}J_1^2(x)\,dx = \tfrac{1}{2}\{1-J_0^2(a)-J_1^2(a)\}$;

(ii) $\displaystyle\int_0^{\pi/2} J_0(x\cos\theta)\cos\theta\,d\theta = \frac{\sin x}{x}$.

8. By expanding $\cos(x\cos\theta)$ in ascending powers of x, obtain *Poisson's integral*

$$J_n(x) = \frac{(x/2)^n}{\Gamma(n+\frac{1}{2})\Gamma(\frac{1}{2})}\int_0^\pi \cos(x\cos\theta)\sin^{2n}\theta\,d\theta,$$

where $n = 0, 1, 2, 3, \ldots$.

9. Show that:

(i) $\displaystyle\int_0^\infty J_1(ax)\,dx = \frac{1}{a}$;

(ii) $\displaystyle\int_0^\infty x^{1-n}J_n(x)\,dx = \frac{2^{1-n}}{\Gamma(n)}$, $(n = 0, 1, 2, 3, \ldots)$.

10. If α_r, $(r = 1, 2, 3, \ldots)$ is a positive root of the equation $J_0(\alpha_r) = 0$, show that

$$\sum_{r=1}^\infty \frac{J_0(\alpha_r x)}{\alpha_r J_1(\alpha_r)} = \frac{1}{2}, \quad (0 < x < 1).$$

11. λ_r, $(r = 1, 2, 3, \ldots)$ is a positive root of the equation

$$hJ_\nu(\lambda_r a)+k\lambda_r aJ_\nu'(\lambda_r a) = 0,$$

h and k being constants. Assuming it is possible to expand a function $f(x)$ in the interval $(0, a)$ in the form

$$f(x) = \sum_{r=1}^\infty c_r J_\nu(\lambda_r x),$$

show that the coefficients c_r are given by

$$c_r = \frac{2k^2\lambda_r^2}{\{k^2\lambda_r^2a^2+h^2-k^2\nu^2\}J_\nu^2(\lambda_r a)}\int_0^a xf(x)J_\nu(\lambda_r x)\,dx.$$

12. Prove that

(i) $I_{1/2}(x) = \left(\dfrac{2}{\pi x}\right)^{1/2}\sinh x$; (ii) $I_{-1/2}(x)-I_{1/2}(x) = \dfrac{2}{\pi}K_{1/2}(x)$;

(iii) $K_{1/2}(x) = \left(\frac{\pi}{2x}\right)^{1/2} e^{-x}$.

13. Use the relation $I_\nu(x) = (i)^{-\nu} J_\nu(ix)$ and the recurrence formulae for $J_\nu(x)$ to show that:

(i) $\frac{2\nu}{x} I_\nu(x) = I_{\nu-1}(x) - I_{\nu+1}(x)$; (ii) $2I_\nu'(x) = I_{\nu-1}(x) + I_{\nu+1}(x)$.

Deduce that $I_0'(x) = I_1(x)$.

14. Using the definition of $K_\nu(x)$ and the results of Exercise 13 above, obtain the recurrence formulae:

(i) $-\frac{2\nu}{x} K_\nu(x) = K_{\nu-1}(x) - K_{\nu+1}(x)$; (ii) $-2K_\nu'(x) = K_{\nu-1}(x) + K_{\nu+1}(x)$.

Deduce that $K_0'(x) = -K_1(x)$.
Given that

$$K_\nu(x) = \tfrac{1}{2}\int_{-\infty}^{\infty} \exp(-x \cosh t - \nu t)\, dt,$$

deduce, by writing $x\, e^t = 2\phi$, that

$$K_\nu(x) = \tfrac{1}{2}(\tfrac{1}{2}x)^\nu \int_0^\infty \exp\left(-\phi - \frac{x^2}{4\phi}\right) \frac{d\phi}{\phi^{\nu+1}}.$$

15. Use the relation $I_\nu(x) = (i)^{-\nu} J_\nu(ix)$ and the result of Exercise 2 above to show that

$$I_\nu(x) I_{-\nu}'(x) - I_\nu'(x) I_{-\nu}(x) = -\frac{2 \sin \nu\pi}{\pi x}.$$

Hence show that

$$I_\nu(x) K_\nu'(x) - I_\nu'(x) K_\nu(x) = -1/x,$$

and deduce that

$$I_\nu(x) K_{\nu+1}(x) + I_{\nu+1}(x) K_\nu(x) = 1/x.$$

16. Show that

$$\operatorname{ker}(x) = (\log_e 2 - \gamma - \log_e x)\operatorname{ber}(x) + \frac{\pi}{4}\operatorname{bei}(x) - \frac{(x/2)^4}{(2!)^2}(1+\tfrac{1}{2}) + \frac{(x/2)^8}{(4!)^2}(1+\tfrac{1}{2}+\tfrac{1}{3}+\tfrac{1}{4}) - \ldots,$$

$$\operatorname{kei}(x) = (\log_e 2 - \gamma - \log_e x)\operatorname{bei}(x) - \frac{\pi}{4}\operatorname{ber}(x) + \frac{(x/2)^2}{(1!)^2} - \frac{(x/2)^6}{(3!)^2}(1+\tfrac{1}{2}+\tfrac{1}{3}) + \ldots.$$

17. Show that the general solution of the differential equation

$$\frac{d^2y}{dt^2} + (k^2 e^{2t} - n^2)y = 0$$

is

$$y = AJ_n(ke^t) + BY_n(ke^t).$$

18. By changing the dependent variable from y to v where $y = v^{-1}(dv/dx)$, transform the equation

$$\frac{dy}{dx}+y^2+x^m = 0$$

into a second order differential equation in v and x.
By writing

$$v = ux^{1/2}, \qquad t = \frac{2}{m+2}x^{1+m/2},$$

show that the transformed equation can itself be transformed into Bessel's equation of order $(m+2)^{-1}$.

19. By means of the substitutions $y = x^{1/2}u$, $z = kx$, reduce the equation

$$\frac{d^2y}{dx^2}+\left(k^2+\frac{1}{4x^2}\right)y = 0$$

to the form

$$\frac{d^2u}{dz^2}+\frac{1}{z}\frac{du}{dz}+u = 0.$$

If $x^{-\frac{1}{2}}y$ is finite when $x = 0$ and $(dy/dx) = a^{1/2}$ when $x = a$, show that the solution is

$$y = \frac{2ax^{1/2}J_0(kx)}{J_0(ka)-2kaJ_1(ka)}. \qquad \text{(L.U.)}$$

20. In a problem on the stability of a tapered strut the displacement y satisfies the equation

$$\frac{d^2y}{dx^2}+\frac{K^2}{4x}y = 0,$$

where a suitable value of K has to be determined. Show that, by writing $y = x^{1/2}u$, $z = Kx^{1/2}$, this equation can be reduced to the form

$$z^2\frac{d^2u}{dz^2}+z\frac{du}{dz}+(z^2-1)u = 0.$$

If $(dy/dx) = 0$ when $x = a$ and when $x = l$, show that the equation for K is

$$J_0(Ka^{1/2})\,Y_0(Kl^{1/2}) = J_0(Kl^{1/2})\,Y_0(Ka^{1/2}). \qquad \text{(L.U.)}$$

6.13 The confluent hypergeometric equation

The second order differential equation

$$x\frac{d^2y}{dx^2}+(\gamma-x)\frac{dy}{dx}-\alpha y = 0, \qquad (1)$$

in which α and γ are constants, is called the *confluent hypergeometric equation*. Its solution is easily found as a series. The indicial equation and the recurrence relation being respectively $\rho(\rho+\gamma-1) = 0$ and

$$a_{r+1} = \frac{\rho+\alpha+r}{(\rho+\gamma+r)(\rho+1+r)}a_r,$$

the first solution (with $\rho = 0$) is given by

$$y_1 = {}_1F_1(\alpha;\gamma;x) \equiv 1+\frac{\alpha}{\gamma}\frac{x}{1!}+\frac{\alpha(\alpha+1)}{\gamma(\gamma+1)}\frac{x^2}{2!}+\dots$$

$$= \sum_{r=0}^{\infty}\frac{(\alpha)_r}{(\gamma)_r}\frac{x^r}{r!}, \qquad (2)$$

the series being convergent for all finite values of x. If γ is not an integer or zero, the second solution is

$$y_2 = x^{1-\gamma}\,{}_1F_1(\alpha+1-\gamma;2-\gamma;x) = \sum_{r=0}^{\infty}\frac{(\alpha+1-\gamma)_r}{(2-\gamma)_r}\frac{x^{r+1-\gamma}}{r!}, \qquad (3)$$

and the details of the calculation are left to the reader. The function ${}_1F_1(\alpha;\gamma;x)$ defined by equation (2) is known as the *confluent hypergeometric function* and is of importance in some physical problems.

Equation (1) is a limiting case of the hypergeometric equation (§5.8 (1)). Writing x/β in place of x, we see that the function ${}_2F_1(\alpha;\beta;\gamma;x/\beta)$ is a solution of the differential equation

$$x\left(1-\frac{x}{\beta}\right)\frac{d^2y}{dx^2}+\left\{\gamma-(\alpha+\beta+1)\frac{x}{\beta}\right\}\frac{dy}{dx}-\alpha y = 0. \qquad (4)$$

This equation has regular singular points at $x = 0$, $x = \beta$ and at the point at infinity, and there is a "confluence" of the second and third of these points as $\beta \to \infty$. When this occurs equation (4) reduces to equation (1) and it can be shown that

$${}_1F_1(\alpha;\gamma;x) = \lim_{\beta\to\infty}\left\{{}_2F_1\left(\alpha,\beta;\gamma;\frac{x}{\beta}\right)\right\}. \qquad (5)$$

The confluent hypergeometric function possesses properties analogous to those of the ordinary hypergeometric function. For example, the integral representation (analogous to equation (7) of §5.9) is

$${}_1F_1(\alpha;\gamma;x) = \frac{\Gamma(\gamma)}{\Gamma(\alpha)\Gamma(\gamma-\alpha)}\int_0^1 t^{\alpha-1}(1-t)^{\gamma-\alpha-1}e^{xt}\,dt, \quad (\gamma>\alpha>0), \qquad (6)$$

and this is sometimes useful in computing values of the function. The derivation of this formula and of other properties of the confluent function are left to the reader (see Exercises 6(*c*)).

6.14 The Jacobi polynomials

The *Jacobi polynomials* $P_n^{(\alpha,\beta)}(x)$ are another special case of the hypergeometric function and are usually defined by

$$P_n^{(\alpha,\beta)}(x) = \frac{(\alpha+1)_n}{n!}\,{}_2F_1(-n,n+\alpha+\beta+1;\alpha+1;\tfrac{1}{2}-\tfrac{1}{2}x), \qquad (1)$$

where n is a non-negative integer and α, β are real numbers greater than -1. Because of the parameter $-n$ in the hypergeometric function, the series terminates at the term in x^n and hence $P_n^{(\alpha,\beta)}(x)$ is a *polynomial of degree n*. These polynomials are more general than those of Legendre to which they reduce when $\alpha = \beta = 0$.

The differential equation satisfied by $P_n^{(\alpha,\beta)}(x)$ is

$$(1-x^2)\frac{d^2y}{dx^2}+\{\beta-\alpha-(\alpha+\beta+2)x\}\frac{dy}{dx}+n(n+\alpha+\beta+1)y = 0. \qquad (2)$$

This can be seen by writing $-n, n+\alpha+\beta+1, \alpha+1, \frac{1}{2}-\frac{1}{2}x$ in place of α, β, γ, x respectively in the hypergeometric equation (§5.8 (1)). It can also be shown that the Jacobi polynomials are given by

$$P_n^{(\alpha,\beta)}(x) = \frac{(-1)^n}{2^n n!}(1-x)^{-\alpha}(1+x)^{-\beta}\frac{d^n}{dx^n}\{(1-x)^{\alpha+n}(1+x)^{\beta+n}\}, \qquad (3)$$

which is clearly the analogue of Rodrigues' formula for the Legendre polynomials. An outline of the method of derivation of this formula will be found in Exercises 6(*c*), 6.

The Jacobi polynomials possess the important property that the functions $(1-x)^{\frac{1}{2}\alpha}(1+x)^{\frac{1}{2}\beta}P_n^{(\alpha,\beta)}(x)$ are orthogonal in the interval $(-1, 1)$. That this is so can be shown as follows. Let $u = P_n^{(\alpha,\beta)}(x)$ so that u satisfies the differential equation (2) and, after multiplication by $(1-x)^\alpha(1+x)^\beta$, this can be written in the form

$$\frac{d}{dx}\left\{(1-x)^{\alpha+1}(1+x)^{\beta+1}\frac{du}{dx}\right\}+n(n+\alpha+\beta+1)(1-x)^\alpha(1+x)^\beta u = 0.$$

Similarly, if $v = P_m^{(\alpha,\beta)}(x)$,

$$\frac{d}{dx}\left\{(1-x)^{\alpha+1}(1+x)^{\beta+1}\frac{dv}{dx}\right\}+m(m+\alpha+\beta+1)(1-x)^\alpha(1+x)^\beta v = 0.$$

Multiplying these equations respectively by v and u and subtracting,

$$(n-m)(n+m+\alpha+\beta+1)(1-x)^\alpha(1+x)^\beta uv$$

$$= u\frac{d}{dx}\left\{(1-x)^{\alpha+1}(1+x)^{\beta+1}\frac{dv}{dx}\right\}-v\frac{d}{dx}\left\{(1-x)^{\alpha+1}(1+x)^{\beta+1}\frac{du}{dx}\right\}$$

$$= \frac{d}{dx}\left\{(1-x)^{\alpha+1}(1+x)^{\beta+1}\left(u\frac{dv}{dx}-v\frac{du}{dx}\right)\right\}.$$

Integrating with respect to x and remembering that α, β are both greater than -1, the right-hand side vanishes if the limits of integration are taken as ± 1. Hence, if $n \neq m$, we find

$$\int_{-1}^{1} (1-x)^{\alpha}(1+x)^{\beta} uv \, dx = 0,$$

and this establishes the required property.

Some authors take the function

$$\mathscr{F}_n(\alpha, \gamma, x) = {}_2F_1(-n, \alpha+n; \gamma; x) \tag{4}$$

where $\gamma > 0, \alpha > \gamma - 1$, as the standard form of the Jacobi polynomials. It is left as an exercise to show that the two different forms of the polynomials are related by

$$(\gamma)_n \mathscr{F}_n(\alpha, \gamma, x) = n! P_n^{(\gamma-1, \alpha-\gamma)}(1-2x), \tag{5}$$

and that $\mathscr{F}_n(\alpha, \gamma, x)$ satisfies the differential equation

$$x(1-x)\frac{d^2y}{dx^2} + \{\gamma - (\alpha+1)x\}\frac{dy}{dx} + n(\alpha+n)y = 0. \tag{6}$$

6.15 The Gegenbauer polynomials

The *Gegenbauer polynomials* $C_n^{\lambda}(x)$ are a sub-class of the Jacobi polynomials and are defined by

$$\left.\begin{aligned} (\lambda+\tfrac{1}{2})_n C_n^{\lambda}(x) &= (2\lambda)_n P_n^{(\lambda-\frac{1}{2}, \lambda-\frac{1}{2})}(x), \quad (\lambda \neq 0), \\ (\tfrac{1}{2})_n C_n^{0}(x) &= 2(n-1)!\, P_n^{(-\frac{1}{2}, -\frac{1}{2})}(x). \end{aligned}\right\} \tag{1}$$

Hence, writing $\alpha = \beta = \lambda - \frac{1}{2}$ in §6.14(1), the Gegenbauer polynomials can be expressed in terms of a hypergeometric function by

$$n!\, C_n^{\lambda}(x) = (2\lambda)_n \, {}_2F_1(-n, n+2\lambda; \lambda+\tfrac{1}{2}; \tfrac{1}{2}-\tfrac{1}{2}x), \tag{2}$$

with the proviso that $(2\lambda)_n$ is replaced by $2(n-1)!$ when $\lambda = 0$.

The differential equation satisfied by these polynomials is, by writing $\alpha = \beta = \lambda - \frac{1}{2}$ in §6.14(2), easily found to be

$$(1-x^2)\frac{d^2y}{dx^2} - (2\lambda+1)x\frac{dy}{dx} + n(n+2\lambda)y = 0. \tag{3}$$

The functions $C_n^{\lambda}(x)$ are, as is easily seen from (2), *polynomials of degree n*. They reduce to the Legendre polynomials when $\lambda = \frac{1}{2}$ and further properties of the Gegenbauer polynomials can, if required, be obtained in a similar way to those established in §6.3 for the Legendre polynomials.

6.16 The Tchebichef polynomials

The *Tchebichef polynomials* $T_n(x)$ and $U_n(x)$ form an important subclass of the Gegenbauer polynomials. They are defined by

$$T_n(x) = \tfrac{1}{2}nC_n{}^0(x) = {}_2F_1(-n, n; \tfrac{1}{2}; \tfrac{1}{2}-\tfrac{1}{2}x), \tag{1}$$

$$U_n(x) = C_n{}^1(x) = (n+1)\,{}_2F_1(-n, n+2; \tfrac{3}{2}; \tfrac{1}{2}-\tfrac{1}{2}x), \tag{2}$$

the representation as hypergeometric functions following from equation (2) of §6.15. From equation (3) of §6.15, it is easily seen that $T_n(x)$ satisfies the differential equation

$$(1-x^2)\frac{d^2y}{dx^2} - x\frac{dy}{dx} + n^2y = 0, \tag{3}$$

while the differential equation satisfied by $U_n(x)$ is

$$(1-x^2)\frac{d^2y}{dx^2} - 3x\frac{dy}{dx} + n(n+2)y = 0. \tag{4}$$

The functions $T_n(x)$ and $U_n(x)$ are both *polynomials of degree n.* They are useful in certain aerodynamical problems and in connection with numerical analysis. Their properties can again be established in a similar way to those of the Legendre polynomials and some of these will be found in Exercises 6(*c*).

6.17 The Laguerre polynomials

The *Laguerre polynomials* $L_n(x)$ are usually defined, in applied mathematics, by

$$L_n(x) = n!\,{}_1F_1(-n; 1; x), \tag{1}$$

where n is a positive integer (or zero) and ${}_1F_1(-n; 1; x)$ is the confluent hypergeometric function. Thus $L_n(x)$ is a *polynomial of degree n* and, by §6.13 (1), it satisfies the differential equation

$$x\frac{d^2y}{dx^2} + (1-x)\frac{dy}{dx} + ny = 0. \tag{2}$$

By using Leibnitz' theorem and comparing the result with equation (1) it follows that

$$L_n(x) = e^x\frac{d^n}{dx^n}(x^n e^{-x}) \tag{3}$$

and this result is often very useful. It can be used, for example, to show that the functions $e^{-\frac{1}{2}x}L_n(x)$ are orthogonal in the interval $(0, \infty)$. Thus, if m is a positive integer,

$$\int_0^\infty e^{-x} x^m L_n(x)\,dx = \int_0^\infty x^m \frac{d^n}{dx^n}(x^n e^{-x})\,dx$$

$$= (-1)^m m! \int_0^\infty \frac{d^{n-m}}{dx^{n-m}}(x^n e^{-x})\,dx, \tag{4}$$

the last step resulting by integrating by parts m times. The integral on the right of (4) is zero when $n > m$ and, since $L_m(x)$ is a polynomial of degree m in x, it follows that

$$\int_0^\infty e^{-x} L_m(x) L_n(x)\,dx = 0, \quad (m \neq n). \tag{5}$$

It is left as an exercise for the reader to apply equation (4) to show that

$$\int_0^\infty e^{-x}\{L_n(x)\}^2\,dx = (n!)^2,$$

and hence that the functions $\{e^{-\frac{1}{2}x}L_n(x)\}/(n!)$ form an orthonormal system.

A generating function for $L_n(x)$ is found in Example 11 below. This result is useful in obtaining further properties of the Laguerre polynomials (see, for example, Exercises 6(*c*), 12).

Example. 11 *Show that*

$$\exp\left(-\frac{xt}{1-t}\right) = (1-t)\sum_{n=0}^{\infty}\frac{L_n(x)}{n!}t^n, \quad (|t| < 1).$$

Expanding the exponential,

$$(1-t)^{-1}\exp\left(-\frac{xt}{1-t}\right) = \sum_{r=0}^{\infty}\frac{(-1)^r x^r t^r}{r!(1-t)^{r+1}}$$

$$= \sum_{r=0}^{\infty}\sum_{s=0}^{\infty}\frac{(-1)^r(r+1)_s}{r!\,s!}x^r t^{r+s},$$

the last step resulting from the expansion of $(1-t)^{-r-1}$. The coefficient of t^n in this double series is found by taking $s = n-r$ and letting r take the values $0, 1, 2, \ldots, n$; it is

$$\sum_{r=0}^{n}\frac{(-1)^r(r+1)_{n-r}}{r!(n-r)!}x^r.$$

Since

$$(r+1)_{n-r} = \frac{n!}{r!} \quad\text{and}\quad \frac{(-1)^r}{(n-r)!} = \frac{(-n)_r}{n!},$$

the coefficient of t^n is

$$\sum_{r=0}^{n}\frac{(-n)_r}{(r!)^2}x^r = {}_1F_1(-n; 1; x) = \frac{L_n(x)}{n!},$$

and the required result has been established.

6.18 The associated Laguerre polynomials

The *associated Laguerre polynomials* $L_n^m(x)$ are defined by

$$L_n^m(x) = \frac{(-1)^m(n!)^2}{m!(n-m)!}\,{}_1F_1(-n+m;m+1;x), \tag{1}$$

where n and m are non-negative integers such that $n \geqq m$. These functions are therefore *polynomials of degree* $n-m$ and, by §6.13 (1), they satisfy the differential equation

$$x\frac{d^2y}{dx^2}+(m+1-x)\frac{dy}{dx}+(n-m)y = 0. \tag{2}$$

They reduce to the ordinary Laguerre polynomials $L_n(x)$ when $m = 0$.

By comparing the coefficients of x^r it is easy to show that

$$\frac{d}{dx}\{{}_1F_1(\alpha;\gamma;x)\} = \frac{\alpha}{\gamma}\,{}_1F_1(\alpha+1;\gamma+1;x)$$

and, by m applications of this result,

$$\frac{d^m}{dx^m}\{{}_1F_1(-n;1;x)\} = \frac{(-1)^m n!}{m!(n-m)!}\,{}_1F_1(-n+m;m+1;x).$$

Using this result in conjunction with equation (1) above and equations (1), (3) of §6.17, it follows that

$$L_n^m(x) = \frac{d^m}{dx^m}\{L_n(x)\} = \frac{d^m}{dx^m}\left\{e^x\frac{d^n}{dx^n}(x^n e^{-x})\right\}, \tag{3}$$

and this result is useful in establishing further properties of the associated polynomials.

One of the important applications of the associated Laguerre polynomials lies in the discussion of the wave functions of the hydrogen atom. This involves the differential equation

$$\frac{d^2R}{dx^2}+\frac{2}{x}\frac{dR}{dx}-\left\{\frac{l(l+1)}{x^2}-\frac{\nu}{x}+\frac{1}{4}\right\}R = 0, \tag{4}$$

and this equation can be obtained from equation (2) by writing

$$m = 2l+1, \quad n = \nu+l, \quad y = e^{\frac{1}{2}x}x^{-l}R. \tag{5}$$

In practice ν and l are integers and a solution of equation (4) is therefore given by

$$R = e^{-\frac{1}{2}x}x^l L_{\nu+l}^{2l+1}(x). \tag{6}$$

A complete discussion of the physical problem involves the evaluation of the integrals $\int_0^\infty x^2 R^2\,dx$ and this can be done by first setting up a generating function for the associated polynomials. This function can be obtained by using equation (3) and the result given in Example 11 (see Exercises 6(*c*), 13).

6.19 The Hermite polynomials

The *Hermite polynomials* $H_n(x)$ are defined by

$$H_n(x) = (-1)^n e^{x^2} \frac{d^n}{dx^n}(e^{-x^2}), \tag{1}$$

where n is a non-negative integer. $H_n(x)$ is a *polynomial of degree n* and the multiplier $(-1)^n$ is convenient as it makes the coefficient of x^n positive for all values of n. We now show that $H_n(x)$ is a solution of the differential equation

$$\frac{d^2y}{dx^2} - 2x\frac{dy}{dx} + 2ny = 0. \tag{2}$$

Using the notation $D \equiv d/dx$, we have

$$H_n(x) = (-1)^n e^{x^2} D^n(e^{-x^2}),$$

$$D\{H_n(x)\} = (-1)^n e^{x^2}(D^{n+1} + 2xD^n)(e^{-x^2}),$$

$$D^2\{H_n(x)\} = (-1)^n e^{x^2}\{D^{n+2} + 4xD^{n+1} + (4x^2+2)D^n\}(e^{-x^2}).$$

Hence,

$$(D^2 - 2xD + 2n)H_n(x) = (-1)^n e^{x^2}\{D^{n+2} + 2xD^{n+1} + 2(n+1)D^n\}(e^{-x^2}). \tag{3}$$

Now $D(e^{-x^2}) = -2xe^{-x^2}$ so that $(D+2x)e^{-x^2} = 0$ and, differentiating this result $(n+1)$ times, we have

$$\{D^{n+2} + 2xD^{n+1} + 2(n+1)D^n\}(e^{-x^2}) = 0.$$

Using this result in conjunction with (3), we see that $H_n(x)$ satisfies the differential equation (2).

The generating function for $H_n(x)$ can be obtained by expanding the function $\exp(2tx - t^2)$ in ascending powers of t by Maclaurin's theorem. We have

$$\begin{aligned}\exp(2tx - t^2) &= \exp(x^2)\exp\{-(x-t)^2\}\\ &= \exp(x^2)\sum_{n=0}^{\infty}\frac{t^n}{n!}\left[\frac{\partial^n}{\partial t^n}\exp\{-(x-t)^2\}\right]_{t=0}.\end{aligned}$$

Now

$$\left[\frac{\partial^n}{\partial t^n}\exp\{-(x-t)^2\}\right]_{t=0} = (-1)^n \frac{d^n}{dx^n}\{\exp(-x^2)\},$$

and hence

$$\exp(2tx-t^2) = \sum_{n=0}^{\infty}(-1)^n e^{x^2}\frac{d^n}{dx^n}(e^{-x^2})\frac{t^n}{n!}$$

$$= \sum_{n=0}^{\infty} H_n(x)\frac{t^n}{n!}. \tag{4}$$

Recurrence formulae can be deduced from equation (4), a particular instance being given below in Example 12.

Example 12. *Show that* $2nH_{n-1}(x) = H_n'(x)$.

Differentiating equation (4) with respect to x,

$$2t\exp(2tx-t^2) = \sum_{n=0}^{\infty} H_n'(x)\frac{t^n}{n!}.$$

Again using (4), this can be written

$$2t\sum_{n=0}^{\infty} H_n(x)\frac{t^n}{n!} = \sum_{n=0}^{\infty} H_n'(x)\frac{t^n}{n!},$$

and the required result follows by equating the coefficients of t^n.

We now consider the differential equation

$$\frac{d^2y}{dx^2} - 2x\frac{dy}{dx} + 2\nu y = 0 \tag{5}$$

which is the generalization of equation (2) when the integer n is replaced by a non-restricted constant ν. Equation (5) is known as *Hermite's equation* and we assume a series solution of the form $y = \sum_{r=0}^{\infty} a_r x^{\rho+r}$. The usual procedure gives the indicial equation $\rho(\rho-1) = 0$ and the recurrence relation

$$a_{r+2} = \frac{2(\rho+r-\nu)}{(\rho+r+2)(\rho+r+1)}a_r.$$

With $\rho = 0$ one solution is therefore given by

$$y_1 = a_0\left\{1 - \frac{2\nu}{2!}x^2 + \frac{2^2\nu(\nu-2)}{4!}x^4 - \frac{2^3\nu(\nu-2)(\nu-4)}{6!}x^6 + \dots\right\}, \tag{6}$$

while $\rho = 1$ leads to the second solution.

$$y_2 = b_0 x\left\{1 - \frac{2(\nu-1)}{3!}x^2 + \frac{2^2(\nu-1)(\nu-3)}{5!}x^4 - \dots\right\}. \tag{7}$$

When ν is a positive integer n, one or other of these solutions reduces to a polynomial, the polynomial solution being y_1 when ν is even and y_2 when ν is odd. When $\nu = n$, it is possible to identify the Hermite polynomial $H_n(x)$ with the solution y_1 by taking

$$a_0 = (-1)^{\frac{1}{2}n}\frac{n!}{(\frac{1}{2}n)!}, \qquad (n \text{ even}),$$

or with the solution y_2 by taking

$$b_0 = (-1)^{\frac{1}{2}n-\frac{1}{2}}\frac{2(n!)}{(\frac{1}{2}n-\frac{1}{2})!}, \qquad (n \text{ odd}).$$

Writing $y = e^{\frac{1}{2}x^2}\psi$, we have

$$\frac{dy}{dx} = e^{\frac{1}{2}x^2}\left(\frac{d\psi}{dx}+x\psi\right), \qquad \frac{d^2y}{dx^2} = e^{\frac{1}{2}x^2}\left\{\frac{d^2\psi}{dx^2}+2x\frac{d\psi}{dx}+(x^2+1)\psi\right\},$$

and Hermite's equation (5) can, after division by $e^{\frac{1}{2}x^2}$ and a slight reduction, be written in the form

$$\frac{d^2\psi}{dx^2}+(2\nu+1-x^2)\psi = 0, \tag{8}$$

which is a reduced form of *Schrödinger's equation.* This equation plays an important part in wave mechanics and we shall now briefly discuss a solution which, as far as physical applications are concerned, has the correct behaviour for large values of x.

It can be shown that the two solutions (6) and (7) of Hermite's equation behave like ce^{x^2} (c constant) when x is large except when ν is a positive integer n. For such values of ν, one solution reduces to the polynomial $H_n(x)$ and this behaves like cx^n for large x. Now the solution of equation (8) is $\psi = ye^{-\frac{1}{2}x^2}$ where y satisfies Hermite's equation and, since y behaves like ce^{x^2} as $x \to \infty$ except when ν is an integer n, it can be shown that equation (8) only has a solution which remains finite as $x \to \infty$ when ν takes such values. In such cases this solution can be taken to be

$$\Psi_n(x) = e^{-\frac{1}{2}x^2}H_n(x), \tag{9}$$

and the function $\Psi_n(x)$ is sometimes called *Hermite's function of order n.* Properties of this function can be deduced from those of $H_n(x)$.

Exercises 6(*c*)

1. One solution of the confluent hypergeometric equation

$$xy''+(\gamma-x)y'-\alpha y = 0$$

is $y_1 = {}_1F_1(\alpha;\gamma;x)$. If γ is neither zero nor an integer, show that a second solution is $x^{1-\gamma}\,{}_1F_1(\alpha-\gamma+1;2-\gamma;x)$. If $\gamma = 1$, show that a second solution is given by $y_2 = y_1 \log_e x+u$ where u is of the form $\sum_{r=1}^{\infty} a_r x^r$.

2. If y satisfies the confluent hypergeometric equation
$$xy''+(\gamma-x)y'-ay = 0,$$
show that
$$W = x^{\gamma/2}e^{-x/2}y$$
satisfies *Whittaker's confluent hypergeometric equation*
$$\frac{d^2W}{dx^2}+\left\{\frac{\frac{1}{4}-m^2}{x^2}+\frac{k}{x}-\frac{1}{4}\right\}W = 0,$$
where $k = \frac{1}{2}\gamma-a$ and $m = \frac{1}{2}(1-\gamma)$.

3. Show that solutions of Whittaker's confluent hypergeometric equation (Exercise 2 above) are given by the *Whittaker functions*
$$M_{k,m}(x) = x^{1/2+m}e^{-x/2}\,{}_1F_1(\tfrac{1}{2}-k+m;\,1+2m;\,x),$$
$$M_{k-m}(x) = x^{1/2-m}e^{-x/2}\,{}_1F_1(\tfrac{1}{2}-k-m;\,1-2m;\,x).$$

4. Show that
(*a*) $\dfrac{d}{dx}\{{}_1F_1(a;\gamma;x)\} = \dfrac{a}{\gamma}\,{}_1F_1(a+1;\gamma+1;x)$;
(*b*) $a\,{}_1F_1(a+1;\gamma+1;x)+(\gamma-a)\,{}_1F_1(a;\gamma+1;x) = \gamma\,{}_1F_1(a;\gamma;x)$.
Show also that
$${}_1F_1(a;a;x) = e^x.$$

5. By expanding the exponential and integrating term by term, show that, if $\gamma > a > 0$,
$${}_1F_1(a;\gamma;x) = \frac{\Gamma(\gamma)}{\Gamma(a)\Gamma(\gamma-a)}\int_0^1 t^{a-1}(1-t)^{\gamma-a-1}e^{xt}\,dt.$$
Deduce *Kummer's relation* ${}_1F_1(a;\gamma;x) = e^x\,{}_1F_1(\gamma-a;\gamma;-x)$.

6. Assuming that ${}_2F_1(a,b;c;z) = (1-z)^{-a}\,{}_2F_1(a,c-b;c;z/(z-1))$, (see §5.9, Example 6), show that
$$\frac{d^n}{dz^n}\{z^{n+c-1}(1-z)^{b-c}\} = (c)_n z^{c-1}(1-z)^{b-c-n}\,{}_2F_1(-n,b;c;z).$$
Deduce that
$$P_n^{(\alpha,\beta)}(x) = \frac{(-1)^n}{2^n n!}(1-x)^{-\alpha}(1+x)^{-\beta}\frac{d^n}{dx^n}\{(1-x)^{\alpha+n}(1+x)^{\beta+n}\}.$$
Show further that $P_n^{(\alpha,\beta)}(-x) = (-1)^n P_n^{(\alpha,\beta)}(x)$.

7. Use the result of Exercise 6 above to show that:
(i) $(1-x)P_n^{(\alpha+1,\beta)}(x)+(1+x)P_n^{(\alpha,\beta+1)}(x) = 2P_n^{(\alpha,\beta)}(x)$;
(ii) $P_n^{(\alpha,\beta-1)}(x)-P_n^{(\alpha-1,\beta)}(x) = P_{n-1}^{(\alpha,\beta)}(x)$.

8. If $C_n^\lambda(x)$ is the Gegenbauer polynomial, show that

$$2^n n!(\lambda+\tfrac{1}{2})_n(1-x^2)^{\lambda-1/2}C_n^\lambda(x) = (-1)^n(2\lambda)_n \frac{d^n}{dx^n}\{(1-x^2)^{n+\lambda-1/2}\},$$

n being a positive integer. Deduce that $C_1^\lambda(x) = 2\lambda x$.

9. By writing $x = \cos\theta$, show that the general solution of the differential equation satisfied by the Tchebichef polynomial $T_n(x)$, viz.

$$(1-x^2)y''-xy'+n^2y = 0$$

is $y = A\cos(n\cos^{-1}x)+B\sin(n\cos^{-1}x)$. Use the result of §5.9, Example 5 to show that this solution becomes $y = T_n(x)$ when $A = 1$, $B = 0$.

10. Use the result of Exercise 6 above to show that

$$\frac{T_n(x)}{(1-x^2)^{1/2}} = \frac{(-1)^n}{1.3.5\ldots(2n-1)}\frac{d^n}{dx^n}\{(1-x^2)^{n-1/2}\}.$$

Deduce that $T_0(x) = 1$, $T_1(x) = x$, $T_2(x) = 2x^2-1$.

11. Show that the first four Laguerre polynomials are given by:

$$L_0(x) = 1, \quad L_1(x) = 1-x, \quad L_2(x) = 2-4x+x^2,$$
$$L_3(x) = 6-18x+9x^2-x^3.$$

12. Prove the following recurrence formulae for the Laguerre polynomials:

(i) $L_{n+1}(x)+(x-2n-1)L_n(x)+n^2L_{n-1}(x) = 0$;

(ii) $L_n'(x)-nL_{n-1}'(x)+nL_{n-1}(x) = 0$.

13. For the associated Laguerre polynomials $L_n^m(x)$, prove that

$$(-1)^m t^m \exp\left(-\frac{xt}{1-t}\right) = (1-t)^{m+1}\sum_{n=m}^{\infty}\frac{L_n^m(x)}{n!}t^n.$$

Deduce from Exercise 12 above that

(i) $L_{n+1}^m(x)+(x-2n-1)L_n^m(x)+mL_n^{m-1}(x)+n^2L_{n-1}^m(x) = 0$;

(ii) $L_n^m(x)-nL_{n-1}^m(x)+nL_{n-1}^{m-1}(x) = 0$.

14. If $H_n(x)$ is the Hermite polynomial, prove that

$$H_{n+1}(x)-2xH_n(x)+2nH_{n-1}(x) = 0.$$

Show that $H_0(x) = 1$, $H_1(x) = 2x$ and deduce that

$$H_2(x) = 4x^2-2, \quad H_3(x) = 8x^3-12x, \quad H_4(x) = 16x^4-48x^2+12.$$

15. If $\Psi_n(x)$ is a Hermite function defined by

$$\Psi_n(x) = e^{-x^2/2}H_n(x)$$

where $H_n(x)$ is the Hermite polynomial, show that

(i) $\displaystyle\int_{-\infty}^{\infty}\Psi_m(x)\Psi_n(x)\,dx = 0, \quad (m \neq n)$; (ii) $\displaystyle\int_{-\infty}^{\infty}x\{\Psi_n(x)\}^2\,dx = 0.$

CHAPTER 7

PARTIAL DIFFERENTIAL EQUATIONS

7.1 Introduction

The physical problems so far considered in this book have led to *ordinary* differential equations. This is because we have been dealing with physical quantities which have depended on a single variable. There are, however, many physical phenomena which can be described by the modern notion of field action. This is based on the two main ideas of (*a*) continuity and (*b*) the limitation of the influence of an element of space to its immediate neighbourhood. In such cases the values of the physical quantities involved depend on more than one variable. These variables are usually the time and quantities which specify the position of the point under discussion in the field. The problem can then often be expressed in terms of a *partial* differential equation together with certain initial and/or boundary conditions.

Partial differential equations play a very important part in modern scientific theories. Their solution is naturally more difficult than that of ordinary differential equations and it would be impossible to give a complete discussion in a book of this size. In this chapter we consider some of the linear partial differential equations which arise in mathematical physics and indicate how solutions to some of the simpler problems can be obtained by the classical method of separating the variables. Another method of solution which, in certain cases, is less dependent on maturity of judgment in assuming at the outset the appropriate form of solution, will be found in the chapter on integral transforms.

7.2 Solution by separation of variables

We commence by illustrating how a solution can be obtained in the relatively simple case of a partial differential equation containing only two independent variables. Consider, for example, the equation

$$\frac{\partial^2 V}{\partial x^2}-\frac{1}{c^2}\frac{\partial^2 V}{\partial t^2}=0, \tag{1}$$

where V is to be found as a function of the two independent variables x, t, and c is a physical constant. It is easy to verify that equation (1) is satisfied by

$$V=f(x-ct)+g(x+ct)$$

where f and g are arbitrary functions of the variables indicated. This is, in fact, the *general solution* of the equation, and it represents two waves of shapes $V = f(x)$, $V = g(x)$ moving respectively in the positive and negative x directions with velocity c. Another way of describing the solution is to say that the initial values of the waves are propagated along the lines $x-ct$ = constant and $x+ct$ = constant situated in the (x, t)-plane. These lines are called the *characteristics* of the equation. In a similar way, the characteristics of the equation

$$\frac{\partial^2 V}{\partial x^2}+\frac{\partial^2 V}{\partial y^2}=0 \tag{2}$$

are the lines $x \pm iy$ = constant. The theory of characteristics* plays an important part in the theory of partial differential equations when the analysis of such equations is taken further than is done here, but it is worth pointing out that equations of the second order can be conveniently classified according to their characteristics. Equations, such as equation (1), with two real distinct sets of characteristics, are said to be *hyperbolic*, those like equation (2) with imaginary characteristics are called *elliptic*, while an equation like

$$\frac{\partial^2 V}{\partial x^2}=\frac{\partial V}{\partial t}$$

has only one family of characteristics and is classified as *parabolic*.

The solution $V = f(x-ct)+g(x+ct)$ to equation (1) is usually too general for use in practical applications and we are concerned here only with finding *particular solutions* which can be fitted to simple boundary and initial conditions. We assume a particular "product" solution of equation (1) of the form

$$V = XT, \tag{3}$$

where X is a function of x only and T a function of t only. Then

$$\frac{\partial^2 V}{\partial x^2}=T\frac{d^2 X}{dx^2}=\frac{V}{X}\frac{d^2 X}{dx^2},$$

and

$$\frac{\partial^2 V}{\partial t^2}=X\frac{d^2 T}{dt^2}=\frac{V}{T}\frac{d^2 T}{dt^2},$$

so that substitution in equation (1) gives, after division by V,

$$\frac{1}{X}\frac{d^2 X}{dx^2}=\frac{1}{c^2 T}\frac{d^2 T}{dt^2}. \tag{4}$$

* A concise treatment can be found in the National Physical Laboratory pamphlet "Modern Computing Methods", *Notes on Applied Science, No. 16*, H.M.S.O., 1957.

With the assumption (3), the left-hand side of equation (4) is independent of t and the right-hand side is independent of x. Hence each side of the equation must equal some constant. Suppose for the moment that this constant, the so-called *constant of separation*, is negative* and take it to be $-\omega^2$. Equation (4) then gives the two ordinary differential equations

$$\frac{1}{X}\frac{d^2X}{dx^2} = -\omega^2, \qquad \frac{1}{c^2T}\frac{d^2T}{dt^2} = -\omega^2,$$

leading to

$$\frac{d^2X}{dx^2}+\omega^2X = 0, \qquad \frac{d^2T}{dt^2}+\omega^2c^2T = 0. \tag{5}$$

The general solutions of these two equations are respectively

$$X = a\cos(\omega x+\varepsilon), \qquad T = b\cos(\omega ct+\eta), \tag{6}$$

where a, b, ε and η are arbitrary constants. Thus a solution of the original equation is

$$V = A\cos(\omega x+\varepsilon)\cos(\omega ct+\eta) \tag{7}$$

where A, ω, ε and η are arbitrary constants. These constants can be determined only when the boundary and initial conditions to be associated with the partial differential equation (1) are given. Suppose, for example, that the boundary conditions are $V = 0$ when both $x = 0$ and $x = l$ for all values of t. Then the solution (7) satisfies these conditions if $\varepsilon = \frac{1}{2}\pi$ and ω is so chosen that $\sin\omega l = 0$, the solution then being

$$V = A\sin\omega x\cos(\omega ct+\eta).$$

Values of ω satisfying the equation $\sin\omega l = 0$ are $\omega = \pi/l, 2\pi/l, 3\pi/l, \ldots$ and these are said to be the *eigenvalues*, the functions $\sin(\pi x/l)$, $\sin(2\pi x/l)$, $\sin(3\pi x/l), \ldots$ being termed the *eigenfunctions*.

It is an important property of a linear differential equation that, if each of a set of functions satisfies the equation, then any linear combination of these functions is also a solution. Thus a solution of the differential equation (1) with the boundary conditions given in the previous paragraph is

$$V = \sum A_r\sin\left(\frac{r\pi x}{l}\right)\cos\left(\frac{cr\pi t}{l}+\eta_r\right),$$

* If this constant is taken as $+\omega^2$, the ordinary differential equations corresponding to equations (5) would lead to exponential solutions of the form

$$V = A\cosh(\omega x+\varepsilon)\cosh(\omega ct+\eta).$$

Such solutions are not wrong but are usually inappropriate in physical problems as these normally require that solutions remain finite for large values of the variables involved.

the summation being over positive integral values of r. The constants A_r and η_r are then determined from the initial conditions of the problem (see, for instance, Example 4, page 203).

Intuition will sometimes indicate the manner in which the solution of a physical problem depends on one of the independent variables. Thus in a vibration problem it will often be apparent that the displacement will be proportional to (say) $\cos \omega t$. In such instances, substitution in the differential equation will lead to a differential equation in one less independent variable and, in the case of an equation in two independent variables, only an ordinary differential equation will then have to be solved. Illustrations are given in the worked examples below.

Example 1. *ϕ is a function of r and t and satisfies the equation*

$$\frac{\partial^2 \phi}{\partial t^2} = c^2\left(\frac{\partial^2 \phi}{\partial r^2}+\frac{2}{r}\frac{\partial \phi}{\partial r}\right).$$

Show that there is a solution $\phi = r^{-1}f\cos(nct+\alpha)$ where f is a function of r only and determine f. (L.U.)

Taking $\phi = r^{-1}f\cos(nct+\alpha)$,

$$\frac{\partial^2 \phi}{\partial t^2} = -\frac{n^2c^2}{r}f\cos(nct+\alpha),$$

and, if primes denote differentiations with respect to r,

$$\frac{\partial \phi}{\partial r} = \left(-\frac{f}{r^2}+\frac{f'}{r}\right)\cos(nct+\alpha),$$

and

$$\frac{\partial^2 \phi}{\partial r^2} = \left(\frac{2}{r^3}f-\frac{2}{r^2}f'+\frac{1}{r}f''\right)\cos(nct+\alpha).$$

Hence

$$c^2\left(\frac{\partial^2 \phi}{\partial r^2}+\frac{2}{r}\frac{\partial \phi}{\partial r}\right) = \frac{c^2f''}{r}\cos(nct+\alpha),$$

and substitution in the given differential equation gives, after division by $c^2r^{-1}\cos(nct+\alpha)$,

$$f'' = -n^2f.$$

There is therefore a solution of the required type provided f satisfies this ordinary differential equation, the general solution of which is $f = A\cos nr+B\sin nr$, A and B being arbitrary constants.

Example 2. *Assuming that $y = X\cos\omega t$ is a solution of*

$$\frac{\partial^4 y}{\partial x^4}+k^2\frac{\partial^2 y}{\partial t^2} = 0,$$

where ω, k are constants and X is a function of x only, find the differential equation satisfied by X and express its general solution in terms of hyperbolic and trigonometrical functions. Given that $y = 0$ and $\partial y/\partial x = 0$ at both $x = 0$ and $x = l$. show that, if X is not identically zero, then $\cosh nl = \sec nl$ where $n = (\omega k)^{1/2}$. (L.U).

Writing $y = X\cos\omega t$, since X depends on x only,

$$\frac{\partial^4 y}{\partial x^4} = \frac{d^4 X}{dx^4}\cos\omega t, \qquad \frac{\partial^2 y}{\partial t^2} = -\omega^2 X\cos\omega t.$$

Substitution in the given partial differential equation and division by $\cos \omega t$ leads to

$$\frac{d^4X}{dx^4} - k^2\omega^2 X = 0,$$

as the required differential equation giving X. Writing $n = (\omega k)^{1/2}$, the auxiliary equation is $m^4 - n^4 = 0$, with roots $\pm n$, $\pm in$. Hence the general solution is

$$X = A\cosh nx + B\sinh nx + C\cos nx + D\sin nx.$$

The conditions $y = 0$, $\partial y/\partial x = 0$ when $x = 0$ are equivalent to $X = 0$, $dX/dx = 0$ when $x = 0$ and these give

$$A + C = 0, \qquad nB + nD = 0.$$

Hence $C = -A$, $D = -B$ and the solution can be written

$$X = A(\cosh nx - \cos nx) + B(\sinh nx - \sin nx).$$

As both X and dX/dx also vanish when $x = l$,

$$A(\cosh nl - \cos nl) + B(\sinh nl - \sin nl) = 0,$$

$$nA(\sinh nl + \sin nl) + nB(\cosh nl - \cos nl) = 0.$$

If y is not to be identically zero, A and B must not both be zero, and

$$\frac{\cosh nl - \cos nl}{\sinh nl + \sin nl} = \frac{\sinh nl - \sin nl}{\cosh nl - \cos nl}.$$

This easily leads to the required condition $\cosh nl = \sec nl$.

The method of separation of variables may also be applied to partial differential equations in more than two independent variables. Consider, for example, the partial differential equation

$$\frac{\partial^2 V}{\partial x^2} + \frac{\partial^2 V}{\partial y^2} + \frac{\partial^2 V}{\partial z^2} = \frac{1}{c^2}\frac{\partial^2 V}{\partial t^2}, \tag{8}$$

where V is a function of the four independent variables x, y, z, t, and c is again a physical constant. Suppose we attempt to find a "product" solution

$$V = XYZT, \tag{9}$$

where X, Y, Z, T are functions respectively of x, y, z, t only. Working as before we find

$$\frac{1}{X}\frac{d^2X}{dx^2} + \frac{1}{Y}\frac{d^2Y}{dy^2} + \frac{1}{Z}\frac{d^2Z}{dz^2} = \frac{1}{c^2T}\frac{d^2T}{dt^2}.$$

Each of the four terms in this equation must be a constant and we can write

$$\frac{1}{X}\frac{d^2X}{dx^2} = -\omega_1^2, \qquad \frac{1}{Y}\frac{d^2Y}{dy^2} = -\omega_2^2, \qquad \frac{1}{Z}\frac{d^2Z}{dz^2} = -\omega_3^2,$$

and

$$\frac{1}{T}\frac{d^2T}{dt^2} = -(\omega_1^2 + \omega_2^2 + \omega_3^2)c^2.$$

The general solutions of these four ordinary differential equations are

$$X = a_1 \cos(\omega_1 x + \varepsilon_1), \quad Y = a_2 \cos(\omega_2 y + \varepsilon_2), \quad Z = a_3 \cos(\omega_3 z + \varepsilon_3)$$

and

$$T = b \cos(\omega c t + \eta),$$

where $\omega^2 = \omega_1^2 + \omega_2^2 + \omega_3^2$ and a_r, ε_r $(r = 1, 2, 3)$, b and η are arbitrary constants. Hence a solution of the partial differential equation (8) is

$$V = \sum A \cos(\omega_1 x + \varepsilon_1) \cos(\omega_2 y + \varepsilon_2) \cos(\omega_3 z + \varepsilon_3) \cos(\omega c t + \eta), \quad (10)$$

where A is a composite constant $(A = a_1 a_2 a_3 b)$, and the summation is over certain values of ω_1, ω_2 and ω_3. It will often be possible to determine suitable values for ω_1, ω_2 and ω_3 from physical considerations.

Example 3. *Find a solution of the equation*

$$\frac{\partial^2 V}{\partial x^2} + \frac{\partial^2 V}{\partial y^2} = \frac{1}{c^2} \frac{\partial^2 V}{\partial t^2},$$

such that $V = 0$ *along the lines* $x = 0$, $y = 0$, $x = a$, $y = a$.

Here a solution will be

$$V = \sum A \cos(\omega_1 x + \varepsilon_1) \cos(\omega_2 y + \varepsilon_2) \cos(\omega c t + \eta)$$

where now $\omega^2 = \omega_1^2 + \omega_2^2$. V can be made to vanish when $x = 0$ by taking $\varepsilon_1 = \frac{1}{2}\pi$ and, similarly, by taking $\varepsilon_2 = \frac{1}{2}\pi$, V will vanish when $y = 0$. Hence

$$V = \sum A \sin \omega_1 x \sin \omega_2 y \cos(\omega c t + \eta).$$

If this is to vanish when $x = a$, then $\sin \omega_1 a = 0$ and hence $\omega_1 = r\pi/a$ where r is an integer. Similarly, since $V = 0$ when $y = a$, $\omega_2 = s\pi/a$ where s is an integer and the required solution can be written

$$V = \sum A \sin\left(\frac{r\pi x}{a}\right) \sin\left(\frac{s\pi x}{a}\right) \cos\left\{\left(r^2 + s^2\right)^{1/2} \frac{\pi c t}{a} + \eta\right\},$$

r and s being integers.

Exercises 7(a)

1. Find a solution of the equation

$$\frac{\partial^2 V}{\partial x^2} = \frac{1}{c^2} \frac{\partial^2 V}{\partial t^2}, \qquad (c \text{ constant}),$$

which is zero when $x = 0$ and when $t = 0$ and which remains finite for all values of x and t.

2. Find a solution of the equation

$$\frac{\partial^2 V}{\partial x^2} + \frac{\partial^2 V}{\partial y^2} = 0$$

which vanishes when y is zero and when x is infinite.

3. Find a solution of the partial differential equation

$$\frac{\partial^2 z}{\partial x^2} = \frac{1}{a^2} \frac{\partial z}{\partial t}, \qquad (a \text{ constant}),$$

which vanishes as $t \to \infty$.

4. If a is a constant, find a solution of the equation

$$\frac{\partial z}{\partial x}+a\frac{\partial z}{\partial y}=0$$

such that z is never infinite and that $\partial z/\partial x = 0$ when $x = y = 0$.

5. If a solution of the equation

$$\frac{\partial^2 V}{\partial r^2}+\frac{1}{r}\frac{\partial V}{\partial r}+\frac{\partial^2 V}{\partial z^2}=0$$

is of the form $V = ARe^{-nz}$ where A, n are constants and R is a function of r only, find the ordinary differential equation satisfied by R. If the solution is to be finite when $r = 0$, show that $R = J_0(nr)$.

6. If a solution of the equation

$$\frac{\partial^2 V}{\partial r^2}+\frac{2}{r}\frac{\partial V}{\partial r}+\frac{1}{r^2 \sin\theta}\frac{\partial}{\partial\theta}\left(\sin\theta\frac{\partial V}{\partial\theta}\right)=0$$

is of the form $V = r^n F(\theta)$ where n is a positive integer, find the ordinary differential equation satisfied by $F(\theta)$. Show that a solution of the partial differential equation is given by $V = r^n P_n(\cos\theta)$.

7. Find a solution of

$$\frac{\partial^2 u}{\partial t^2}=a^2\frac{\partial^2 u}{\partial x^2}, \qquad (a \text{ constant}),$$

in the form $u = f(x)\sin nt$ such that $\partial u/\partial t = k$ (constant) when $x = 0$ and when $t = 0$ and also such that $\partial u/\partial x = 0$ when $x = 0$ for all t.

8. Find a solution of the equation

$$\frac{\partial^2 V}{\partial x^2}+\frac{\partial^2 V}{\partial y^2}=0$$

such that when $y = n\pi/a$ (where n is an integer and a is a constant), $V = 0$ and $\partial V/\partial y = e^{2ax}$.

9. Assuming that $u = r^{-1}F(r)\cos(\omega t+\alpha)$ is a solution of the partial differential equation

$$\frac{\partial^2 u}{\partial r^2}+\frac{2}{r}\frac{\partial u}{\partial r}=\frac{1}{c^2}\frac{\partial^2 u}{\partial t^2}$$

where ω, α and c are constants and $F(r)$ is a function of r only, obtain the ordinary differential equation satisfied by $F(r)$ and give the general solution for $F(r)$. Given that, for all values of t, (i) u is finite at $r = 0$, (ii) $\partial u/\partial r = 0$ at $r = a$ and that u is not identically zero, prove that $(\omega a/c) = \beta$ must satisfy the equation $\beta = \tan\beta$. (L.U.)

10. If u satisfies the differential equation

$$\frac{1}{r}\frac{\partial}{\partial r}\left(r\frac{\partial u}{\partial r}\right)+\frac{1}{r^2}\frac{\partial^2 u}{\partial\theta^2}+k^2 u=0$$

where k is a constant, show that $u = r^{-1/2}f(r)\cos\frac{1}{2}\theta$ is a solution, where $f(r)$ is a function of r only, provided that $f(r)$ satisfies a certain differential equation. Write down the general solution for $f(r)$. Obtain the solution for u which is of the above form and satisfies the two conditions (i) $u = 0$ when $r = a$, (ii) $u = \cos \frac{1}{2}\theta$ when $r = b$ for all values of θ. (L.U.)

11. Find a solution of the equation

$$\frac{\partial}{\partial r}\left(r^2 \frac{\partial \phi}{\partial r}\right)+\frac{1}{\sin\theta}\frac{\partial}{\partial \theta}\left(\sin\theta \frac{\partial \phi}{\partial \theta}\right) = 0$$

of the form $\phi = f(r)\cos\theta$ such that $\partial\phi/\partial r \to 0$ as $r \to \infty$ and $\partial\phi/\partial r = -\cos\theta$ when $r = a$.

12. If the equation

$$x^2 \frac{\partial^2 u}{\partial x^2}+x\frac{\partial u}{\partial x}+\frac{\partial^2 u}{\partial y^2} = 0$$

has a solution of the form $u = XY$ where X, Y are respectively functions of x and y only, find the differential equations satisfied by X and Y and solve them when Y involves trigonometrical functions only. If $\partial u/\partial x = -\cos 2y$ when $x = a$ and u tends to zero as $x \to \infty$, find u. (L.U.)

13. Show that $u = Ae^{mx}\cos(\omega t+mx)+Be^{-mx}\cos(\omega t-mx)$ is a solution of

$$\frac{\partial^2 u}{\partial x^2} = 2\frac{\partial u}{\partial t},$$

where A, B, m and ω are constants, provided that $m^2 = \omega$. Find the values of the constants, given the conditions (i) $m > 0$, (ii) u remains finite as $x \to \infty$, (iii) $u = \cos t$ when $x = 0$. (L.U.)

14. If $v = e^{-k\alpha^2 t}f(r)\sin m\pi\theta$ is a solution of the partial differential equation

$$\frac{\partial v}{\partial t} = k\left(\frac{\partial^2 v}{\partial r^2}+\frac{1}{r}\frac{\partial v}{\partial r}+\frac{1}{r^2}\frac{\partial^2 v}{\partial \theta^2}\right), \qquad (k \text{ constant}),$$

for all values of m and α, find the ordinary differential equation satisfied by $f(r)$. Hence show that $v = e^{-k\alpha^2 t}J_{m\pi}(\alpha r)\sin m\pi\theta$ is a possible solution of the partial differential equation.

15. Find X, a function of x only, such that $v = X\cos ct$ is a solution of

$$\frac{\partial^2 v}{\partial x^2} = \frac{\partial^2 v}{\partial t^2}+n^2 v$$

for the cases (a) $c < n$, (b) $c > n$, (c) $c = n$ where c and n are constants. Determine the solution of this equation when $n = 2{\cdot}5$, given that $v = \cos 1{\cdot}5t$ when $x = 0$ and when $x = 0{\cdot}5$. Verify that the greatest value of v when $x = 1$ is $e+e^{-1}-1$. (L.U.)

16. Obtain the solution of the differential equation

$$\frac{\partial^2 V}{\partial x^2}+\frac{\partial^2 V}{\partial t^2}+6\frac{\partial V}{\partial x}+9V = 0$$

in the form $V = f(x)g(t)$ satisfying the following conditions: (i) V is periodic in t, (ii) $V = 0$ when $x = 0$ for all values of t, (iii) $(\partial^2 V/\partial x\,\partial t) = 6\cos 3t$ when $x = 0$ for all values of t. (L.U.)

17. Find a solution of the equation

$$\frac{\partial^2 z}{\partial x^2}+2k\frac{\partial z}{\partial x}-\frac{\partial^2 z}{\partial t^2} = 0, \quad \text{where } k \text{ is positive,}$$

of the form $z = f(x)\sin kt$, given that $z = 0$ when $x = 0$ for all values of t, and $z = 1/(ke)$ when $x = 1/k$ and $t = \pi/2k$. Show that as $x \to \infty$, $z \to 0$ for all values of t. (L.U.)

7.3 Some practical applications

In this section we consider some practical problems which lead to simple partial differential equations. As with problems involving ordinary differential equations, the solution is in two parts:

(i) setting up the differential equation and appropriate boundary and initial conditions:

(ii) finding a solution of the differential equation which satisfies these conditions.

In the course of finding a complete solution, it is often necessary to express the values taken on one boundary (or, in a problem involving the time, the initial value) as trigonometrical series. It is out of place to give a complete discussion of such series here but the following brief survey should be useful.

Suppose a periodic function $f(x)$, of period $2l$, is approximated to by the *finite* series of sines and cosines whose sum $s_n(x)$ is given by

$$s_n(x) = \tfrac{1}{2}a_0+a_1\cos(\pi x/l)+a_2\cos(2\pi x/l)+\ldots+a_n\cos(n\pi x/l)$$
$$+b_1\sin(\pi x/l)+b_2\sin(2\pi x/l)+\ldots+b_n\sin(n\pi x/l).$$

The error of this approximation is given by $\varepsilon_n(x) = f(x)-s_n(x)$ and the approximation is the "best possible", in the sense of least squares, if we make the mean value of the square of the error over the interval $-l < x < l$ to be a minimum. This requires that the value of the integral

$$\frac{1}{2l}\int_{-l}^{l}\{f(x)-s_n(x)\}^2\,dx$$

should be made a minimum by a proper choice of the a's and b's and this condition can be shown to imply that the a's and b's are given by the formulae

$$a_r = \frac{1}{l}\int_{-l}^{l} f(x)\cos\left(\frac{r\pi x}{l}\right)dx, \qquad (r = 0, 1, 2, \ldots, n),$$

$$b_r = \frac{1}{l}\int_{-l}^{l} f(x)\sin\left(\frac{r\pi x}{l}\right)dx, \qquad r = 1, 2, \ldots, n).$$

All that has been said so far is that $s_n(x)$ is a good approximation to $f(x)$ when the coefficients are calculated in this way and it is natural to enquire if modifications can be made to provide an exact representation of $f(x)$. It turns out that this can be done, providing $f(x)$ satisfies certain conditions, by allowing the number of terms in $s_n(x)$ to become infinite. In such cases we can write

$$f(x) = \tfrac{1}{2}a_0 + \sum_{r=1}^{\infty} \{a_r \cos(r\pi x/l) + b_r \sin(r\pi x/l)\},$$

the coefficients a_r, b_r being calculated, for all values of r, from the formulae already given. The infinite series is called the *Fourier expansion* of $f(x)$ and a_r, b_r are said to be its *Fourier coefficients.*

It should be noticed that if $f(x)$ is an *even* function, that is, one for which $f(x) = f(-x)$, its Fourier coefficients are given by

$$a_r = \frac{2}{l}\int_0^l f(x)\cos\left(\frac{r\pi x}{l}\right)dx, \qquad b_r = 0,$$

as is easily seen by dividing the range of integration in the formulae giving the coefficients into sub-ranges $(-l, 0)$ and $(0, l)$. Similarly, if $f(x)$ is an *odd* function in that $f(x) = -f(-x)$,

$$a_r = 0, \qquad b_r = \frac{2}{l}\int_0^l f(x)\sin\left(\frac{r\pi x}{l}\right)dx.$$

We are led then to the following results, which should prove useful in what follows:

(i) If $f(x)$ is an even function possessing the Fourier cosine expansion

$$\tfrac{1}{2}a_0 + a_1\cos(\pi x/l) + a_2\cos(2\pi x/l) + a_3\cos(3\pi x/l) + \ldots,$$

then the coefficients a_r ($r = 0, 1, 2, 3, \ldots$) are given by

$$a_r = \frac{2}{l}\int_0^l f(x)\cos\left(\frac{r\pi x}{l}\right)dx.$$

(ii) If $f(x)$ is an odd function possessing the Fourier sine expansion

$$b_1\sin(\pi x/l) + b_2\sin(2\pi x/l) + b_3\sin(3\pi x/l) + \ldots,$$

then the coefficients b_r($r = 1, 2, 3, \ldots$) are given by

$$b_r = \frac{2}{l}\int_0^l f(x)\sin\left(\frac{r\pi x}{l}\right)dx.$$

These series represent the function $f(x)$ over the interval $-l$ to l. Irrespective of whether $f(x)$ be even or odd, it can be represented in the interval 0 to l either by the cosine or by the sine series. There is thus a choice of series to represent a function in the interval 0 to l.

The examples and exercises which follow have been chosen so that only two independent variables are involved. More general problems will be considered later in this chapter.

(a) *Transverse vibrations of a stretched string*

Suppose a stretched string of length l and mass m per unit length is fixed at its two ends and that the coordinates of the end-points are taken as $(0, 0)$ and $(l, 0)$. In its undisturbed position, the string lies along the x-axis and we assume it to be vibrating with small amplitude in the plane containing the x and y axes. Neglecting gravity, the only forces acting on an element PQ (of length δx) of the string are the tensions T, $T+\delta T$ at P and Q. The direction of the tension at P makes an angle ψ (say) with the x-axis while that at Q makes an angle $\psi+\delta\psi$, and the total force on the element in the direction of the y-axis is

$$(T+\delta T)\sin(\psi+\delta\psi)-T\sin\psi = T\delta\psi = T\frac{\partial\psi}{\partial x}\delta x, \text{ approximately,}$$

if ψ is small. If y is the transverse displacement at time t of the point of the string whose coordinates were $(x, 0)$ in the undisturbed position, $\psi = \tan\psi = (\partial y/\partial x)$ approximately. Hence the force in the positive direction of the y-axis can be written

$$T\frac{\partial^2 y}{\partial x^2}\delta x.$$

The acceleration of the element PQ in the y direction is $(\partial^2 y/\partial t^2)$ and, since the mass of the element is $m\delta x$, we have

$$\frac{m\delta x}{g}\frac{\partial^2 y}{\partial t^2} = T\frac{\partial^2 y}{\partial x^2}\delta x,$$

leading to the partial differential equation

$$\frac{\partial^2 y}{\partial x^2} = \frac{1}{c^2}\frac{\partial^2 y}{\partial t^2}, \tag{1}$$

where

$$c^2 = \frac{gT}{m}. \tag{2}$$

Since the transverse vibrations are small, the extension of the string will, to the first order of small quantities, remain constant. Hence T, and therefore c, in equation (1) can be taken as constant. As the displacement is zero at both ends of the string, $y = 0$ when $x = 0$ and when $x = l$ for all values of t. As shown in §7.2, the solution of equation (1) satisfying these two boundary conditions can be taken as

$$y = \sum_{r=1}^{\infty} A_r \sin\left(\frac{r\pi x}{l}\right) \cos\left(\frac{cr\pi t}{l} + \eta_r\right), \tag{3}$$

where A_r and η_r are arbitrary constants.

Equation (3) shows that the motion is, in general, made up of a number of vibrations. The amplitudes of these vibrations can only be determined when the initial configuration of the string is given but the frequencies of vibration can be determined without a knowledge of the amplitudes. The vibration with the lowest frequency arises from the term in equation (3) in which $r = 1$ and is called the *fundamental* vibration. If only this vibration were present, the string would move

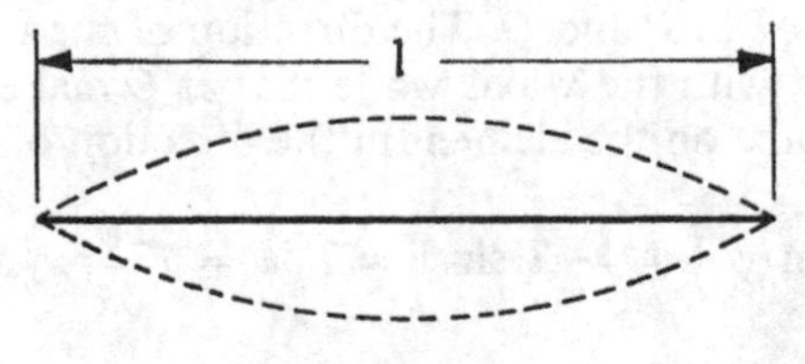

FIG. 14

between two extreme positions as shown in Fig. 14, the period in this case being $2l/c$, that is, $2l(m/gT)^{\frac{1}{2}}$. The vibrations with higher frequencies are called *harmonics* and, for that given by $r = 2$, the vibration

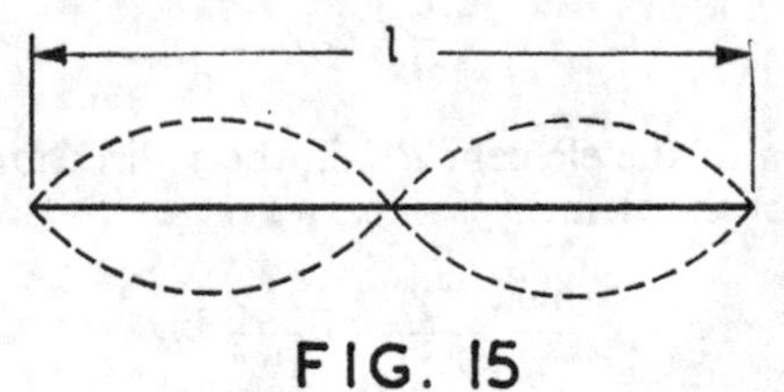

FIG. 15

would be one with extreme positions shown in Fig. 15. This harmonic has a node at $x = \frac{1}{2}l$, and its period is $l(m/gT)^{\frac{1}{2}}$. Similarly, when $r = 3$,

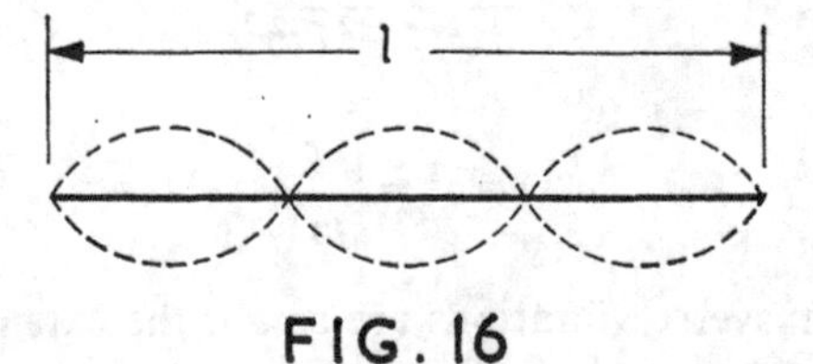

FIG. 16

the extreme positions are as shown in Fig. 16; there are now nodes at $x = \frac{1}{3}l$ and at $x = \frac{2}{3}l$ and the period of this harmonic is $\{2l(m/gT)^{\frac{1}{2}}\}/3$.

In general, the motion is made up of a fundamental mode together with harmonics of increasing frequencies and diminishing amplitudes. Example 4 below shows how these amplitudes can be calculated in a typical case.

Example 4. *A stretched string of length l and mass m per unit length lies along the x-axis with its ends fixed at the points* $(0, 0)$ *and* $(l, 0)$. *The tension of the string is T and its middle point is displaced by a small distance b measured perpendicular to the equilibrium position of the string and then let go from rest. Find an expression for the displacement y at time t of a point in the string which was at distance x from one end in the equilibrium position.*

The differential equation governing the motion is equation (1) above and the solution of this equation such that $y = 0$ when $x = 0$ and when $x = l$ is given by equation (3), that is,

$$y = \sum_{r=1}^{\infty} A_r \sin\left(\frac{r\pi x}{l}\right) \cos\left(\frac{cr\pi t}{l}+\eta_r\right).$$

This gives

$$\frac{\partial y}{\partial t} = -\frac{cr\pi}{l} \sum_{r=1}^{\infty} A_r \sin\left(\frac{r\pi x}{l}\right) \sin\left(\frac{cr\pi t}{l}+\eta_r\right).$$

Since the initial velocity of the string is zero, $\partial y/\partial t = 0$ when $t = 0$ so that $\eta_r = 0$, $(r = 1, 2, 3, \ldots)$ and we have

$$y = \sum_{r=1}^{\infty} A_r \sin\left(\frac{r\pi x}{l}\right) \cos\left(\frac{cr\pi t}{l}\right), \tag{4}$$

giving, as the initial configuration of the string,

$$y = \sum_{r=1}^{\infty} A_r \sin\left(\frac{r\pi x}{l}\right). \tag{5}$$

We now have to choose A_r so that the value of y given by equation (5) is consistent with that resulting when the middle point of the string is displaced by a distance b, that is,

$$\left.\begin{aligned} y &= 2bx/l && \text{when } 0 < x < l/2, \\ y &= 2b(l-x)/l && \text{when } l/2 < x < l. \end{aligned}\right\} \tag{6}$$

This can be done by taking A_r to be the coefficient of $\sin(r\pi x/l)$ in the sine series for the function given by equations (6). Hence, using the formula given on page 200,

$$\begin{aligned} \frac{lA_r}{2} &= \int_0^{l/2} \frac{2bx}{l} \sin\left(\frac{r\pi x}{l}\right) dx + \int_{l/2}^{l} \frac{2b(l-x)}{l} \sin\left(\frac{r\pi x}{l}\right) dx \\ &= \frac{4bl}{r^2\pi^2} \sin\left(\frac{r\pi}{2}\right) \end{aligned}$$

when the integrations are performed. Thus $A_{2r} = 0$ and

$$A_{2r+1} = (-1)^r \frac{8b}{(2r+1)^2\pi^2}, \qquad (r = 0, 1, 2, \ldots).$$

The required expression for y is therefore

$$y = \frac{8b}{\pi^2}\left\{\sin\left(\frac{\pi x}{l}\right)\cos\left(\frac{c\pi t}{l}\right) - \frac{1}{3^2}\sin\left(\frac{3\pi x}{l}\right)\cos\left(\frac{3c\pi t}{l}\right) + \frac{1}{5^2}\sin\left(\frac{5\pi x}{l}\right)\cos\left(\frac{5c\pi t}{l}\right) - \ldots\right\},$$

where $c^2 = gT/m$.

(b) Vibrations of a uniform beam

Suppose one end of a uniform horizontal beam of mass m per unit length is fixed at the origin O. Take the x and y axes as being respectively horizontal and vertical through O and assume that small vibrations of the beam take place in the plane containing the coordinate axes. If an element $m\delta x$ at distance x from O has displacement y at time t, the effective force on the element is $(m\delta x/g)(\partial^2 y/\partial t^2)$, that is, the effect of the vibration is that of loading w where

$$w = -\frac{m}{g}\frac{\partial^2 y}{\partial t^2}.$$

If EI is the flexural rigidity of the beam, the deflexion equation is

$$EI\frac{\partial^4 y}{\partial x^4} = w$$

and, combining these two equations, the partial differential equation governing the vibrations is

$$EI\frac{\partial^4 y}{\partial x^4} + \frac{m}{g}\frac{\partial^2 y}{\partial t^2} = 0. \tag{7}$$

Harmonic vibrations of period $2\pi/\omega$ are given by writing $y = X\cos\omega t$ where X is a function of x only. Substitution in equation (7) then gives

$$EI\frac{d^4 X}{dx^4} - \frac{m\omega^2}{g}X = 0$$

as the differential equation for X. Writing

$$\alpha = \left(\frac{m\omega^2}{gEI}\right)^{\frac{1}{4}}, \tag{8}$$

this can be written

$$\frac{d^4 X}{dx^4} - \alpha^4 X = 0, \tag{9}$$

and the general solution of this equation is

$$X = A\cosh\alpha x + B\sinh\alpha x + C\cos\alpha x + D\sin\alpha x, \tag{10}$$

where A, B, C and D are arbitrary constants.

The periods of the modes of vibration depend on the manner in which the beam is fixed at its ends. If the beam is of length l and is simply supported at each end, we have

$$X = 0, \qquad \frac{d^2X}{dx^2} = 0 \quad \text{when } x = 0 \text{ and when } x = l, \tag{11}$$

for then there is zero displacement and bending moment at each end. If both ends of the beam are clamped, the conditions are

$$X = 0, \qquad \frac{dX}{dx} = 0 \quad \text{when } x = 0 \text{ and when } x = \ . \tag{12}$$

For a cantilever, clamped at $x = 0$,

$$\left.\begin{aligned} X = 0, &\qquad \frac{dX}{dx} = 0 \quad \text{when } x = 0, \\ \frac{d^2X}{dx^2} = 0, &\qquad \frac{d^3X}{dx^3} = 0 \quad \text{when } x = l, \end{aligned}\right\} \tag{13}$$

since in this case there is no bending moment or shearing force at the free end. When these conditions are used in conjunction with equation (10), a transcendental equation for α is obtained. The periods $(2\pi/\omega)$ of the vibrations can then be found from the roots of this transcendental equation and equation (8). An illustration is given in Example 5 below.

Example 5. *A uniform beam of length l, flexural rigidity EI and mass m per unit length is simply supported at its ends. Find the fundamental mode of vibration.*

Inserting the conditions $X = 0$, $X'' = 0$ when $x = 0$ in equation (10), $A+C = 0 = \alpha^2(A-C)$ so that $A = C = 0$ and equation (10) becomes

$$X = B \sinh \alpha x + D \sin \alpha x.$$

The remaining two conditions in (11) are $X = 0$, $X'' = 0$ when $x = l$ and these now give

$$B \sinh \alpha l + D \sin \alpha l = 0,$$
$$\alpha^2 B \sinh \alpha l - \alpha^2 D \sin \alpha l = 0.$$

Hence $B = 0$ and, if D is not zero, $\sin \alpha l = 0$. Thus $\alpha l = r\pi$ where r is an integer and equation (8) gives

$$\omega = \left(\frac{gEI}{m}\right)^{1/2} \frac{r^2\pi^2}{l^2},$$

the period being $2\pi/\omega$. The fundamental mode is that which corresponds to the lowest frequency ($r = 1$). This mode coexists with modes of higher frequencies ($r = 2, 3, \ldots$) as in the case of the vibrating string.

(c) *Heat conduction along a bar*

Consider a bar of uniform cross-section A whose sides are insulated so that the flow of heat can be assumed to be entirely along the length of the bar. Let V be the temperature at time t at a point P of the bar at distance x from one end. The flow of heat across a plane through P at right angles to the length of the rod is given by

$$f(x) = -KA\frac{\partial V}{\partial x},$$

where K is the thermal conductivity of the material of the bar. The flow of heat across a plane through Q at distance $x+\delta x$ from the end is

$$\begin{aligned} f(x+\delta x) &= f(x)+f'(x)\,\delta x \\ &= -KA\frac{\partial V}{\partial x}-KA\frac{\partial^2 V}{\partial x^2}\delta x, \end{aligned}$$

to the first order of small quantities. The total gain of heat to the element bounded by planes through P and Q is therefore

$$f(x)-f(x+\delta x) = KA\frac{\partial^2 V}{\partial x^2}\delta x.$$

But, if ρ is the density and c the specific heat of the material of the bar, the gain of heat to the element is

$$\rho c A\delta x\frac{\partial V}{\partial t}.$$

By equating these quantities,

$$\frac{\partial^2 V}{\partial x^2} = \frac{1}{k}\frac{\partial V}{\partial t}, \tag{14}$$

where $k = K/\rho c$ is the so-called diffusivity (or thermometric conductivity) of the bar.

Equation (14) can also be used in other cases in which the flow of heat is entirely in one direction. For example, it can be used to find the temperature in a slab of material bounded by two parallel planes provided that the thickness of the slab is small compared with the linear dimensions of the bounding planes.

A product solution $V = XT$, where X and T are respectively functions of x and t only, can be found in the usual way. Substituting in (14),

$$\frac{1}{X}\frac{d^2X}{dx^2} = \frac{1}{kT}\frac{dT}{dt},$$

and, taking the constant of separation to be $-\alpha^2$,

$$\frac{d^2X}{dx^2}+\alpha^2 X = 0, \qquad \frac{dT}{dt}+k\alpha^2 T = 0.$$

These give $X = a\cos(\alpha x+\varepsilon)$, $T = be^{-k\alpha^2 t}$ so that a solution of equation (14) is

$$V = \sum A\, e^{-k\alpha^2 t}\cos(\alpha x+\varepsilon), \tag{15}$$

where A, ε and α are at present undetermined constants. These constants can be found when the boundary and initial conditions are given. A typical case is given in Example 6.

Example 6. *The ends of a uniform bar of length l whose sides are heat insulated are kept at zero temperature. The initial temperature of the bar is V_0 (constant) and its diffusivity is k. Find an expression for the average temperature of the bar at time t.*

Taking the bar to lie along the x-axis with one end at the origin, the differential equation for the temperature V is equation (14) with solution (15), that is

$$V = \sum A e^{-k\alpha^2 t}\cos(\alpha x+\varepsilon).$$

The conditions $V = 0$ when $x = 0$ and $x = l$ for all t can be satisfied by taking $\varepsilon = \pi/2$, $\alpha = r\pi/l$ where $r = 1, 2, 3, \ldots$, and hence the initial temperature of the bar is given by

$$V = \sum_{r=1}^{\infty} A_r \sin\left(\frac{r\pi x}{l}\right).$$

The coefficients A_r have now to be chosen so that this is equal to the constant V_0 for all values of x such that $0 < x < l$. This can be done by taking A_r to be the coefficient of $\sin(r\pi x/l)$ in the sine series representing V_0. Hence

$$\tfrac{1}{2}lA_r = \int_0^l V_0 \sin\left(\frac{r\pi x}{l}\right)dx$$

$$= -\frac{lV_0}{r\pi}\left[\cos\left(\frac{r\pi x}{l}\right)\right]_0^l,$$

so that $A_{2r} = 0$ and

$$A_{2r+1} = \frac{4V_0}{(2r+1)\pi}, \qquad (r = 0, 1, 2, 3, \ldots).$$

Thus

$$V = \frac{4V_0}{\pi}\sum_{r=0}^{\infty}\frac{1}{2r+1}\exp\left\{-k(2r+1)^2\frac{\pi^2 t}{l^2}\right\}\sin\frac{(2r+1)\pi x}{l},$$

and the mean value $\bar{V}$ of V is given by

$$\bar{V} = \frac{1}{l}\int_0^l V\,dx$$

$$= \frac{8V_0}{\pi^2}\sum_{r=0}^{\infty}\frac{1}{(2r+1)^2}\exp\left\{-k(2r+1)^2\frac{\pi^2 t}{l^2}\right\}.$$

(d) *The equation of telegraphy*

Suppose L, R, C and G are respectively the inductance, resistance, capacitance and leakage conductance per unit length of an electric transmission line. Then if i is the current flowing and V the potential at distance x from some fixed point in the line, the equation for an element of length δx is

$$(L\delta x)\frac{\partial i}{\partial t}+(R\delta x)i = -\delta V,$$

$-\delta V$ being the potential drop over the element. This can be written

$$L\frac{\partial i}{\partial t}+Ri = -\frac{\partial V}{\partial x}. \tag{16}$$

In time δt, a charge $i\delta t$ enters the element and a charge

$$\left(i+\frac{\partial i}{\partial x}\delta x\right)\delta t$$

leaves at the other end. The difference between these quantities represents the charge taken by the element and the charge that has leaked away through the insulation. Hence

$$(C\delta x)\delta V+(G\delta x)V\delta t = -\frac{\partial i}{\partial x}\delta x\,\delta t$$

giving

$$C\frac{\partial V}{\partial t}+GV = -\frac{\partial i}{\partial x}. \tag{17}$$

Equations (16) and (17) are a pair of simultaneous linear partial differential equations of the first order. Either i or V can be eliminated to give a single second order equation. Thus, from (17),

$$\begin{aligned}\left(L\frac{\partial}{\partial t}+R\right)\left(C\frac{\partial V}{\partial t}+GV\right) &= -\left(L\frac{\partial}{\partial t}+R\right)\frac{\partial i}{\partial x}\\ &= -\frac{\partial}{\partial x}\left(L\frac{\partial i}{\partial t}+Ri\right)\\ &= \frac{\partial^2 V}{\partial x^2}, \quad \text{using (16).}\end{aligned}$$

Hence

$$LC\frac{\partial^2 V}{\partial t^2}+(RC+LG)\frac{\partial V}{\partial t}+RGV = \frac{\partial^2 V}{\partial x^2}, \tag{18}$$

and this equation is often known as the *equation of telegraphy*. The elimination of V between equations (16) and (17) leads in a similar way to

$$LC\frac{\partial^2 i}{\partial t^2}+(RC+LG)\frac{\partial i}{\partial t}+RGi = \frac{\partial^2 i}{\partial x^2}. \tag{19}$$

Example 7. *Show that $V = \sum Ae^{-\alpha x} \sin \omega(t-x/u)$ is a solution of the equation of telegraphy provided that the constants α, ω and u are related by*

$$2\alpha = u(RC+LG) \quad \textit{and} \quad u^{-2} = LC+(\alpha^2-RG)\omega^{-2}.$$

Deduce that the waves represented by this solution are propagated without distortion if $RC = LG$.

If $V = \sum Ae^{-\alpha x} \sin \omega(t-x/u)$, it follows that

$$\frac{\partial V}{\partial t} = \sum \omega A e^{-\alpha x} \cos \omega\left(t-\frac{x}{u}\right), \qquad \frac{\partial^2 V}{\partial t^2} = -\sum \omega^2 A e^{-\alpha x} \sin \omega\left(t-\frac{x}{u}\right),$$

$$\frac{\partial V}{\partial x} = \sum Ae^{-\alpha x}\left\{-\alpha \sin \omega\left(t-\frac{x}{u}\right)-\frac{\omega}{u}\cos \omega\left(t-\frac{x}{u}\right)\right\},$$

$$\frac{\partial^2 V}{\partial x^2} = \sum Ae^{-\alpha x}\left\{\left(\alpha^2-\frac{\omega^2}{u^2}\right)\sin \omega\left(t-\frac{x}{u}\right)+\frac{2\alpha\omega}{u}\cos \omega\left(t-\frac{x}{u}\right)\right\}.$$

Substituting in the equation of telegraphy (18) and equating the coefficients of $\sin \omega(t-x/u)$ and $\cos \omega(t-x/u)$,

$$-\omega^2 LC+RG = \alpha^2-\frac{\omega^2}{u^2},$$

$$\omega(RC+LG) = \frac{2\alpha\omega}{u},$$

and these, with trifling rearrangements, give the required relations between the constants.

The solution is such that, in general, waves of different frequencies travel at different rates. If, however, $\alpha^2 = RG$, the first of the above relations gives $u = 1/(LC)^{1/2}$ and all the waves travel with the same velocity u. All the waves also then contain the same constant α in the factor $e^{-\alpha x}$ governing the decrease of their amplitudes. Hence with these values of α and u, the waves are propagated without distortion. Finally, writing $\alpha = (RG)^{1/2}$, $u = 1/(LC)^{1/2}$ in the second of the relations between the constants, we have $RC+LG = 2(RGLC)^{1/2}$. This gives $\{(RC)^{1/2}-(LG)^{1/2}\}^2 = 0$, leading to $RC = LG$.

Exercises 7(b)

1. A stretched string of length l and mass m per unit length lies along the x-axis with its ends fixed at the points $(0, 0)$ and $(l, 0)$. The tension of the string is T and it is initially displaced into the curve $y = d \sin^3 (\pi x/l)$ where d is small and then let go from rest. Show that the displacement y at time t of a point in the string which was at distance x from the origin in the equilibrium position is given by

$$4\frac{y}{d} = 3 \sin\left(\frac{\pi x}{l}\right)\cos\left(\frac{c\pi t}{l}\right)-\sin\left(\frac{3\pi x}{l}\right)\cos\left(\frac{3c\pi t}{l}\right),$$

where $c^2 = gT/m$.

2. A uniform string of length l and line density ρ is stretched between two fixed points $(0, 0)$ and $(l, 0)$ to tension $\rho c^2/g$. It is plucked a small distance d at a point distant a from the origin and released at $t = 0$. Show that its subsequent displacement is

$$\frac{2dl^2}{\pi^2 a(l-a)}\sum_{r=1}^{\infty}\frac{1}{r^2}\sin\left(\frac{r\pi a}{l}\right)\sin\left(\frac{r\pi x}{l}\right)\cos\left(\frac{rc\pi t}{l}\right).$$

3. A string of line density ρ is stretched tightly between the points (0, 0) and $(l, 0)$. Both its ends are given small displacements $y = a \sin pt$ perpendicular to the length of the string. Show that, in the steady state, the displacements of points of the string are given by

$$y = a \sec \frac{pl}{2c} \cos \frac{p}{c}(x - \tfrac{1}{2}l) \sin pt.$$

4. A uniform cantilever of length l, mass m per unit length and flexural rigidity EI is vibrating in a vertical plane. Show that the period of free oscillation is $2\pi/\omega$ where $m\omega^2 = gEI\alpha^4$ and $\cosh \alpha l \cos \alpha l = -1$.

5. A vibrating uniform beam of flexural rigidity EI, mass m per unit length and length l is clamped horizontally at each end. Show that the period of free oscillation is $2\pi/\omega$ where $m\omega^2 = gEI\alpha^4$ and α is given by the transcendental equation $\cosh \alpha l \cos \alpha l = 1$. Find graphically the least root of this equation and hence show that the fundamental period is

$$\left(\frac{m}{gEI}\right)^{1/2} \frac{l^2}{1{\cdot}135\pi}, \text{ approximately.}$$

Show also that, for the vibrations of higher frequencies, ω is given approximately by

$$\omega = \left(\frac{gEI}{m}\right)^{1/2} \left\{\frac{(2r+1)\pi}{2l}\right\}^2, \qquad (r = 2, 3, 4, \ldots).$$

6. A uniform beam of flexural rigidity EI, mass m per unit length and length l, is clamped horizontally at one end and freely supported at the other end. Show that the modes of vibration of the beam are given by the equation $\tan \alpha l = \tanh \alpha l$ where $gEI\alpha^4 = m\omega^2$. Show also that the frequency of the fundamental mode of vibration is approximately

$$\left(\frac{gEI}{m}\right)^{1/2} \frac{(3{\cdot}93)^2}{2l^2\pi}.$$

7. The second moments of area of a steel joist about principal axes through the centroid of a section are 35 in.^{4} and 7·93 in.^{4} and the weight is 24 lb. per ft. If the joist is 10 ft. long and the ends are fixed but direction free, find the frequencies of the fundamental modes of vibration in the directions of the principal axes. (Take $E = 30 \times 10^6$ lb./in.^{2}).

8. A slab of conducting material of diffusivity k is bounded by the planes $x = 0$ and $x = \pi$. These planes are kept at zero temperature and the heat flow can be assumed to be entirely in the direction of the x-axis. If the initial temperature V_0 of the slab is given by

$$V_0 = \frac{\pi}{4}x \text{ when } 0 < x < \frac{\pi}{2} \text{ and } V = \frac{\pi}{4}(\pi - x) \text{ when } \frac{\pi}{2} < x < \pi,$$

show that the temperature V at time t is given by

$$V = e^{-kt} \sin x - \tfrac{1}{9} e^{-9kt} \sin 3x + \tfrac{1}{25} e^{-25kt} \sin 5x - \ldots.$$

9. A thin uniform bar of material of diffusivity k is heat insulated along its sides. It lies along the x-axis with its ends at the points (0, 0) and $(\pi, 0)$. It is initially at a constant temperature V_0 and the end at the

origin is also heat insulated. If the other end of the bar is kept at zero temperature, find an expression giving the temperature at the point $(x, 0)$ of the bar at time t.

10. The equation of heat conduction in a rod of length l is

$$\frac{\partial^2 V}{\partial x^2} = \frac{1}{k}\frac{\partial V}{\partial t}.$$

If $V = V_0 x$ when $t = 0$ and also $\partial V/\partial x = 0$ when $x = 0$ and when $x = l$ for all values of $t(>0)$, show that

$$V = \frac{1}{2}V_0 l - \frac{4V_0 l}{\pi^2}\sum_{n=1}^{\infty}\frac{\exp\{-k(2n-1)^2\pi^2 t/l^2\}}{(2n-1)^2}\cos(2n-1)\frac{\pi x}{l}. \quad \text{(L.U.)}$$

11. A large sheet of metal 5 cm. thick is initially at a temperature of 50° C. The temperature of its two parallel faces is suddenly dropped to, and thereafter kept at, 0° C., and the heat flow can be assumed to take place entirely in a direction perpendicular to the faces of the sheet. Taking the diffusivity of the metal to be 0·04 c.g.s. units, calculate the temperature in the middle plane of the sheet 25 secs. after the temperature of the faces has been dropped to 0° C.

12. The current i in a cable satisfies the equation

$$\frac{\partial^2 i}{\partial x^2} = \frac{2}{k}\frac{\partial i}{\partial t} + i, \qquad (k \text{ constant}).$$

By assuming a solution of the type $i = XT$, where X is a function of x alone and T is a function of t alone, show that if $i = 0$ when $x = l$ and $\partial i/\partial x = -a\,e^{-kt}$ when $x = 0$, the current is given by

$$i = ae^{-kt}\frac{\sin(l-x)}{\cos l}. \quad \text{(L.U.)}$$

13. In a leaky telegraph wire the resistance R and the leakage conductance G are very large in comparison with the capacitance and inductance. The wire is of length l and the voltage at the sending end is V_0 (constant). If the other end of the wire is earthed, show that the voltage V and the current i at distance x from the sending end are given approximately by

$$V = V_0\frac{\sinh \alpha(l-x)}{\sinh \alpha l}, \qquad i = V_0\left(\frac{G}{R}\right)^{1/2}\frac{\cosh \alpha(l-x)}{\sinh \alpha l},$$

where $\alpha = (RG)^{1/2}$.

14. A distortionless transmission line ($RC = LG$) of length l is initially charged to unit potential and the end $x = l$ is insulated. If, at time $t = 0$, the end $x = 0$ is earthed, show that the potential at time t and distance x from the earthed end is given by

$$\frac{4}{\pi}e^{-kt}\sum_{r=0}^{\infty}\frac{1}{2r+1}\sin\frac{(2r+1)\pi x}{2l}\cos\frac{(2r+1)\pi ut}{2l},$$

where $u = (LC)^{-1/2}$ and $k = R/L$.

15. In a submarine cable the inductance and leakage conductance can be taken to be negligible. The sending end ($x = 0$) of such a cable is raised

to potential V_0 and the receiving end ($x = l$) is earthed. Show that, when the steady state has been reached, the potential at distance x from the sending end is $V_0(1-x/l)$. When this steady state has been reached the sending end is suddenly earthed. Show that the potential along the cable is given by

$$V = \frac{2V_0}{\pi}\sum_{r=1}^{\infty}\frac{1}{r}\exp\{-r^2\pi^2 t/CRl^2\}\sin\left(\frac{r\pi x}{l}\right),$$

where t is reckoned from the instant at which the sending end is earthed. (C and R are respectively the capacitance and resistance per unit length of cable.)

7.4 The equations of mathematical physics

We now consider four partial differential equations which occur in a great variety of physical problems. In this section we merely state these equations and indicate the various contexts in which each equation is appropriate.

(a) Laplace's equation

This is the partial differential equation

$$\frac{\partial^2 V}{\partial x^2}+\frac{\partial^2 V}{\partial y^2}+\frac{\partial^2 V}{\partial z^2} = 0. \tag{1}$$

If (x, y, z) be the cartesian coordinates of a point in space, examples of quantities which can be represented by the function V are as follows:

(i) the gravitational potential in a region devoid of attracting matter;
(ii) the electrostatic potential in a uniform dielectric;
(iii) the magnetic potential;
(iv) the velocity potential in the irrotational motion of a homogeneous fluid;
(v) the steady state temperature in a uniform solid.

(b) Poisson's equation

This is the equation

$$\frac{\partial^2 V}{\partial x^2}+\frac{\partial^2 V}{\partial y^2}+\frac{\partial^2 V}{\partial z^2} = f(x, y, z), \tag{2}$$

where $f(x, y, z)$ is a specified function of x, y and z. Examples of quantities which can be represented by the function V are:

(i) the gravitational potential in a region in which $f(x, y, z)$ is proportional to the density of the material at any point of the region;
(ii) the electrostatic potential in a region in which $f(x, y, z)$ is proportional to the charge distribution;

(iii) in the two-dimensional (z absent) form of the equation when $f(x, y)$ is a constant, V is a measure of the shear stress entailed by twisting a long bar of specified cross-section.

(*c*) *The equation of heat conduction*

The temperature V at time t at a point (x, y, z) in a homogeneous isotropic body satisfies the equation

$$\frac{\partial^2 V}{\partial x^2}+\frac{\partial^2 V}{\partial y^2}+\frac{\partial^2 V}{\partial z^2} = \frac{1}{k}\frac{\partial V}{\partial t}, \tag{3}$$

where k is the diffusivity (here assumed constant) of the material of the body. The same equation can, in certain circumstances, be used in diffusion problems, V then being the concentration of the diffusing substance. Many such problems require, however, additional terms in the equation for a realistic representation of the physical data. It is worth noticing that the one-dimensional form of equation (3) arises in calculating the voltage (or current) in an electrical transmission line in which both the inductance and leakance are negligible (equations (18), (19) of §7.3).

The temperature in a heat-conducting solid becomes *steady* when $\partial V/\partial t = 0$. Inserting this condition in equation (3), the steady temperature is therefore given by Laplace's equation (see §7.4, (*a*), (v) above).

(*d*) *The wave equation*

The equation

$$\frac{\partial^2 V}{\partial x^2}+\frac{\partial^2 V}{\partial y^2}+\frac{\partial^2 V}{\partial z^2} = \frac{1}{c^2}\frac{\partial^2}{\partial t^2} \tag{4}$$

arises in investigations of waves propagated with velocity c independent of wavelength. Typical examples of quantities which can be represented by the function V are as follows:

(i) each component of the displacement in vibrating systems;
(ii) the velocity potential of a gas in the theory of sound;
(iii) each component of the electric or magnetic vector in the electromagnetic theory of light.

The above equations are those appropriate to general problems in three dimensions. In certain problems the quantity represented by V can be assumed to be independent of one (or two) of the cartesian coordinates. For example, in a long thin bar which lies along the x-axis and which is heat insulated along its sides, the flow of heat can be taken to be entirely along the length of the bar and the temperature V is then independent of y and z. In this case, equation (3) takes the so-called one-dimensional form

$$\frac{\partial^2 V}{\partial x^2} = \frac{1}{k}\frac{\partial V}{\partial t}.$$

Again, consider the electrostatic potential in the field of a charged cylindrical body whose axis lies along the z-axis. If the length of the cylinder is large compared with the dimensions of its cross-section, the potential V will be approximately independent of z at points in the field far removed from the ends of the cylinder. At such points V will satisfy

$$\frac{\partial^2 V}{\partial x^2}+\frac{\partial^2 V}{\partial y^2}=0,$$

the two-dimensional form of Laplace's equation.

7.5 The physical significance of $\nabla^2 V$

All the equations discussed in the last section contain the expression

$$\nabla^2 V \equiv \frac{\partial^2 V}{\partial x^2}+\frac{\partial^2 V}{\partial y^2}+\frac{\partial^2 V}{\partial z^2}, \tag{1}$$

the symbol ∇^2, commonly referred to as "del-squared", being a convenient notation for the differential operator $(\partial^2/\partial x^2+\partial^2/\partial y^2+\partial^2/\partial z^2)$ and we now briefly investigate the physical significance of this expression.

Suppose the quantity V is a function of the three space coordinates x, y, z and that it takes the value V_0 at a point P with coordinates (x_0, y_0, z_0). The average value $\bar{V}$ of V inside a small cube whose centre is at P and whose sides are parallel to the coordinate axes and of length a is given by

$$a^3\bar{V}=\iiint V\,dx\,dy\,dz, \tag{2}$$

where each integration is from $-a/2$ to $a/2$. The Taylor expansion of V gives, if suffix 0 denotes values at P,

$$\begin{aligned} V = V_0 &+ x\left(\frac{\partial V}{\partial x}\right)_0+y\left(\frac{\partial V}{\partial y}\right)_0+z\left(\frac{\partial V}{\partial z}\right)_0 \\ &+\frac{1}{2}\left\{x^2\left(\frac{\partial^2 V}{\partial x^2}\right)_0+y^2\left(\frac{\partial^2 V}{\partial y^2}\right)_0+z^2\left(\frac{\partial^2 V}{\partial z^2}\right)_0\right\} \\ &+xy\left(\frac{\partial^2 V}{\partial x\,\partial y}\right)_0+yz\left(\frac{\partial^2 V}{\partial y\,\partial z}\right)_0+zx\left(\frac{\partial^2 V}{\partial z\,\partial x}\right)_0+\ldots . \end{aligned}$$

Inserting this in equation (2) and noting that the terms which are odd in x, y or z vanish while the other terms give

$$\iiint dx\,dy\,dz = a^3, \qquad \iiint x^2\,dx\,dy\,dz=\frac{a^5}{12}, \quad \text{etc.,}$$

we have

$$a^3\bar{V} = a^3V_0 + \frac{a^5}{24}\left(\frac{\partial^2 V}{\partial x^2} + \frac{\partial^2 V}{\partial y^2} + \frac{\partial^2 V}{\partial z^2}\right)_0 + \ldots.$$

Neglecting terms of order a^4 and above, this gives

$$\bar{V} - V_0 = \frac{a^2}{24}\nabla^2 V_0. \tag{3}$$

Hence *the quantity* $\nabla^2 V$ *is a measure of the difference between the value of the quantity* V *at a point* P *and the average value of* V *in an infinitesimal neighbourhood of* P.

An explanation can now be given of the occurrence of the expression $\nabla^2 V$ in the equations of mathematical physics. In certain instances the value of V at any point is equal to the value in the neighbourhood, in which case $\nabla^2 V = 0$ and Laplace's equation is appropriate. Simple examples are the gravitational field in a region devoid of attracting matter and the electrostatic field in a uniform dielectric. Again, in a gravitational field containing attracting matter, the value of the potential V at a point differs from its average value in the neighbourhood and, in this case, Poisson's equation applies.

There are physical processes in which the value of V at a point differs from its average value at some time. In such cases there is a tendency towards equalization with time and the smaller the difference between V and its average the greater is its rate of increase with time. This can be expressed by writing

$$\nabla^2 V = \lambda \frac{\partial V}{\partial t},$$

where λ is the constant of proportionality. An example of such a process is the flow of heat in a conducting medium. Finally, the difference between the value of V and its average value has a similar effect to the displacement from a position of equilibrium in a vibrating system. In such systems, forces are set up which tend to give the displacement its average value and thus to restore the system to the equilibrium position. In such cases $\nabla^2 V$ is proportional to a force and hence to an acceleration and the differential equation will take the form of the wave equation

$$\nabla^2 V = \lambda \frac{\partial^2 V}{\partial t^2},$$

where again λ is a constant of proportionality.

7.6 Transformation of coordinates

To obtain the solution of a partial differential equation satisfying given boundary conditions, it is almost essential that the boundaries shall be capable of being described by the constancy of one of the coordinates in terms of which the equation is expressed. Thus, for problems involving rectangular parallel epipeds, cartesian coordinates would be chosen, while cylindrical polar coordinates would be used in cases where the boundaries were the surfaces of a right circular cylinder. It is useful therefore to be able to transform the expression $\nabla^2 V$ from cartesians to other coordinate systems* and, as examples, we consider below the transformations to cylindrical polar and spherical polar coordinates.

(a) Cylindrical polar coordinates

The relations between the cartesian coordinates (x, y, z) of a point P and its cylindrical polar coordinates (ρ, ϕ, z) are (see Fig. 17),

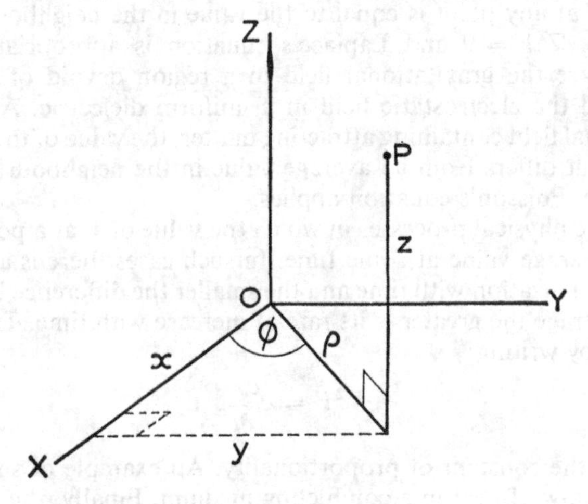

FIG. 17

$$x = \rho \cos \phi, \quad y = \rho \sin \phi, \quad z = z. \tag{1}$$

Hence

$$\frac{\partial V}{\partial \rho} = \frac{\partial V}{\partial x}\frac{\partial x}{\partial \rho} + \frac{\partial V}{\partial y}\frac{\partial y}{\partial \rho} = \cos \phi \frac{\partial V}{\partial x} + \sin \phi \frac{\partial V}{\partial y},$$

$$\frac{\partial V}{\partial \phi} = \frac{\partial V}{\partial x}\frac{\partial x}{\partial \phi} + \frac{\partial V}{\partial y}\frac{\partial y}{\partial \phi} = -\rho \sin \phi \frac{\partial V}{\partial x} + \rho \cos \phi \frac{\partial V}{\partial y},$$

the values of $\partial x/\partial \rho$, $\partial y/\partial \rho$, $\partial x/\partial \phi$, $\partial y/\partial \phi$ being found from equations (1). Solving the above equations for $\partial V/\partial x$ and $\partial V/\partial y$,

* It is worth pointing out that the operator ∇^2 is invariant when the coordinate axes are translated to a new origin and when they are rotated. This is another reason why the operator occurs naturally in the equations of mathematical physics.

$$\frac{\partial V}{\partial x} = \cos\phi\frac{\partial V}{\partial \rho} - \frac{\sin\phi}{\rho}\frac{\partial V}{\partial \phi}, \qquad \frac{\partial V}{\partial y} = \sin\phi\frac{\partial V}{\partial \rho} + \frac{\cos\phi}{\rho}\frac{\partial V}{\partial \phi}, \tag{2}$$

so that, since $\cos\phi \pm i\sin\phi = e^{\pm i\phi}$,

$$\frac{\partial V}{\partial x} + i\frac{\partial V}{\partial y} = e^{i\phi}\left(\frac{\partial V}{\partial \rho} + \frac{i}{\rho}\frac{\partial V}{\partial \phi}\right), \qquad \frac{\partial V}{\partial x} - i\frac{\partial V}{\partial y} = e^{-i\phi}\left(\frac{\partial V}{\partial \rho} - \frac{i}{\rho}\frac{\partial V}{\partial \phi}\right).$$

Consequently

$$\begin{aligned}\frac{\partial^2 V}{\partial x^2} + \frac{\partial^2 V}{\partial y^2} &= \left(\frac{\partial}{\partial x} + i\frac{\partial}{\partial y}\right)\left(\frac{\partial V}{\partial x} - i\frac{\partial V}{\partial y}\right) \\ &= e^{i\phi}\left(\frac{\partial}{\partial \rho} + \frac{i}{\rho}\frac{\partial}{\partial \phi}\right)\left\{e^{-i\phi}\left(\frac{\partial V}{\partial \rho} - \frac{i}{\rho}\frac{\partial V}{\partial \phi}\right)\right\} \\ &= \frac{\partial^2 V}{\partial \rho^2} + \frac{i}{\rho^2}\frac{\partial V}{\partial \phi} - \frac{i}{\rho}\frac{\partial^2 V}{\partial \rho\,\partial \phi} + \frac{1}{\rho}\frac{\partial V}{\partial \rho} + \frac{i}{\rho}\frac{\partial^2 V}{\partial \phi\,\partial \rho} - \frac{i}{\rho^2}\frac{\partial V}{\partial \phi} + \frac{1}{\rho^2}\frac{\partial^2 V}{\partial \phi^2} \\ &= \frac{\partial^2 V}{\partial \rho^2} + \frac{1}{\rho}\frac{\partial V}{\partial \rho} + \frac{1}{\rho^2}\frac{\partial^2 V}{\partial \phi^2}. \end{aligned} \tag{3}$$

Thus, in cylindrical polar coordinates,

$$\nabla^2 V = \frac{\partial^2 V}{\partial \rho^2} + \frac{1}{\rho}\frac{\partial V}{\partial \rho} + \frac{1}{\rho^2}\frac{\partial^2 V}{\partial \phi^2} + \frac{\partial^2 V}{\partial z^2}. \tag{4}$$

(b) Spherical polar coordinates

The relations between the cartesian coordinates (x, y, z) of a point P and its spherical polar coordinates (r, θ, ϕ) are (see Fig. 18),

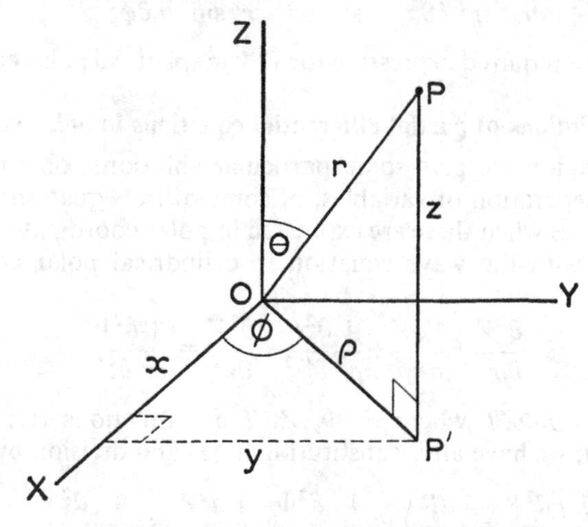

FIG. 18

$$x = r\sin\theta\cos\phi, \quad y = r\sin\theta\sin\phi, \quad z = r\cos\theta. \tag{5}$$

If P' is the projection of P on the plane XOY and $OP' = \rho$, we have

$$\left.\begin{aligned} z &= r\cos\theta, \quad \rho = r\sin\theta, \\ x &= \rho\cos\phi, \quad y = \rho\sin\phi, \end{aligned}\right\} \tag{6}$$

so that the relations between z, ρ and r, θ are similar to those between x, y and ρ, ϕ. By suitable interchange of symbols we can therefore make use of formulae obtained in (*a*) above.

Equation (4) is

$$\nabla^2 V = \frac{\partial^2 V}{\partial \rho^2} + \frac{1}{\rho}\frac{\partial V}{\partial \rho} + \frac{1}{\rho^2}\frac{\partial^2 V}{\partial \phi^2} + \frac{\partial^2 V}{\partial z^2}, \tag{7}$$

and writing z for x, ρ for y, r for ρ, θ for ϕ, the second of equations (2) and equation (3) give

$$\frac{\partial V}{\partial \rho} = \sin\theta\frac{\partial V}{\partial r} + \frac{\cos\theta}{r}\frac{\partial V}{\partial \theta},$$

$$\frac{\partial^2 V}{\partial z^2} + \frac{\partial^2 V}{\partial \rho^2} = \frac{\partial^2 V}{\partial \rho^2} + \frac{1}{\rho}\frac{\partial V}{\partial \rho} + \frac{1}{\rho^2}\frac{\partial^2 V}{\partial \theta^2}.$$

Substituting in (7) and remembering that $\rho = r\sin\theta$,

$$\begin{aligned} \nabla^2 V &= \frac{\partial^2 V}{\partial r^2} + \frac{1}{r}\frac{\partial V}{\partial r} + \frac{1}{r^2}\frac{\partial^2 V}{\partial \theta^2} + \frac{1}{r\sin\theta}\left(\sin\theta\frac{\partial V}{\partial r} + \frac{\cos\theta}{r}\frac{\partial V}{\partial \theta}\right) + \frac{1}{r^2\sin^2\theta}\frac{\partial^2 V}{\partial \phi^2} \\ &= \frac{\partial^2 V}{\partial r^2} + \frac{2}{r}\frac{\partial V}{\partial r} + \frac{1}{r^2}\frac{\partial^2 V}{\partial \theta^2} + \frac{\cot\theta}{r^2}\frac{\partial V}{\partial \theta} + \frac{1}{r^2\sin^2\theta}\frac{\partial^2 V}{\partial \phi^2}, \end{aligned} \tag{8}$$

and this is the required expression for $\nabla^2 V$ in spherical polar coordinates.

7.7 Some solutions of partial differential equations in polar coordinates

In this section we give some particular solutions, obtained by the method of separation of variables, of some of the equations of mathematical physics when these are expressed in polar coordinates.

We start with the wave equation in cylindrical polar coordinates, that is,

$$\frac{\partial^2 V}{\partial \rho^2} + \frac{1}{\rho}\frac{\partial V}{\partial \rho} + \frac{1}{\rho^2}\frac{\partial^2 V}{\partial \phi^2} + \frac{\partial^2 V}{\partial z^2} = \frac{1}{c^2}\frac{\partial^2 V}{\partial t^2}. \tag{1}$$

Writing $V = R\Phi ZT$ where R, Φ, Z, T are functions respectively of ρ, ϕ, z and t, we have after substitution in (1) and division by V

$$\frac{1}{R}\left(\frac{d^2 R}{d\rho^2} + \frac{1}{\rho}\frac{dR}{d\rho}\right) + \frac{1}{\rho^2\Phi}\frac{d^2\Phi}{d\phi^2} + \frac{1}{Z}\frac{d^2 Z}{dz^2} = \frac{1}{c^2 T}\frac{d^2 T}{dt^2}.$$

Let

$$\frac{1}{\Phi}\frac{d^2\Phi}{d\phi^2} = -m^2, \qquad \frac{1}{Z}\frac{d^2Z}{dz^2} = -n^2, \qquad \frac{1}{T}\frac{d^2T}{dt^2} = -c^2p^2, \tag{2}$$

where m, n and p are constants, then

$$\frac{1}{R}\left(\frac{d^2R}{d\rho^2}+\frac{1}{\rho}\frac{dR}{d\rho}\right)-\frac{m^2}{\rho^2} = n^2-p^2 = -\alpha^2 \quad \text{(say)},$$

giving

$$\frac{d^2R}{d\rho^2}+\frac{1}{\rho}\frac{dR}{d\rho}+\left(\alpha^2-\frac{m^2}{\rho^2}\right)R = 0. \tag{3}$$

The solutions of equations (2) are

$$\Phi = A_1\cos m\phi+B_1\sin m\phi, \qquad Z = A_2\cos nz+B_2\sin nz,$$
$$T = A_3\cos cpt+B_3\sin cpt,$$

and equation (3), being Bessel's equation of order m with argument $\alpha\rho$, has solution

$$R = A_4J_m(\alpha\rho)+B_4Y_m(\alpha\rho),$$

the A's and B's being arbitrary constants. Hence a solution of the wave equation in these coordinates is given by an expression of the type

$$V = \sum AJ_m(\alpha\rho)\cos m\phi\cos nz\cos cpt, \tag{4}$$

where $\alpha^2 = -n^2+p^2$. A solution which has axial symmetry and is therefore independent of ϕ is given by taking $m = 0$ and is

$$V = \sum AJ_0(\alpha\rho)\cos nz\cos cpt, \tag{5}$$

and, if the wave is also independent of z,

$$V = \sum AJ_0(p\rho)\cos cpt, \tag{6}$$

for now $n = 0$ and $\alpha = p$.

Solutions of the equation of heat conduction in cylindrical coordinates

$$\frac{\partial^2V}{\partial\rho^2}+\frac{1}{\rho}\frac{\partial V}{\partial\rho}+\frac{1}{\rho^2}\frac{\partial^2V}{\partial\phi^2}+\frac{\partial^2V}{\partial z^2} = \frac{1}{k}\frac{\partial V}{\partial t} \tag{7}$$

can be obtained in a similar way. A typical result is

$$V = \sum A\,e^{-kp^2t}J_m(\alpha\rho)\cos m\phi\cos nz, \tag{8}$$

where $\alpha^2 = -n^2+p^2$, with the simplifications

$$V = \sum A\,e^{-kp^2t}J_0(\alpha\rho)\cos nz \tag{9}$$

in the case of axial symmetry, and

$$V = \sum A\, e^{-kp^2 t} J_0(p\rho) \tag{10}$$

when the temperature V is independent of both ϕ and z.

Example 8. *The material of a long circular cylinder of radius a is of diffusivity k. The cylinder is initially at constant temperature V_0 and its curved surface is maintained at zero temperature. Find an expression for the temperature V at time t of a point in the cylinder at distance ρ from its axis.*

Here there is axial symmetry, and as the cylinder is long, the temperature will be independent of z. The temperature V satisfies the equation

$$\frac{\partial^2 V}{\partial \rho^2} + \frac{1}{\rho}\frac{\partial V}{\partial \rho} = \frac{1}{k}\frac{\partial V}{\partial t}$$

and a particular solution of this equation is, equation (10) above,

$$V = \sum A e^{-kp^2 t} J_0(p\rho),$$

where A and p are, as yet, undetermined constants. By taking $p = p_n (n = 1, 2, 3, \ldots)$ where $J_0(p_n a) = 0$, the condition $V = 0$ on the surface $\rho = a$ of the cylinder is satisfied for all t. Hence we can write

$$V = \sum_{n=1}^{\infty} A_n e^{-kp_n^2 t} J_0(p_n \rho),$$

where p_n is a positive root of the equation $J_0(p_n a) = 0$ and the value of V when $t = 0$ is given by

$$V = \sum_{n=1}^{\infty} A_n J_0(p_n \rho).$$

The coefficients A_n have now to be chosen so that this is equal to the constant V_0 for all values of ρ such that $0 < \rho < a$. This can be done by taking A_n to be the coefficient of $J_0(p_n\rho)$ in the Fourier-Bessel series representing V_0. By equation (20) of §6.9, A_n is therefore given by

$$\begin{aligned} A_n &= \frac{2}{a^2 J_1^2(p_n a)} \int_0^a \rho V_0 J_0(p_n \rho)\, d\rho \\ &= \frac{2V_0}{p_n^2 a^2 J_1^2(p_n a)} \int_0^{p_n a} t J_0(t)\, dt, \quad (t = p_n \rho), \\ &= \frac{2V_0}{p_n a J_1(p_n a)}, \end{aligned}$$

the last step following from the result (12) of §6.9. Hence the required expression for the temperature V is given by

$$V = \frac{2V_0}{a} \sum_{n=1}^{\infty} e^{-kp_n^2 t} \frac{J_0(p_n \rho)}{p_n J_1(p_n a)}, \tag{11}$$

where p_n is a positive root of $J_0(p_n a) = 0$.

Solutions of Laplace's equation

$$\frac{\partial^2 V}{\partial \rho^2} + \frac{1}{\rho}\frac{\partial V}{\partial \rho} + \frac{1}{\rho^2}\frac{\partial^2 V}{\partial \phi^2} + \frac{\partial^2 V}{\partial z^2} = 0, \tag{12}$$

can, since V is now independent of t, be obtained by writing $p = 0$ in either (4) or (8). It is generally more convenient to use in in place of n and a typical solution in this case is

$$V = \sum AJ_m(n\rho)\cos m\phi \cosh nz. \tag{13}$$

Similar solutions can be obtained when the equations are expressed in spherical polar coordinates. For example, the wave equation is

$$\frac{\partial^2 V}{\partial r^2}+\frac{2}{r}\frac{\partial V}{\partial r}+\frac{1}{r^2}\frac{\partial^2 V}{\partial \theta^2}+\frac{\cot\theta}{r^2}\frac{\partial V}{\partial \theta}+\frac{1}{r^2\sin^2\theta}\frac{\partial^2 V}{\partial \phi^2}=\frac{1}{c^2}\frac{\partial^2 V}{\partial t^2}, \tag{14}$$

and if $V = R\Theta\Phi T$ where R, Θ, Φ and T are respectively functions of r, θ, ϕ and t only, we have

$$\frac{1}{R}\left\{\frac{d^2R}{dr^2}+\frac{2}{r}\frac{dR}{dr}\right\}+\frac{1}{r^2\Theta}\left\{\frac{d^2\Theta}{d\theta^2}+\cot\theta\frac{d\Theta}{d\theta}\right\}+\frac{1}{r^2\sin^2\theta}\frac{1}{\Phi}\frac{d^2\Phi}{d\phi^2}=\frac{1}{c^2T}\frac{d^2T}{dt^2}.$$

Let

$$\frac{1}{\Phi}\frac{d^2\Phi}{d\phi^2} = -m^2, \quad \frac{1}{c^2T}\frac{d^2T}{dt^2} = -p^2, \tag{15}$$

and

$$\frac{1}{\Theta}\left\{\frac{d^2\Theta}{d\theta^2}+\cot\theta\frac{d\Theta}{d\theta}\right\} = \frac{m^2}{\sin^2\theta}-n(n+1), \tag{16}$$

where again m, n and p are constants, then

$$\frac{1}{R}\left\{\frac{d^2R}{dr^2}+\frac{2}{r}\frac{dR}{dr}\right\} = \frac{n(n+1)}{r^2}-p^2. \tag{17}$$

Solutions of the ordinary differential equations (15) are

$$\left.\begin{aligned}\Phi &= \cos m\phi \quad \text{or} \quad \Phi = \sin m\phi,\\ T &= \cos cpt \quad \text{or} \quad T = \sin cpt.\end{aligned}\right\} \tag{18}$$

Equation (16) can be written in the form

$$\frac{1}{\sin\theta}\frac{d}{d\theta}\left(\sin\theta\frac{d\Theta}{d\theta}\right)+\left\{n(n+1)-\frac{m^2}{\sin^2\theta}\right\}\Theta = 0.$$

Writing $x = \cos\theta$ so that

$$\frac{1}{\sin\theta}\frac{d}{d\theta} = -\frac{d}{dx},$$

this becomes

$$\frac{d}{dx}\left\{(1-x^2)\frac{d\Theta}{dx}\right\}+\left\{n(n+1)-\frac{m^2}{1-x^2}\right\}\Theta = 0,$$

or

$$(1-x^2)\frac{d^2\Theta}{dx^2}-2x\frac{d\Theta}{dx}+\left\{n(n+1)-\frac{m^2}{1-x^2}\right\}\Theta = 0.$$

This is Legendre's associated equation (§6.4) with solution

$$\begin{aligned}\Theta &= AP_n{}^m(x)+BQ_n{}^m(x)\\ &= AP_n{}^m(\cos\theta)+BQ_n{}^m(\cos\theta),\end{aligned} \tag{19}$$

$P_n{}^m(\cos\theta)$ and $Q_n{}^m(\cos\theta)$ being the associated Legendre functions. Finally, equation (17) can be written in the form

$$r^2\frac{d^2R}{dr^2}+2r\frac{dR}{dr}+\{p^2r^2+\tfrac{1}{4}-(n+\tfrac{1}{2})^2\}R = 0, \tag{20}$$

and this equation, see equation (3) of §6.12, has the solution

$$R = r^{-\frac{1}{2}}\{AJ_{n+\frac{1}{2}}(pr)+BY_{n+\frac{1}{2}}(pr)\}. \tag{21}$$

Thus a solution of the wave equation in spherical polar coordinates is of the form

$$V = \sum Ar^{-\frac{1}{2}}J_{n+\frac{1}{2}}(pr)P_n{}^m(\cos\theta)\cos m\phi\cos cpt. \tag{22}$$

A solution of the wave equation which has spherical symmetry and is independent of θ and ϕ is found by putting $m = n = 0$ in (22). Such a solution is

$$\begin{aligned}V &= \sum Ar^{-\frac{1}{2}}J_{\frac{1}{2}}(pr)\cos cpt\\ &= \sum Br^{-1}\sin pr\cos cpt,\end{aligned} \tag{23}$$

where the constants A and B are related by $B = A(2/\pi)^{\frac{1}{2}}$. For a solution which possesses axial symmetry and which is therefore independent of ϕ, we write $m = 0$ in (22) and have

$$V = \sum Ar^{-\frac{1}{2}}J_{n+\frac{1}{2}}(pr)P_n(\cos\theta)\cos cpt, \tag{24}$$

where now $P_n(\cos\theta)$ is the Legendre polynomial.

When the equation of heat conduction is expressed in spherical polar coordinates, its solution can be found in the same way. The solution corresponding to (22) is

$$V = \sum A\,e^{-kp^2t}r^{-\frac{1}{2}}J_{n+\frac{1}{2}}(pr)P_n{}^m(\cos\theta)\cos m\phi. \tag{25}$$

For Laplace's equation, we work similarly but with $p = 0$. In this case equation (20) becomes

$$r^2\frac{d^2R}{dr^2}+2r\frac{dR}{dr}-n(n+1)R = 0,$$

and it is easy to verify that the solution of this equation is

$$R = Ar^n + Br^{-n-1}.$$

Hence solutions of Laplace's equation are of the form

$$V = \sum Ar^n P_n^m(\cos\theta)\cos m\phi \quad \text{or} \quad V = \sum Ar^{-n-1} P_n^m(\cos\theta)\cos m\phi. \tag{26}$$

Example 9. *Find the gravitational potential of a thin uniform wire of mass M bent into a circle of radius a.*

Take the centre of the circle as the origin O and the perpendicular through O to the plane of the circle as the z-axis. Then, in suitable units, the potential V_0 of the wire at a point $(0, 0, z)$ on the z-axis is given by

$$V_0 = \frac{M}{(a^2+z^2)^{1/2}}$$

$$= \frac{M}{a}\left(1 - \frac{1}{2}\frac{z^2}{a^2} + \frac{1.3}{2.4}\frac{z^4}{a^4} - \dots\right) \quad \text{when } z < a, \tag{27}$$

$$= \frac{M}{z}\left(1 - \frac{1}{2}\frac{a^2}{z^2} + \frac{1.3}{2.4}\frac{a^4}{z^4} - \dots\right) \quad \text{when } z > a. \tag{28}$$

The potential V at a point with spherical polar coordinates (r, θ, ϕ) satisfies Laplace's equation. From symmetry, V will be independent of ϕ and, from (26) with $m = 0$, possible solutions are

$$V = \sum_{n=0}^{\infty} A_n r^n P_n(\cos\theta), \tag{29}$$

and

$$V = \sum_{n=0}^{\infty} A_n r^{-n-1} P_n(\cos\theta). \tag{30}$$

The first of these solutions is suitable for points for which $r < a$ and the second for those at which $r > a$, for no infinities are then involved. The coefficients A_n in equations (29) and (30) have to be chosen so that these solutions reduce to (27) and (28) respectively when $r = z$ and $\theta = 0$, for the point (r, θ, ϕ) then becomes the point $(0, 0, z)$. Since $P_n(1) = 1$, it should be clear that

$$A_1 = A_3 = \dots = 0$$

and that in (29)

$$A_0 = \frac{M}{a}, \qquad A_2 = -\frac{1}{2}\frac{M}{a^3}, \qquad A_4 = \frac{1.3}{2.4}\frac{M}{a^5}, \qquad \dots,$$

while in (30),

$$A_0 = M, \qquad A_2 = -\frac{1}{2}Ma^2, \qquad A_4 = \frac{1.3}{2.4}Ma^4, \qquad \dots.$$

Hence the required results are

$$V = \frac{M}{a}\left\{P_0(\cos\theta) - \frac{1}{2}\frac{r^2}{a^2}P_2(\cos\theta) + \frac{1.3}{2.4}\frac{r^4}{a^4}P_4(\cos\theta) - \dots\right\} \quad \text{when } r < a,$$

$$V = \frac{M}{r}\left\{P_0(\cos\theta) - \frac{1}{2}\frac{a^2}{r^2}P_2(\cos\theta) + \frac{1.3}{2.4}\frac{a^4}{r^4}P_4(\cos\theta) - \dots\right\} \quad \text{when } r > a.$$

Exercises 7(c)

1. A long bar of square cross-section has the faces $x = 0, x = a$ and $y = 0$ maintained at zero temperature and the face $y = a$ at a constant temperature V_0. Under steady state conditions, the temperature V at a point (x, y) in the cross-section satisfies

$$\frac{\partial^2 V}{\partial x^2}+\frac{\partial^2 V}{\partial y^2} = 0.$$

Show that V is given by

$$V = \frac{4V_0}{\pi}\sum_{n=0}^{\infty}\left\{\frac{\operatorname{cosech}(2n+1)\pi}{2n+1}\sinh(2n+1)\frac{\pi y}{a}\sin(2n+1)\frac{\pi x}{a}\right\}. \quad \text{(L.U.)}$$

2. A square plate is bounded by the lines $x = 0, x = a, y = 0$ and $y = a$. The edges $x = 0, y = a$ are maintained at zero temperature, the edge $y = 0$ is heat insulated and the edge $x = a$ is kept at temperature V_0. Under steady state conditions, show that the temperature V at the point (x, y) is given by the formula

$$V = \frac{4V_0}{\pi}\sum_{n=0}^{\infty}\left\{\frac{(-1)^n}{2n+1}\frac{\sinh(2n+1)\dfrac{\pi x}{2a}}{\sinh(2n+1)\dfrac{\pi}{2}}\cos(2n+1)\frac{\pi y}{2a}\right\}.$$

3. A semi-infinite plate is bounded by the lines $x = 0, x = \pi$ and $y = 0$. The edges $x = 0, x = \pi$ are both maintained at zero temperature and the edge $y = 0$ is kept at unit temperature. Show that, in the steady state, the temperature V at the point (x, y) is given by

$$\pi V = 4(e^{-y}\sin x+\tfrac{1}{3}e^{-3y}\sin 3x+\tfrac{1}{5}e^{-5y}\sin 5x+\ldots).$$

4. If $x = ce^{-z}\sin\theta$ and $y = ce^{-z}\cos\theta$, change the variables from (x, y) to (z, θ) in the equation

$$y^2\frac{\partial^2 u}{\partial x^2}-2xy\frac{\partial^2 u}{\partial x\,\partial y}+x^2\frac{\partial^2 u}{\partial y^2} = 0.$$

Hence show that the equation has a solution of the form

$$u = \sum A_n e^{-n^2 z}\cos(n\theta+\alpha)$$

where A, α and n are constants. (L.U.)

5. If $x = \log_e u$, $y = \log_e v$, show that by changing the independent variables from (u, v) to (x, y) the partial differential equation

$$u^2\frac{\partial^2 z}{\partial u^2}+v^2\frac{\partial^2 z}{\partial v^2}+u\frac{\partial z}{\partial u}+v\frac{\partial z}{\partial v} = 0$$

is transformed into Laplace's equation.

6. For the transformations $x = \rho\cos\phi, y = \rho\sin\phi$ and $\rho = a\cosh u\cos v$, $z = a\sinh u\sin v$ applied to the function $F(x, y, z)$,

(i) express $\dfrac{\partial^2 F}{\partial x^2}+\dfrac{\partial^2 F}{\partial y^2}$ in terms of $\dfrac{\partial^2 F}{\partial \rho^2}, \dfrac{\partial^2 F}{\partial \phi^2}$ and $\dfrac{\partial F}{\partial \rho}$;

(ii) show that the equation $\dfrac{\partial^2 F}{\partial x^2}+\dfrac{\partial^2 F}{\partial y^2}+\dfrac{\partial^2 F}{\partial z^2} = 0$ transforms into

$$\frac{\partial^2 F}{\partial u^2}+\frac{\partial^2 F}{\partial v^2}+\tanh u\frac{\partial F}{\partial u}-\tan v\frac{\partial F}{\partial v}+\frac{\sinh^2 u+\sin^2 v}{\cosh^2 u\cos^2 v}\frac{\partial^2 F}{\partial \phi^2} = 0.$$

Hence find a solution of this equation which is a function of u only.

(L.U.)

7. Transform Laplace's equation in cylindrical coordinates (ρ, ϕ, z) to one in which the variables are u, v, ϕ where $\rho = (uv)^{1/2}$, $z = \frac{1}{2}(u-v)$. Hence show that a solution of Laplace's equation is

$$V = F_n(u)F_{-n}(v)\cos\omega\phi$$

where $F_n(u)$ satisfies the equation

$$u^2\frac{d^2F_n}{du^2}+u\frac{dF_n}{du}+\left(nu-\frac{\omega^2}{4}\right)F_n = 0.$$

8. The velocity potential V of a homogeneous fluid streaming with general velocity U past a fixed long circular cylinder of radius a satisfies the two-dimensional form of Laplace's equation and the boundary conditions

$$\frac{\partial V}{\partial \rho} = 0 \text{ when } \rho = a, \quad V \to -U\rho\cos\phi \text{ as } \rho \to \infty,$$

ρ and ϕ being cylindrical polar coordinates. Show that

$$V = -U\left(\rho+\frac{a^2}{\rho}\right)\cos\phi.$$

9. A thin semi-circular plate of radius a has its bounding diameter kept at temperature zero and its circumference is kept at temperature unity. Show that the temperature V at the point with cylindrical polar coordinates $(\rho, \phi, 0)$ is, under steady state conditions, given by

$$V = \frac{4}{\pi}\sum_{n=0}^{\infty}\frac{(\rho/a)^{2n+1}}{2n+1}\sin(2n+1)\phi.$$

10. A thin metal plate lies in the plane $z = 0$ and is bounded by $\rho = a$, $\rho = b$ $(a < b)$, $\phi = 0$ and $\phi = \frac{1}{2}\pi$, (ρ, ϕ, z) being cylindrical polar coordinates. The boundary $\rho = a$ is maintained at temperature $\phi(\frac{1}{2}\pi-\phi)$ and the other boundaries are kept at zero temperature. Show that, in the steady state, the temperature V is given by

$$V = \frac{2}{\pi}\sum_{n=0}^{\infty}\frac{(\rho/b)^{4n+2}-(b/\rho)^{4n+2}}{(a/b)^{4n+2}-(b/a)^{4n+2}}\cdot\frac{\sin(4n+2)\phi}{(2n+1)^3}.$$

11. A thin membrane is stretched tightly over a circular ring of unit radius and its tension T and surface density m are uniform over its whole area. Given that the smallest root of the equation $J_0(p) = 0$ is $p = 2{\cdot}405$, show that, when the membrane vibrates, the frequency of the fundamental mode is

$$\frac{2{\cdot}405}{2\pi}\left(\frac{Tg}{m}\right)^{1/2}.$$

12. A semi-infinite circular cylinder occupies the region $0 \leqq \rho < a$, $z > 0$ where ρ and z are cylindrical polar coordinates. The surface $\rho = a$ is kept at zero temperature while $z = 0$ is maintained at temperature $f(\rho)$. Show that the steady temperature V at the point (ρ, ϕ, z) is given by

$$V = \frac{2}{a^2} \sum_{n=1}^{\infty} \frac{J_0(\rho\alpha_n)}{J_1^2(a\alpha_n)} e^{-\alpha_n z} \int_0^a \rho f(\rho) J_0(\rho\alpha_n)\, d\rho,$$

where the α_n are the positive roots of the equation $J_0(a\alpha_n) = 0$.

13. A conducting sphere of radius a is placed with its centre at the origin in an originally uniform electric field of strength E_0 extending along the z-axis. The potential V at the point with spherical polar coordinates (r, θ, ϕ) satisfies Laplace's equation and the boundary conditions $V = 0$ at $r = a$, $V \to -E_0 r \cos\theta$ as $r \to \infty$. Show that

$$V = -E_0 r \cos\theta \left\{1 - \left(\frac{a}{r}\right)^3\right\}.$$

14. When air vibrates freely in a rigid sphere of radius a, the velocity potential V at a point P satisfies the wave equation and the boundary condition $\partial V/\partial r = 0$ when $r = a$, r being the distance of P from the centre of the sphere. Show that if P is the point (r, θ, ϕ),

$$V = \sum A r^{-1/2} J_{n+1/2}(pr) P_n^m(\cos\theta) \cos m\phi \cos cpt$$

where p is chosen so that

$$\frac{d}{dr}\{r^{-1/2} J_{n+1/2}(pr)\}_{r=a} = 0,$$

A is a constant and the summation is over m, n and p. Deduce that for radial vibrations, in which V is independent of θ and ϕ, the equation satisfied by p reduces to $\tan pa = pa$.

15. A spherical cap occupying the region $r = a$, $0 \leqq \theta < \alpha$ where r and θ are spherical polar coordinates carries an electric charge of uniform surface density σ. If V_+ and V_- are the electrostatic potentials at the point (r, θ, ϕ) when $r > a$ and $r < a$ respectively, V_+, V_- satisfy Laplace's equation and the boundary conditions on $r = a$ are $V_+ = V_-$ and

$$\frac{\partial V_+}{\partial r} - \frac{\partial V_-}{\partial r} = -4\pi\sigma, \quad (0 < \theta < \alpha), \qquad \frac{\partial V_+}{\partial r} - \frac{\partial V_-}{\partial r} = 0, \quad (\alpha < \theta < \pi).$$

Show that

$$V_+ = 2\pi a\sigma \left\{(1-\cos\alpha)\left(\frac{a}{r}\right) + \sum_{n=1}^{\infty} \frac{P_{n-1}(\cos\alpha) - P_{n+1}(\cos\alpha)}{2n+1} \left(\frac{a}{r}\right)^{n+1} P_n(\cos\theta)\right\},$$

$$V_- = 2\pi a\sigma \left\{1 - \cos\alpha + \sum_{n=1}^{\infty} \frac{P_{n-1}(\cos\alpha) - P_{n+1}(\cos\alpha)}{2n+1} \left(\frac{r}{a}\right)^{n} P_n(\cos\theta)\right\}.$$

7.8 Maxwell's equations

Maxwell's equations for the electromagnetic field in a uniform isotropic medium are

$$\left.\begin{aligned}\frac{\partial H_z}{\partial y}-\frac{\partial H_y}{\partial z} &= \frac{4\pi}{c}j_x+\frac{K}{c}\frac{\partial E_x}{\partial t}\\ \frac{\partial H_x}{\partial z}-\frac{\partial H_z}{\partial x} &= \frac{4\pi}{c}j_y+\frac{K}{c}\frac{\partial E_y}{\partial t}\\ \frac{\partial H_y}{\partial x}-\frac{\partial H_x}{\partial y} &= \frac{4\pi}{c}j_z+\frac{K}{c}\frac{dE_z}{dt}\end{aligned}\right\}, \tag{1}$$

$$\left.\begin{aligned}\frac{\partial E_z}{\partial y}-\frac{\partial E_y}{\partial z} &= -\frac{\mu}{c}\frac{\partial H_x}{\partial t}\\ \frac{\partial E_x}{\partial z}-\frac{\partial E_z}{\partial x} &= -\frac{\mu}{c}\frac{\partial H_y}{\partial t}\\ \frac{\partial E_y}{\partial x}-\frac{\partial E_x}{\partial y} &= -\frac{\mu}{c}\frac{\partial H_z}{\partial t}\end{aligned}\right\}, \tag{2}$$

$$\frac{\partial E_x}{\partial x}+\frac{\partial E_y}{\partial y}+\frac{\partial E_z}{\partial z} = \frac{4\pi}{K}\rho, \tag{3}$$

$$\frac{\partial H_x}{\partial x}+\frac{\partial H_y}{\partial y}+\frac{\partial H_z}{\partial z} = 0. \tag{4}$$

In the above E_x, E_y, E_z, H_x, H_y, H_z and j_x, j_y, j_z are the components parallel to the coordinate axes of the electric field strength **E**, the magnetic field strength **H** and the current density **j** respectively, K and μ are respectively the dielectric constant and permeability of the medium, ρ is the volume density of free electricity and t is the time. The reader who is familiar with vectors will see that the equations can be written compactly in the form

$$\operatorname{curl}\mathbf{H} = \frac{4\pi}{c}\mathbf{j}+\frac{K}{c}\frac{\partial \mathbf{E}}{\partial t}, \tag{1}$$

$$\operatorname{curl}\mathbf{E} = -\frac{\mu}{c}\frac{\partial \mathbf{H}}{\partial t}, \tag{2}$$

$$\operatorname{div}\mathbf{E} = \frac{4\pi}{K}\rho, \tag{3}$$

$$\operatorname{div}\mathbf{H} = 0, \tag{4}$$

and it should be noted that the electrical quantities ($\mathbf{E}$, $\mathbf{j}$, K, ρ) are in electrostatic units while the magnetic quantities ($\mathbf{H}$, μ) are in electromagnetic units, c being the factor relating the two sets of units. The above equations are in c.g.s. units; in m.k.s. units they become

$$\text{curl}\,\mathbf{H} = \mathbf{j}+KK_0(\partial\mathbf{E}/\partial t), \qquad \text{curl}\,\mathbf{E} = -\mu\mu_0(\partial\mathbf{H}/\partial t),$$
$$\text{div}\,\mathbf{E} = \rho/KK_0, \qquad \text{div}\,\mathbf{H} = 0,$$

where K_0 and μ_0 are respectively the absolute permittivity and permeability of free space.

In an isotropic medium the components of the current density are proportional to the corresponding components of the electric field strength and we have, by Ohm's law,

$$j_x = \frac{E_x}{\tau}, \qquad j_y = \frac{E_y}{\tau}, \qquad j_z = \frac{E_z}{\tau}, \tag{5}$$

where τ is the specific resistance of the medium in electrostatic units. Using the relations (5), equations (1) can be written

$$\frac{\partial H_z}{\partial y}-\frac{\partial H_y}{\partial z} = \frac{4\pi}{c\tau}E_x+\frac{K}{c}\frac{\partial E_x}{\partial t} \tag{6}$$

and two similar equations. Differentiating these equations with respect to x, y, z respectively, multiplying by c and adding

$$\left(\frac{4\pi}{\tau}+K\frac{\partial}{\partial t}\right)\left(\frac{\partial E_x}{\partial x}+\frac{\partial E_y}{\partial y}+\frac{\partial E_z}{\partial z}\right) = 0,$$

and this, when use is made of equation (3), gives

$$K\frac{\partial\rho}{\partial t}+\frac{4\pi}{\tau}\rho = 0.$$

If ρ_0 is the value of ρ at time $t = 0$, the solution of this differential equation is

$$\rho = \rho_0 \exp\left(-\frac{4\pi}{K\tau}t\right),$$

showing that ρ decreases exponentially and any initial charge in a conducting material will therefore rapidly disappear. Hence, for practical purposes, we can take $\rho = 0$ both for a charge-free dielectric and for a conductor, and equation (3) can in these cases be replaced by

$$\frac{\partial E_x}{\partial x}+\frac{\partial E_y}{\partial y}+\frac{\partial E_z}{\partial z} = 0. \tag{7}$$

Multiplying equation (6) by μ/c, differentiating with respect to t and using equations (2)

$$\frac{4\pi\mu}{c^2\tau}\frac{\partial E_x}{\partial t}+\frac{K\mu}{c^2}\frac{\partial^2 E_x}{\partial t^2}=\frac{\partial}{\partial y}\left(\frac{\mu}{c}\frac{\partial H_z}{\partial t}\right)-\frac{\partial}{\partial z}\left(\frac{\mu}{c}\frac{\partial H_y}{\partial t}\right)$$

$$=\frac{\partial}{\partial z}\left(\frac{\partial E_x}{\partial z}-\frac{\partial E_z}{\partial x}\right)-\frac{\partial}{\partial y}\left(\frac{\partial E_y}{\partial x}-\frac{\partial E_x}{\partial y}\right)$$

$$=-\frac{\partial}{\partial x}\left(\frac{\partial E_y}{\partial y}+\frac{\partial E_z}{\partial z}\right)+\frac{\partial^2 E_x}{\partial y^2}+\frac{\partial^2 E_x}{\partial z^2}$$

$$=\left(\frac{\partial^2}{\partial x^2}+\frac{\partial^2}{\partial y^2}+\frac{\partial^2}{\partial z^2}\right)E_x, \tag{8}$$

when equation (7) is used in the last step. The other components of the electric field strength and those of the magnetic field strength can be shown to satisfy similar equations. Hence in such cases, Maxwell's equations can be written compactly in the form

$$\nabla^2\mathbf{E}=\frac{K\mu}{c^2}\frac{\partial^2\mathbf{E}}{\partial t^2}+\frac{4\pi\mu}{c^2\tau}\frac{\partial\mathbf{E}}{\partial t},$$

$$\nabla^2\mathbf{H}=\frac{K\mu}{c^2}\frac{\partial^2\mathbf{H}}{\partial t^2}+\frac{4\pi\mu}{c^2\tau}\frac{\partial\mathbf{H}}{\partial t}.$$

In many practical applications the constant τ is either small or large, and the above equations then approximate respectively to the equation of heat conduction and the wave equation.

Example 10. *Show that in a dielectric of permeability μ and dielectric constant K, electromagnetic waves are propagated with velocity $c/(K\mu)^{1/2}$.*

In a dielectric, the specific resistance $\tau \to \infty$ and equation (8) reduces to

$$\left(\frac{\partial^2}{\partial x^2}+\frac{\partial^2}{\partial y^2}+\frac{\partial^2}{\partial z^2}\right)E_x=\frac{K\mu}{c^2}\frac{\partial^2 E_x}{\partial t^2}.$$

This is the wave equation already discussed and the velocity of propagation is $c/(K\mu)^{1/2}$ (see §7.2). In free space, $\mu = K = 1$ and the velocity of propagation reduces to c. Now c is the ratio of the two sets of electrical units and its value has been determined experimentally as $2{\cdot}998\times 10^{10}$ cm. per sec. It is also known that the velocity of light in free space has precisely this value and it can be deduced that light waves are electromagnetic in nature.

Example 11. *A long metallic solid circular cylinder of specific resistance τ and unit permeability carries an alternating current of period $2\pi/p$. Show that the current density at time t and at distance r from the axis of the cylinder is given by $A(\operatorname{ber} x+i \operatorname{bei} x)e^{ipt}$ where $x = 2(r/c)(\pi p/\tau)^{1/2}$ and A is a constant.*

Taking the axis of the cylinder to coincide with the x-axis, the only non-vanishing component of the current density is $j_x = E_x/\tau$. Since the cylinder is long, we can take j_x to be independent of x and hence equation (8) can be written

$$\left(\frac{\partial^2}{\partial y^2}+\frac{\partial^2}{\partial z^2}\right)j_x = \frac{4\pi}{c^2\tau}\frac{\partial j_x}{\partial t}+\frac{K}{c^2}\frac{\partial^2 j_x}{\partial t^2}, \tag{9}$$

for here $\mu = 1$. Since there is axial symmetry we can write

$$j_x = \Omega(r)e^{ipt}, \tag{10}$$

where $\Omega(r)$ is a function of r only and $y = r\cos\theta$, $z = r\sin\theta$. It is easy to show that

$$\frac{\partial^2\Omega}{\partial y^2}+\frac{\partial^2\Omega}{\partial z^2} \quad \text{transforms into} \quad \frac{d^2\Omega}{dr^2}+\frac{1}{r}\frac{d\Omega}{dr},$$

so that substitution from (10) in (9) and division by e^{ipt} gives

$$\frac{d^2\Omega}{dr^2}+\frac{1}{r}\frac{d\Omega}{dr} = \left(\frac{4\pi}{c^2\tau}ip-\frac{K}{c^2}p^2\right)\Omega.$$

In metallic conductors τ is small and the expression $(K\tau p/4\pi)$ is small unless we are dealing with currents whose frequency is very large. Hence the second term on the right-hand side can be neglected in comparison with the first and the equation satisfied by Ω is

$$\frac{d^2\Omega}{dr^2}+\frac{1}{r}\frac{d\Omega}{dr}-\frac{4\pi}{c^2\tau}ip\Omega = 0.$$

From §6.11 the solution of this equation, finite at $r = 0$, is

$$\Omega = AI_0(i^{1/2}x) = A(\text{ber}\, x+i\,\text{bei}\, x)$$

where $x = 2(r/c)(\pi p/\tau)^{1/2}$ and the required result is obtained by substitution in (10).

7.9 Schrödinger's equation

In wave mechanics, Schrödinger's equation for the wave function ψ associated with a particle of mass m moving in a field of force is

$$\nabla^2\psi+\frac{8\pi^2 m}{h^2}(W-V)\psi = 0, \tag{1}$$

where W is the total and V the potential energy of the particle, h is Planck's constant and ∇^2 is the usual operator $\partial^2/\partial x^2+\partial^2/\partial y^2+\partial^2/\partial z^2$. The physical significance of the function ψ is that $|\psi|^2\,dv$ represents the probability that the particle whose wave function is ψ is to be found in a small volume dv centred at the point (x, y, z). It is convenient to make the total probability unity, and for this to be so

$$\int|\psi|^2\,dv = 1, \tag{2}$$

where the integral is taken throughout the whole space: when this is done, the wave function is said to be "normalized".

A simple but important example is that of the "linear oscillator" in which a particle of mass m is attracted to a fixed point P by a restoring force $-\mu x$ where x is its displacement from P and μ is a constant. By classical mechanics, the particle vibrates about P with frequency $(\mu/4\pi^2 m)^{\frac{1}{2}}$ and with arbitrary amplitude and energy. For such an oscillator, the potential energy is given by $V = \frac{1}{2}\mu x^2$ and, since ψ is here dependent only on x, Schrödinger's equation (1) becomes

$$\frac{d^2\psi}{dx^2}+\frac{8\pi^2 m}{h^2}(W-\tfrac{1}{2}\mu x^2)\psi = 0. \tag{3}$$

Writing

$$x = \left(\frac{h^2}{4\pi^2 m\mu}\right)^{\frac{1}{4}}\xi, \tag{4}$$

equation (3) becomes

$$\frac{d^2\psi}{d\xi^2}+\left\{\frac{4\pi}{h}\left(\frac{m}{\mu}\right)^{\frac{1}{2}}W-\xi^2\right\}\psi = 0. \tag{5}$$

This is of the form of equation (8) of §6.19. Since, on physical grounds ψ must tend to zero as $|\xi| \to \infty$, we can conclude from the last paragraph of §6.19 that acceptable solutions only exist when

$$\frac{4\pi}{h}\left(\frac{m}{\mu}\right)^{\frac{1}{2}}W = 2n+1, \quad \text{where } n \text{ is a positive integer,} \tag{6}$$

and the values of W obtained from this equation are known as the energy levels. Furthermore these solutions are given by

$$\psi_n = C_n\Psi_n = C_n e^{-\frac{1}{2}\xi^2}H_n(\xi) \tag{7}$$

where Ψ_n is Hermite's function, H_n the Hermite polynomial of order n and C_n is a constant. C_n is chosen so that ψ_n is normalized, i.e. so that

$$\int_{-\infty}^{\infty}|\psi_n|^2\,dx = 1 \quad \text{or} \quad \int_{-\infty}^{\infty}|\psi_n|^2\,d\xi = \left(\frac{4\pi^2 m\mu}{h^2}\right)^{\frac{1}{4}}, \tag{8}$$

and substitution from (7) in (8) leads to the actual value of C_n.

Another important problem is the determination of the energy levels for a hydrogen-like atom. Such an atom consists of a nucleus and a single electron and, as a first approximation, the mass of the nucleus can be considered to be infinite. If the electron is of mass m and charge $-\varepsilon$ and the nucleus is of charge $Z\varepsilon$, the potential of the field can be taken as $V = -Z\varepsilon^2/r$ where r is the distance between electron and nucleus. In this case, Schrödinger's equation (1) becomes

$$\nabla^2\psi+\frac{8\pi^2 m}{h^2}\left(W+\frac{Z\varepsilon^2}{r}\right)\psi = 0, \tag{9}$$

where $\nabla^2\psi$ is now expressed in spherical polar coordinates (see equation (8) of §7.6). Writing $\psi = R\Theta\Phi$ where R, Θ, Φ are respectively functions of r, θ and ϕ only and, separating the variables as in §7.7, we find

$$\frac{d^2R}{dr^2}+\frac{2}{r}\frac{dR}{dr}-\left\{\frac{l(l+1)}{r^2}-\frac{8\pi^2 m}{h^2}\left(W+\frac{Z\varepsilon^2}{r}\right)\right\}R = 0, \qquad (10)$$

$$\frac{d^2\Theta}{d\theta^2}+\cot\theta\frac{d\Theta}{d\theta}+\left\{l(l+1)-\frac{\alpha^2}{\sin^2\theta}\right\}\Theta = 0, \qquad (11)$$

$$\frac{d^2\Phi}{d\phi^2}+\alpha^2\Phi = 0, \qquad (12)$$

where l and α are constants of separation. Suitable solutions of equations (11) and (12) are $P_l^\alpha(\cos\theta)$ and $e^{i\alpha\phi}$ where $P_l^\alpha(\cos\theta)$ is the associated Legendre function, l and α are integers or zero and $l \geqq |\alpha|$. Writing

$$x = \left(-\frac{32\pi^2 mW}{h^2}\right)^{\frac{1}{2}} r, \qquad \nu = \frac{2\pi Z\varepsilon^2}{h}\left(-\frac{m}{2W}\right)^{\frac{1}{2}}, \qquad (13)$$

equation (10) transforms into

$$\frac{d^2R}{dx^2}+\frac{2}{x}\frac{dR}{dx}-\left\{\frac{l(l+1)}{x^2}-\frac{\nu}{x}+\frac{1}{4}\right\}R = 0.$$

This is equation (4) of §6.18 and a solution is given by equation (6) of that section as

$$R = e^{-\frac{1}{2}x}x^l L_{\nu+l}^{2l+1}(x),$$

where $L_{\nu+l}^{2l+1}(x)$ is the associated Laguerre polynomial. For physical reasons R remains finite as x tends to zero and R tends to zero as x tends to infinity. Using arguments similar to those used for the Hermite function in §6.19, this only occurs when ν is an integer n greater than l. Putting $\nu = n$, equation (13) then shows that the energy levels are given by

$$n = \frac{2\pi Z\varepsilon^2}{h}\left(-\frac{m}{2W}\right)^{\frac{1}{2}},$$

that is,

$$W = -\frac{2\pi^2 Z^2\varepsilon^4 m}{h^2 n^2}, \qquad (n = 1, 2, 3, \ldots). \qquad (14)$$

The wave function corresponding to the constants n, l, α above is given by

$$\psi_{nl\alpha}(r,\theta,\phi) = C_{nl\alpha}\, e^{-\frac{1}{2}x}x^l L_{\nu+l}^{2l+1}(x) P_l^\alpha(\cos\theta)\, e^{i\alpha\phi}, \qquad (15)$$

where x is given in terms of r by the first of equations (13) and $C_{nl\alpha}$

is a normalizing constant. In spherical polar coordinates the element of volume $dv = r^2 \sin\theta\, dr\, d\theta\, d\phi$ and equation (2) gives

$$\int_0^\infty \int_0^\pi \int_0^{2\pi} |\psi_{nl\alpha}|^2 r^2 \sin\theta\, dr\, d\theta\, d\phi = 1 \tag{16}$$

as the relation to determine $C_{nl\alpha}$.

Example 12. *Show that the wave function $\psi_{100}(r, \theta, \phi)$ for a hydrogen-like atom of one electron of charge ε and mass m and a heavy nucleus of charge $Z\varepsilon$ is given by*

$$\psi_{100}(r, \theta, \phi) = \frac{1}{\pi^{1/2}}\left(\frac{Z}{a}\right)^{3/2} e^{-Zr/a},$$

where $a = h^2/(4\pi^2 m\varepsilon^2)$.

With $n = 1$, equation (14) gives $W = (-2\pi^2 Z^2 \varepsilon^4 m)/h^2$ and substitution in (13) yields

$$x = \left(\frac{64\pi^4 Z^2 \varepsilon^4 m^2}{h^4}\right)^{1/2} r = \left(\frac{8\pi^2 Z \varepsilon^2 m}{h^2}\right) r = \frac{2Zr}{a},$$

$$\nu = \frac{2\pi Z\varepsilon^2}{h}\left(\frac{mh^2}{4\pi^2 Z^2 \varepsilon^4 m}\right)^{1/2} = 1.$$

Substituting for x, ν and writing $n = 1$, $l = \alpha = 0$ in (15),

$$\psi_{100}(r, \theta, \phi) = C_{100}\, e^{-Zr/a} L_1^{1}(2Zr/a),$$

since $P_0(\cos\theta) = 1$. From equation (3) of §6.18,

$$L_1^{1}(x) = \frac{d}{dx}\left\{e^x \frac{d}{dx}(xe^{-x})\right\} = \frac{d}{dx}\{e^x(e^{-x} - xe^{-x})\} = \frac{d}{dx}(1-x) = -1,$$

and hence

$$\psi_{100}(r, \theta, \phi) = -C_{100}\, e^{-Zr/a}.$$

Substituting in (16),

$$C_{100}^2 \int_0^\infty \int_0^\pi \int_0^{2\pi} r^2 e^{-2Zr/a} \sin\theta\, dr\, d\theta\, d\phi = 1,$$

and the value of the triple integral is easily found to be $\pi a^3/Z^3$. It follows that the value of C_{100} can be taken as $-\pi^{-1/2}(Z/a)^{3/2}$ and hence

$$\psi_{100}(r, \theta, \phi) = \frac{1}{\pi^{1/2}}\left(\frac{Z}{a}\right)^{3/2} e^{-Zr/a}.$$

Exercises 7(d)

1. A long circular conducting cylinder of radius a, small specific resistance τ and unit permeability is initially free from electric and magnetic fields. The axis of the cylinder coincides with the x-axis, and at time $t = 0$ a constant magnetic field H_0 is established outside the cylinder and parallel to its axis. If the axial distance of a point P inside the cylinder is r, show that the magnetic field at P at time t is given by

$$H_x = H_0 - \frac{2H_0}{a} \sum_{n=1}^{\infty} e^{-\lambda p_n^2 t} \frac{J_0(p_n r)}{p_n J_1(p_n a)},$$

where $\lambda = c^2\tau/4\pi$ and p_n is a positive root of $J_0(p_n a) = 0$.

2. Prove that in an electromagnetic field in free space independent of the coordinate y, where the axes Ox, Oy, Oz are rectangular cartesian, the expressions

$$E_x = \frac{\partial S}{\partial z}, \qquad E_y = 0, \qquad E_z = -\frac{\partial S}{\partial x},$$

$$H_x = 0, \qquad H_y = -\frac{1}{c}\frac{\partial S}{\partial t}, \qquad H_z = 0,$$

satisfy Maxwell's equations provided that S (a function of x, z and t) satisfies the equation

$$\frac{\partial^2 S}{\partial x^2}+\frac{\partial^2 S}{\partial y^2} = \frac{1}{c^2}\frac{\partial^2 S}{\partial t^2}. \qquad \text{(L.U.)}$$

3. Verify that there is a solution of Maxwell's equations in free space of the form

$$E_x = \frac{\partial^2 \psi}{\partial z\,\partial x}, \qquad E_y = \frac{\partial^2 \psi}{\partial z\,\partial y}, \qquad E_z = \frac{\partial^2 \psi}{\partial z^2} - \frac{1}{c^2}\frac{\partial^2 \psi}{\partial t^2},$$

$$H_x = \frac{1}{c}\frac{\partial^2 \psi}{\partial t\,\partial y}, \qquad H_y = -\frac{1}{c}\frac{\partial^2 \psi}{\partial t\,\partial x}, \qquad H_z = 0,$$

provided that ψ satisfies the wave equation.

4. The components E_x, E_y, E_z of the electric field strength **E** and those H_x, H_y, H_z of the magnetic field strength **H** are expressed in terms of a *scalar potential* ϕ and the components A_x, A_y, A_z of a *vector potential* **A** by the equations

$$E_x = -\frac{\partial \phi}{\partial x}-\frac{1}{c}\frac{\partial A_x}{\partial t}, \qquad E_y = -\frac{\partial \phi}{\partial y}-\frac{1}{c}\frac{\partial A_y}{\partial t}, \qquad E_z = -\frac{\partial \phi}{\partial z}-\frac{1}{c}\frac{\partial A_z}{\partial t},$$

$$\mu H_x = \frac{\partial A_z}{\partial y}-\frac{\partial A_y}{\partial z}, \qquad \mu H_y = \frac{\partial A_x}{\partial z}-\frac{\partial A_z}{\partial x}, \qquad \mu H_z = \frac{\partial A_y}{\partial x}-\frac{\partial A_x}{\partial y},$$

while A_x, A_y, A_z and ϕ are related by the equation

$$\frac{\partial A_x}{\partial x}+\frac{\partial A_y}{\partial y}+\frac{\partial A_z}{\partial z}+\frac{K\mu}{c}\frac{\partial \phi}{\partial t} = 0.$$

Show that Maxwell's equations are satisfied if

$$\nabla^2\phi - \frac{K\mu}{c^2}\frac{\partial^2 \phi}{\partial t^2} = -\frac{4\pi}{K}\rho,$$

and

$$\nabla^2 A_x - \frac{K\mu}{c^2}\frac{\partial^2 A_x}{\partial t^2} = -\frac{4\pi\mu}{c}j_x,$$

with similar equations for A_y and A_z.

5. Show that, in free space, there is a solution for the vector and scalar potentials of Exercise 4 above of the form

$$A_x = 0, \qquad A_y = 0, \qquad A_z = \frac{f'(u)}{cr},$$

$$\phi = \frac{z}{r}\left\{\frac{f'(u)}{cr}+\frac{f(u)}{r^2}\right\},$$

where $r^2 = x^2+y^2+z^2$ and $f(u)$ is an arbitrary function of $u = t-r/c$. (L.U.)

6. Show that the wave function ψ_1 for a linear oscillator of mass m and frequency ν is given by

$$\psi_1 = 2\left(\frac{\pi m\nu}{h}\right)^{1/4} \xi\, e^{-\xi^2/2} \quad \text{where} \quad \xi = 2\pi\left(\frac{m\nu}{h}\right)^{1/2} x.$$

7. Show that for a hydrogen-like atom of one electron of charge ε and mass m and a heavy nucleus of charge $Z\varepsilon$,

$$\psi_{200}(r, \theta, \phi) = \frac{1}{4(2\pi)^{1/2}}\left(\frac{Z}{a}\right)^{3/2} (2-\rho)\, e^{-\rho/2},$$

$$\psi_{210}(r, \theta, \phi) = \frac{1}{4(2\pi)^{1/2}}\left(\frac{Z}{a}\right)^{3/2} \rho\, e^{-\rho/2} \cos\theta,$$

$$\psi_{211}(r, \theta, \phi) = \frac{1}{8(\pi)^{1/2}}\left(\frac{Z}{a}\right)^{3/2} \rho\, e^{-\rho/2} \sin\theta\, e^{i\phi},$$

where $a = h^2/(4\pi^2 m\varepsilon^2)$ and $\rho = Zr/a$.

8. Show that in a hydrogen-like atom with a heavy nucleus, the probability that the electron is at radial distance between $r-\frac{1}{2}\delta r$ and $r+\frac{1}{2}\delta r$ from the nucleus is $F(r)\delta r$ where

$$F(r) = r^2 \int_0^{\pi}\int_0^{2\pi} |\psi(r, \theta, \phi)|^2 \sin\theta\, d\theta\, d\phi$$

and $\psi(r, \theta, \phi)$ is the normalized wave function corresponding to the state of the atom. Show also that the mean distance of the electron from the nucleus is given by $\int_0^{\infty} rF(r)\, dr$.

9. A hydrogen-like atom consists of a single electron of mass m and charge ε and its heavy nucleus carries a charge $Z\varepsilon$. Use the results of Exercise 8 above and that of Example 12 of §7.9 to show that when the atom is in a state for which the wave function is ψ_{100}, the mean distance of the electron from the nucleus is $3a/2Z$ where $a = h^2/(4\pi^2 m\varepsilon^2)$.

10. Write down Schrödinger's equation in cylindrical polar coordinates (ρ, ϕ, z) when the potential energy of the field is $2\pi^2 m(a^2\rho^2+b^2z^2)$ where a and b are constants.
Writing $\alpha h = 4\pi^2 ma$, $\beta h = 4\pi^2 mb$, $\lambda = (8\pi^2 mW/h^2)-(2n+1)\beta$, where n is a positive integer, show that a solution of the equation is

$$\exp\{il\phi-\tfrac{1}{2}\alpha\rho^2-\tfrac{1}{2}\beta z^2\} f(\rho) H_n(\beta^{1/2}z),$$

where l is a constant, H_n is the Hermite polynomial and $f(\rho)$ satisfies the ordinary differential equation

$$\frac{d^2f}{d\rho^2}+\left(\frac{1}{\rho}-2\alpha\rho\right)\frac{df}{d\rho}+\left(\lambda-2\alpha-\frac{l^2}{\rho^2}\right)f = 0.$$

CHAPTER 8

INTEGRAL TRANSFORMS

8.1 Introduction

The integral transform $\bar{f}(p)$ of a function $f(x)$ of x in the range (a, b) is defined by the relation

$$\bar{f}(p) = \int_a^b f(x)K(p, x)\,dx, \qquad (1)$$

where $K(p, x)$ is a known function of p and x, called the *kernel* of the transform, and it is assumed that the integral on the right-hand side of equation (1) exists. Such transforms have been used increasingly in recent years to obtain the solution of both ordinary and partial linear differential equations and it is the purpose of this chapter to outline the procedure.

A particularly important special case of (1) is that in which $K(p, x) = e^{-px}$ and the limits of integration are 0 and ∞; in this case $\bar{f}(p)$ is said to be the *Laplace* transform of $f(x)$. This transform has been very widely used and is, in fact, particularly suitable for the solution of ordinary differential equations with given initial conditions, but other integral transforms are often equally useful in the solution of the differential equations of mathematical physics.

In what follows we show that a similar technique can be employed whatever the kernel or range of integration of the transform. Once this technique has been mastered, this method of solving differential equations is often found to be more direct and straightforward than the classical methods already described. These methods often require a good deal of experience and judgment whereas the procedure to be followed in using an integral transform can be reduced almost to a "drill".

8.2 The Laplace transform

As already stated, the Laplace transform $\bar{f}(p)$ of a function $f(x)$ is defined by

$$\bar{f}(p) = \int_0^\infty e^{-px} f(x)\,dx, \qquad (1)$$

and it is a comparatively simple matter to draw up a table of the transforms of given functions of x. For example, if $f(x) = 1$,

$$\bar{f}(p) = \int_0^\infty e^{-px}\,dx = \left[-\frac{e^{-px}}{p}\right]_0^\infty = \frac{1}{p},$$

it being assumed here that $p > 0$ in order that the integral may exist. Again, if $f(x) = x^n$, where $n > -1$,

$$\bar{f}(p) = \int_0^\infty x^n e^{-px}\,dx = \frac{\Gamma(n+1)}{p^{n+1}},$$

when use is made of equation (9) of §5.5.

Next consider the Laplace transform of $e^{ax}f(x)$ where a is a constant. This is given by

$$\int_0^\infty e^{ax} f(x)\, e^{-px}\,dx = \int_0^\infty e^{-(p-a)x} f(x)\,dx, \qquad (p > a),$$

and we can conclude that if $\bar{f}(p)$ is the Laplace transform of $f(x)$, then the transform of $e^{ax}f(x)$ is $\bar{f}(p-a)$. Taking $f(x)$ to be unity and x^n in turn we therefore see that the transforms of e^{ax} and $x^n e^{ax}$ are respectively

$$\frac{1}{p-a} \quad \text{and} \quad \frac{\Gamma(n+1)}{(p-a)^{n+1}}.$$

Writing ib in place of a in the first of these results, we have

$$f(x) = e^{ibx}, \qquad \bar{f}(p) = \frac{1}{p-ib} = \frac{p+ib}{p^2+b^2}.$$

Since $e^{ibx} = \cos bx + i \sin bx$, it follows by equating real and imaginary parts that

$$f(x) = \cos bx, \qquad \bar{f}(p) = \frac{p}{p^2+b^2},$$

$$f(x) = \sin bx, \qquad \bar{f}(p) = \frac{b}{p^2+b^2}.$$

The transforms of $e^{ax} \cos bx$ and $e^{ax} \sin bx$ are then given by writing $p-a$ in place of p.

The above results are summarized in the short table below and this table should be comprehensive enough for present purposes. It will be noticed that some of the entries given are particular cases of others: these occur, however, so frequently in practical problems that it seems worth while to record them separately.

Short table of Laplace transforms

$$\bar{f}(p) = \int_0^\infty e^{-px} f(x)\, dx.$$

	$f(x)$	$\bar{f}(p)$	*Remarks*
1	1	$1/p$	
2	x^n	$\Gamma(n+1)/p^{n+1}$	$n > -1$
3	e^{ax}	$1/(p-a)$	a constant
4	$x^n e^{ax}$	$\Gamma(n+1)/(p-a)^{n+1}$	$n > -1$, a constant
5	$\cos bx$	$p/(p^2+b^2)$	b constant
6	$e^{ax} \cos bx$	$(p-a)/\{(p-a)^2+b^2\}$	a, b constant
7	$\sin bx$	$b/(p^2+b^2)$	b constant
8	$e^{ax} \sin bx$	$b/\{(p-a)^2+b^2\}$	a, b constant

We shall require the Laplace transforms of the derivatives of $f(x)$. These can be obtained by integrating by parts as follows.

$$\int_0^\infty e^{-px} \frac{df}{dx}\, dx = \Big[e^{-px} f \Big]_0^\infty + p \int_0^\infty e^{-px} f(x)\, dx = -f(0) + p\bar{f}(p) \qquad (2)$$

where $f(0)$ denotes the value of $f(x)$ when $x = 0$ and we have assumed, as is often the case in physical problems, that $\lim_{x \to \infty} (e^{-px} f) = 0$. Again,

$$\int_0^\infty e^{-px} \frac{d^2 f}{dx^2}\, dx = \left[e^{-px} \frac{df}{dx} \right]_0^\infty + p \int_0^\infty e^{-px} \frac{df}{dx}\, dx$$

$$= -f'(0) + p\{-f(0) + p\bar{f}(p)\} = -f'(0) - pf(0) + p^2 \bar{f}(p), \qquad (3)$$

where $f'(0)$ denotes the value of df/dx when $x = 0$. In obtaining (3) we have used equation (2) and assumed further that $e^{-px}(df/dx)$ tends to zero as x tends to infinity. It can be shown similarly that, provided $e^{-px}(d^{r-1}f/dx^{r-1})$ vanishes as x tends to infinity for $r = 1, 2, 3, \ldots, n$,

$$\int_0^\infty e^{-px} \frac{d^n f}{dx^n}\, dx = -f^{(n-1)}(0) - pf^{(n-2)}(0) - \ldots - p^{n-1} f(0) + p^n \bar{f}(p). \quad (4)$$

Equation (4) permits the Laplace transform of the derivatives of a function to be expressed in terms of the Laplace transform of the function, its initial value and the initial values of its lower derivatives. This result, together with the table of transforms, provides the necessary equipment for the solution of linear differential equations.

8.3 The solution of ordinary differential equations by the Laplace transform

The "drill" to be followed in the solution of an ordinary linear differential equation (in which the independent and dependent variables are respectively taken as x and y) is:

(i) multiply throughout by e^{-px} and integrate with respect to x between the limits 0 and ∞, thus forming the Laplace transform of the equation;

(ii) apply the result (4) using the given initial values of y and its derivatives and use the table of transforms as is necessary: the result of these operations is an algebraic equation (called the *subsidiary equation*) which is then solved to give the transform $\bar{y}$ of the wanted function y in terms of p;

(iii) find y as a function of x corresponding to the transform $\bar{y}$ by using the table inversely: some algebraic manipulation is often required to enable this step to be carried through.

The method is illustrated in detail in the examples given below.

Example 1. *Find the current i at time t in a circuit consisting of an inductance L and a resistance R when a constant electromotive force E is applied at time $t = 0$.*

The current is given by the first order equation

$$L\frac{di}{dt}+Ri = E,$$

and the initial condition is $i = 0$ when $t = 0$. The independent variable here being t, we multiply by e^{-pt} and integrate with respect to t between 0 and ∞. Thus

$$L\int_0^\infty e^{-pt}\frac{di}{dt}\,dt+R\int_0^\infty e^{-pt}\,i\,dt = E\int_0^\infty e^{-pt}\,dt. \tag{1}$$

Denote the Laplace transform of i by $\bar{i}$ so that

$$\bar{i} = \int_0^\infty e^{-pt}\,i\,dt,$$

and, since $i = 0$ when $t = 0$, equation (2) of §8.2 shows that

$$\int_0^\infty e^{-pt}\frac{di}{dt}\,dt = p\bar{i}.$$

The term on the right-hand side of (1) is E times the transform of unity and, using entry 1 of the table, this is E/p. Hence equation (1) can be written

$$Lp\bar{i}+R\bar{i} = E/p,$$

and this is the subsidiary equation giving $\bar{\imath}$ in terms of p; solving for $\bar{\imath}$,

$$\bar{\imath} = \frac{E}{p(Lp+R)}.$$

Before using the table inversely to find i from this value of $\bar{\imath}$, we express $\bar{\imath}$ in partial fractions in the form

$$\bar{\imath} = \frac{E}{R}\left(\frac{1}{p} - \frac{1}{p+R/L}\right).$$

From the first and third entries of the table we find that $1/p$ and $1/(p+R/L)$ are respectively the transforms of unity and $e^{-Rt/L}$, since here we are using t in place of x and $-R/L$ in place of a. Hence the required value of the current i is given by

$$i = \frac{E}{R}\left(1-e^{-Rt/L}\right).$$

Many of the intermediate steps used in the above example can be omitted when the user has gained experience with the method. In the examples which follow, less detail is given and the advantages of the method become more apparent. In particular it should be noticed that whereas in the classical method it was first necessary to find the general solution of the differential equation and then to determine the arbitrary constants in it from the initial conditions, these two distinct steps in the solution are no longer involved.

Example 2. *Solve the equation $y''+y = 0$ given that $y = 1$, $y' = 0$ when $x = 0$.*

The subsidiary equation is

$$-p+p^2\bar{y}+\bar{y} = 0,$$

so that

$$\bar{y} = \frac{p}{p^2+1}.$$

Inverting (using entry 5 of the table), $y = \cos x$.

Example 3. *Solve the equation $y''-5y'+6y = 0$ given that $y = 0$, $y' = 1$ when $x = 0$.*

The subsidiary equation is

$$-1+p^2\bar{y}-5p\bar{y}+6\bar{y} = 0,$$

giving

$$\bar{y} = \frac{1}{p^2-5p+6} = \frac{1}{p-3} - \frac{1}{p-2}.$$

Inverting (by entry 3), $y = e^{3x}-e^{2x}$.

Example 4. *Solve the differential equation $\ddot{x}+m^2x = a\cos nt$, given that $\dot{x} = x = 0$ when $t = 0$ and that a, m, n are constants.*

The subsidiary equation is

$$(p^2+m^2)\bar{x} = \frac{ap}{p^2+n^2},$$

the term on the right-hand side being the transform of $a\cos nt$. Hence

$$\bar{x} = \frac{ap}{(p^2+m^2)(p^2+n^2)} = \frac{a}{m^2-n^2}\left(\frac{p}{p^2+n^2} - \frac{p}{p^2+m^2}\right).$$

Inverting,

$$x = \frac{a}{m^2-n^2}(\cos nt - \cos mt).$$

Example 5. *Solve the equation* $y'''+y = \frac{1}{2}x^2e^x$ *given that* $y = y' = y'' = 0$ *when* $x = 0$.

The subsidiary equation is

$$(p^3+1)\bar{y} = \frac{1}{(p-1)^3},$$

the term on the right coming from entry 4 of the table. Hence

$$\bar{y} = \frac{1}{(p^3+1)(p-1)^3}$$

$$= \frac{1}{2(p-1)^3} - \frac{3}{4(p-1)^2} + \frac{3}{8(p-1)} - \frac{1}{24(p+1)} - \frac{p-2}{3(p^2-p+1)}.$$

The last term on the right can be written in the form

$$\frac{\left(p-\frac{1}{2}\right)-\sqrt{3}\left(\frac{\sqrt{3}}{2}\right)}{3\left\{\left(p-\frac{1}{2}\right)^2+\left(\frac{\sqrt{3}}{2}\right)^2\right\}},$$

and the inversion can now be carried out from the table. This gives

$$y = \frac{1}{4}x^2e^x - \frac{3}{4}xe^x + \frac{3}{8}e^x - \frac{1}{24}e^{-x} - \frac{1}{3}e^{x/2}\cos\frac{\sqrt{3}}{2}x + \frac{\sqrt{3}}{3}e^{x/2}\sin\frac{\sqrt{3}}{2}x$$

$$= \frac{1}{4}\left(x^2-3x+\frac{3}{2}\right)e^x - \frac{1}{24}e^{-x} - \frac{1}{3}e^{x/2}\left(\cos\frac{\sqrt{3}}{2}x - \sqrt{3}\sin\frac{\sqrt{3}}{2}x\right).$$

A similar method can be followed in the case of simultaneous differential equations. We consider a pair of such equations involving dependent variables x, y and independent variable t. The Laplace transform of each equation is formed in the usual way and this yields a pair of simultaneous algebraic equations which can be solved to give $\bar{x}$ and $\bar{y}$ as functions of p. x and y can then be found as functions of t by using the table inversely. The method can clearly be extended when three or more differential equations are involved.

Example 6. *Solve the simultaneous differential equations*

$$5\dot{x}-2\dot{y}+4x-y = e^{-t}, \qquad \dot{x}+8x-3y = 5e^{-t},$$

given that $x = y = 0$ *when* $t = 0$.

The subsidiary equations are

$$(5p+4)\bar{x}-(2p+1)\bar{y} = \frac{1}{p+1},$$

$$(p+8)\bar{x}-3\bar{y} = \frac{5}{p+1}.$$

Solving these simultaneous algebraic equations for $\bar{x}$, $\bar{y}$ we find

$$\bar{x} = \frac{5p+1}{(p+1)(p^2+p-2)}, \qquad \bar{y} = \frac{12p+6}{(p+1)(p^2+p-2)},$$

and expressing these results in partial fractions,

$$\bar{x} = \frac{2}{p+1}+\frac{1}{p-1}-\frac{3}{p+2}, \qquad \bar{y} = \frac{3}{p+1}+\frac{3}{p-1}-\frac{6}{p+2}.$$

Inversion from the table of transforms then gives

$$x = 2e^{-t}+e^t-3e^{-2t}, \qquad y = 3e^{-t}+3e^t-6e^{-2t}.$$

Exercises 8(*a*)

Use the method of the Laplace transform to solve the following differential equations under the conditions given:

1. $y'+y = 0$ with $y = 1$ when $x = 0$.
2. $y'-2y = 1-2x$ with $y = 3$ when $x = 0$.
3. $\dot{x}-ax = e^{at}$, a constant, with $x = 0$ when $t = 0$.
4. $\dot{x}-2x = 4\sin 2t$ with $x = 0$ when $t = 0$.
5. $y'-y = 4e^x\cos x$ with $y = 3$ when $x = 0$.
6. $\ddot{y}+8\dot{y}+15y = 0$ with $\dot{y} = 3$, $y = 1$ when $t = 0$.
7. $y''+10y'+25y = 0$ with $y' = y = 1$ when $x = 0$.
8. $\ddot{y}+2\dot{y}+10y = 0$ with $\dot{y} = 3$, $y = 0$ when $t = 0$.
9. $\ddot{x}+2k\dot{x}+(k^2+m^2)x = 0$, ($k$, m constants) with $\dot{x} = 0$, $x = 1$ when $t = 0$.
10. $y''-5y'+6y = e^{2x}$ with $y' = 0$, $y = 1$ when $x = 0$.
11. $\ddot{\theta}+4\dot{\theta}+8\theta = 1$ with $\dot{\theta} = 1$, $\theta = 0$ when $t = 0$.
12. $4\ddot{y}+4\dot{y}+37y = 99\cos t-12\sin t$, with $\dot{y} = y = 1$ when $t = 0$. (C.U.)
13. $y''+4y'+13y = 5\cos 3x$ with $y' = 2$, $y = \frac{1}{4}$ when $x = 0$. (C.U.)
14. $y'''+y = 1$ with $y'' = y' = y = 0$ when $x = 0$.
15. $y^{iv}+4y'''+4y'' = 0$ with $y''' = y'' = y' = 0$, $y = 1$ when $x = 0$.
16. $y'''-2y''-y'+2y = 4x^2$ with $y'' = 8$, $y' = 4$, $y = 6$ when $x = 0$.
17. $\left.\begin{aligned}\dot{x}+2x+y &= 0\\ \dot{y}+x+2y &= 0\end{aligned}\right\}$, with $x = 1$, $y = 0$ when $t = 0$. (L.U.)
18. $\left.\begin{aligned}\dot{x}-x+2y &= 0\\ \dot{y}-5x-3y &= 0\end{aligned}\right\}$, with $x = y = 1$ when $t = 0$. (L.U.)
19. $\left.\begin{aligned}3\dot{x}+2x-y &= t\\ 2\dot{y}-x+y &= 5e^{-t}\end{aligned}\right\}$, with $x = y = 0$ when $t = 0$.
20. $\left.\begin{aligned}6\ddot{x}-3\dot{x}+6x+\dot{y} &= t\\ \ddot{x}+8x+y &= 0\end{aligned}\right\}$, with $x = \frac{11}{36}$, $y = -\frac{22}{9}$, $\dot{x} = \frac{1}{6}$ when $t = 0$. (L.U.)
21. Show that $(p^2+b^2)^{-2}$ is the Laplace transform of
$$\frac{1}{2b^3}(\sin bx-bx\cos bx).$$
Hence solve the differential equation $y''+y = x\cos 2x$ given that y and y' are both zero when $x = 0$.
22. The output θ_0 at time t from a control mechanism corresponding to an input θ_i is given by the differential equation
$$\frac{d^2\theta_0}{dt^2}+2\zeta\omega\frac{d\theta_0}{dt} = \omega^2(\theta_i-\theta_0),$$

where ζ is the damping ratio and ω is the undamped natural frequency. If $\zeta = 1$, $\theta_i = \Omega t$ (Ω constant) and $\dot{\theta}_0 = \theta_0 = 0$ when $t = 0$, use the method of the Laplace transform to show that

$$\theta_0 = \Omega\left\{t-\frac{2}{\omega}+\left(t+\frac{2}{\omega}\right)e^{-\omega t}\right\}.$$

23. A condenser of capacity C charged to voltage E is discharged through a circuit containing an inductance L and a resistance R. If

$$\mu = \frac{R}{2L} \quad \text{and} \quad n^2 = \frac{1}{LC}-\frac{R^2}{4L^2} > 0,$$

use the method of the Laplace transform to show that the charge q at time t on the condenser is given by

$$q = CEe^{-\mu t}\left(\cos nt+\frac{\mu}{n}\sin nt\right).$$

24. Two flywheels of moments of inertia I, $2I$ are connected by a light shaft of length l and torsional rigidity μ and the whole system is rotating at constant angular velocity ω. At time $t = 0$, a constant retarding couple P is applied to the wheel with the smaller moment of inertia. Use the Laplace transform to show that the angular velocity of the other wheel at time t is

$$\omega-\frac{Pt}{3I}+\frac{P}{3\lambda I}\sin \lambda t,$$

where $2\lambda^2 = 3\mu/lI$.

8.4 Fourier's integral formula

Integral transforms can be used in the solution of many of the partial differential equations discussed in the last chapter. Here we shall consider only transforms in which the kernel is $\sin px$, $\cos px$ or $J_n(px)$ and, in these cases, it is possible to give simple formulae, called *inversion formulae*, which enable a function to be found when its transform is known. For this purpose Fourier's integral formula is required and a formal derivation of this is given below.

If a function $f(x)$, of period $2\pi\lambda$, is expressed by its Fourier series

$$f(x) = \frac{1}{2}a_0+\sum_{n=1}^{\infty}\left\{a_n \cos\frac{nx}{\lambda}+b_n \sin\frac{nx}{\lambda}\right\}, \tag{1}$$

then the coefficients a_n, b_n are given by the relations

$$\pi\lambda a_n = \int_{-\lambda\pi}^{\lambda\pi} f(u)\cos\frac{nu}{\lambda}\,du, \qquad (n = 0, 1, 2, 3, \ldots),$$

$$\pi\lambda b_n = \int_{-\lambda\pi}^{\lambda\pi} f(u)\sin\frac{nu}{\lambda}\,du, \qquad (n = 1, 2, 3, \ldots).$$

Substituting for a_n, b_n in equation (1), we have

$$f(x) = \frac{1}{2\pi\lambda}\int_{-\lambda\pi}^{\lambda\pi} f(u)\,du + \frac{1}{\pi\lambda}\sum_{n=1}^{\infty}\int_{-\lambda\pi}^{\lambda\pi} f(u)\cos\frac{n(x-u)}{\lambda}\,du.$$

Writing $n/\lambda = p$, $1/\lambda = \delta p$ and letting λ tend to infinity, the sum passes formally into an integral and thus

$$\pi f(x) = \int_0^{\infty} dp \int_{-\infty}^{\infty} f(u)\cos p(x-u)\,du, \tag{2}$$

which is *Fourier's integral formula.* It is emphasized that the above derivation is purely formal and, for a rigorous discussion together with a precise statement of the conditions under which (2) holds, reference should be made to a standard text.*

8.5 Fourier transforms and their inversion formulae

Fourier's integral formula can be written in the form

$$\pi f(x) = \int_0^{\infty}\left\{\int_{-\infty}^{\infty} f(u)\cos pu\,du\right\}\cos xp\,dp$$
$$+\int_0^{\infty}\left\{\int_{-\infty}^{\infty} f(u)\sin pu\,du\right\}\sin xp\,dp. \tag{1}$$

If $f(x)$ is an odd function of x,

$$\int_{-\infty}^{\infty} f(u)\cos pu\,du = 0, \qquad \int_{-\infty}^{\infty} f(u)\sin pu\,du = 2\int_0^{\infty} f(u)\sin pu\,du,$$

and equation (1) becomes

$$f(x) = \frac{2}{\pi}\int_0^{\infty}\left\{\int_0^{\infty} f(u)\sin pu\,du\right\}\sin xp\,dp. \tag{2}$$

If therefore we define the *Fourier sine transform* $\bar{f}_s(p)$ by

$$\bar{f}_s(p) = \int_0^{\infty} f(x)\sin px\,dx, \tag{3}$$

equation (2) then gives, as the required *inversion formula,*

$$f(x) = \frac{2}{\pi}\int_0^{\infty} \bar{f}_s(p)\sin xp\,dp, \tag{4}$$

since the variable x in equation (3) can be replaced by u.

* For example, E. C. Titchmarsh, *Theory of Fourier Integrals*, Oxford (1937), §1.9.

Similarly, if $f(x)$ is an even function of x and its *Fourier cosine transform* $\bar{f}_c(p)$ is defined by

$$\bar{f}_c(p) = \int_0^\infty f(x) \cos px \, dx, \tag{5}$$

the *inversion formula* is

$$f(x) = \frac{2}{\pi}\int_0^\infty \bar{f}_c(p) \cos xp \, dp. \tag{6}$$

These inversion formulae will be of great use later in the solution of partial differential equations. They are also of use in evaluating certain definite integrals involving trigonometrical functions and an example is given below.

Example 7. *Find the Fourier cosine transform of the function $f(x)$ defined by $f(x) = 1$ when $x^2 < a^2$, $f(x) = 0$ when $x^2 > a^2$. Deduce the value of the definite integral*

$$\int_0^\infty \frac{\sin ap \cos xp}{p} \, dp.$$

Equation (5) can be written

$$\bar{f}_c(p) = \int_0^a f(x) \cos px \, dx + \int_a^\infty f(x) \cos px \, dx,$$

and here $f(x) = 1$ in the first integral and $f(x) = 0$ in the second. Hence

$$\bar{f}_c(p) = \int_0^a \cos px \, dx = \left[\frac{\sin px}{p}\right]_0^a = \frac{\sin ap}{p},$$

giving the required cosine transform. Inserting this result in (6),

$$\frac{2}{\pi}\int_0^\infty \frac{\sin ap \cos xp}{p} \, dp = f(x) = \begin{rcases} 1, & (x^2 < a^2) \\ 0, & (x^2 > a^2) \end{rcases},$$

and the value of the integral follows immediately.

8.6 The Hankel transform and its inversion formula

Fourier's integral formula (2) of §8.4 can be written in the form

$$2\pi f(x) = \int_{-\infty}^\infty dp \int_{-\infty}^\infty f(u) \cos p(x-u) \, du,$$

and it is clear that

$$\int_{-\infty}^\infty dp \int_{-\infty}^\infty f(u) \sin p(x-u) \, du = 0.$$

Multiplying the second of these results by i and adding to the first,

$$2\pi f(x) = \int_{-\infty}^\infty e^{ipx} \, dp \int_{-\infty}^\infty f(u) \, e^{-ipu} \, du. \tag{1}$$

Writing

$$(p) = \int_{-\infty}^{\infty} f(u)\, e^{ipu}\, du = \int_{-\infty}^{\infty} f(x)\, e^{ipx}\, dx, \tag{2}$$

(for the variable u can be replaced by x), equation (1) can be written

$$2\pi f(x) = \int_{-\infty}^{\infty} \bar{f}(p)\, e^{-ixp}\, dp. \tag{3}$$

Equations (2) and (3) can be modified to cover functions of two variables. Thus if

$$\bar{f}(s,t) = \int_{-\infty}^{\infty}\int_{-\infty}^{\infty} f(x,y) \exp\{i(sx+ty)\}\, dx\, dy, \tag{4}$$

then

$$4\pi^2 f(x,y) = \int_{-\infty}^{\infty}\int_{-\infty}^{\infty} \bar{f}(s,t) \exp\{-i(xs+yt)\}\, ds\, dt. \tag{5}$$

The result expressed by equations (4) and (5) can be used to obtain an inversion formula for a transform, known as the *Hankel transform*, in which the kernel is a Bessel function. Writing

$$x = \rho\cos\theta, \qquad y = \rho\sin\theta, \qquad s = p\cos\alpha, \qquad t = p\sin\alpha,$$

equations (4) and (5) become

$$\bar{f}(p,\alpha) = \int_0^{\infty} \rho\, d\rho \int_0^{2\pi} f(\rho,\theta) \exp\{ip\rho\cos(\theta-\alpha)\}\, d\theta, \tag{6}$$

$$4\pi^2 f(\rho,\theta) = \int_0^{\infty} p\, dp \int_0^{2\pi} \bar{f}(p,\alpha) \exp\{-ip\rho\cos(\theta-\alpha)\}\, d\alpha. \tag{7}$$

We now take for $f(\rho, \theta)$ the special function $f(\rho)e^{-in\theta}$ and equation (6) gives

$$\bar{f}(p,\alpha) = \int_0^{\infty} \rho f(\rho)\, d\rho \int_0^{2\pi} \exp[i\{-n\theta + p\rho\cos(\theta-\alpha)\}]\, d\theta. \tag{8}$$

Writing $\phi = \alpha - \theta - \frac{1}{2}\pi$, the integral in θ can be written

$$\exp\{in(\tfrac{1}{2}\pi-\alpha)\} \int_0^{2\pi} \exp\{i(n\phi - p\rho\sin\phi)\}\, d\phi$$

and the integral in ϕ is Bessel's integral. By equation (10) of §6.9, its value is $2\pi J_n(p\rho)$, and hence equation (8) can be written

$$\begin{aligned} \bar{f}(p,\alpha) &= 2\pi \exp\{in(\tfrac{1}{2}\pi-\alpha)\} \int_0^{\infty} \rho f(\rho) J_n(p\rho)\, d\rho \\ &= 2\pi \exp\{in(\tfrac{1}{2}\pi-\alpha)\} \bar{f}_H(p), \end{aligned} \tag{9}$$

where

$$\bar{f}_H(p) = \int_0^\infty \rho f(\rho) J_n(p\rho)\, d\rho. \tag{10}$$

Writing $f(\rho, \theta) = f(\rho)e^{-in\theta}$ and substituting from (9) in (7), we obtain after division by 2π,

$$2\pi f(\rho)\, e^{-in\theta} = \int_0^\infty p\bar{f}_H(p)\, dp \int_0^{2\pi} \exp[i\{n(\tfrac{1}{2}\pi - \alpha) - p\rho\cos(\theta - \alpha)\}]\, d\alpha. \tag{11}$$

By writing $\phi = \theta - \alpha + \frac{1}{2}\pi$ and again using Bessel's integral, the integral in α can be expressed as $2\pi e^{-in\theta} J_n(\rho p)$, and hence equation (11) gives

$$f(\rho) = \int_0^\infty p\bar{f}_H(p) J_n(\rho p)\, dp. \tag{12}$$

The *Hankel transform* $\bar{f}_H(p)$ (*of order n*) of a function $f(\rho)$ is defined by equation (10), and equation (11) provides an *inversion formula* giving $f(\rho)$ when $\bar{f}_H(p)$ is known. As a simple example, take $f(\rho) = 1$ when $0 < \rho < a$ and $f(\rho) = 0$ when $\rho > a$. Then its Hankel transform of order zero is given by equation (10) as

$$\begin{aligned} \bar{f}_H(p) &= \int_0^\infty \rho f(\rho) J_0(p\rho)\, d\rho = \int_0^a \rho J_0(p\rho)\, d\rho \\ &= \frac{1}{p^2}\int_0^{ap} x J_0(x)\, dx, \qquad \text{putting } x = p\rho, \\ &= \frac{a}{p} J_1(ap), \end{aligned}$$

using equation (12) of §6.9. The inversion formula (12), besides being of use later in the solution of partial differential equations, can again be used to evaluate certain definite integrals. Thus, in the above example

$$f(\rho) = 1, \quad (0 < \rho < a); \quad f(\rho) = 0, \quad (\rho > a); \quad \bar{f}_H(p) = (a/p)J_1(ap);$$

hence equation (12) gives

$$\int_0^\infty J_1(ap) J_0(\rho p)\, dp = \left.\begin{matrix} a^{-1}, & (0 < \rho < a), \\ 0, & (\rho > a). \end{matrix}\right\}$$

8.7 Fourier and Hankel transforms of derivatives

As is the case when the Laplace transform is applied to the solution of ordinary differential equations, the application of Fourier and Hankel transforms to the solution of partial differential equations requires formulae for the transforms of certain derivatives. In physical applications it is usually possible to assume a certain behaviour of the functions involved for specific values (usually 0 and ∞) of the variable, and the formulae given below are obtained with this type of assumption.

The sine transform of the second derivative of a function of x, which in the work in view will also be a function of one or more other variables as well, is given by

$$\int_0^\infty \frac{\partial^2 f}{\partial x^2} \sin px\, dx = \left[\frac{\partial f}{\partial x} \sin px\right]_0^\infty - p\int_0^\infty \frac{\partial f}{\partial x} \cos px\, dx.$$

The first term on the right vanishes at the upper limit if we assume $(\partial f/\partial x)$ vanishes as $x \to \infty$, and it vanishes at the lower limit through the term $\sin px$, and a second integration by parts gives

$$\int_0^\infty \frac{\partial^2 f}{\partial x^2} \sin px\, dx = -p\left[f \cos px\right]_0^\infty - p^2\int_0^\infty f \sin px\, dx.$$

The first term on the right again vanishes at the upper limit if we assume f vanishes as $x \to \infty$ and hence

$$\int_0^\infty \frac{\partial^2 f}{\partial x^2} \sin px\, dx = pf(0) - p^2 \bar{f}_s(p), \tag{1}$$

where $f(0)$ is the value of f when $x = 0$ and $\bar{f}_s(p)$ is the sine transform of f. In a similar way, the cosine transform of $(\partial^2 f/\partial x^2)$ is given by

$$\int_0^\infty \frac{\partial^2 f}{\partial x^2} \cos px\, dx = -f'(0) - p^2 \bar{f}_c(p), \tag{2}$$

where $f'(0)$ is the value of $(\partial f/\partial x)$ when $x = 0$.

To obtain a similar result for the Hankel transform, we have

$$\int_0^\infty \rho\left(\frac{\partial^2 f}{\partial \rho^2} + \frac{1}{\rho}\frac{\partial f}{\partial \rho}\right) J_n(p\rho)\, d\rho = \int_0^\infty \frac{\partial}{\partial \rho}\left(\rho \frac{\partial f}{\partial \rho}\right) J_n(p\rho)\, d\rho$$

$$= \left[\rho \frac{\partial f}{\partial \rho} J_n(p\rho)\right]_0^\infty - p\int_0^\infty \rho \frac{\partial f}{\partial \rho} J_n'(p\rho)\, d\rho.$$

The first term on the right vanishes if we assume that $\rho(\partial f/\partial \rho)$ vanishes as ρ tends to zero and to infinity and a second integration by parts then gives

$$\int_0^\infty \rho\left(\frac{\partial^2 f}{\partial \rho^2} + \frac{1}{\rho}\frac{\partial f}{\partial \rho}\right) J_n(p\rho)\, d\rho = -p\left[\rho f J_n'(p\rho)\right]_0^\infty$$

$$+ p\int_0^\infty f\{p\rho J_n''(p\rho) + J_n'(p\rho)\}\, d\rho.$$

The first term on the right now vanishes if we assume that ρf vanishes as ρ tends to zero and to infinity. With this assumption and noting that

$$p\rho J_n''(p\rho)+J_n'(p\rho) = \left(\frac{n^2}{p\rho}-p\rho\right)J_n(p\rho),$$

(Bessel's equation (1) of §6.5 with $\nu = n$, $x = p\rho$) we obtain, after a slight rearrangement,

$$\int_0^\infty \rho\left(\frac{\partial^2 f}{\partial\rho^2}+\frac{1}{\rho}\frac{\partial f}{\partial\rho}-\frac{n^2}{\rho^2}f\right)J_n(p\rho)\,d\rho = -p^2\int_0^\infty \rho f J_n(p\rho)\,d\rho = -p^2\bar{f}_H(p) \quad (3)$$

where $\bar{f}_H(p)$ is the Hankel transform of order n of $f(\rho)$.

8.8 The solution of partial differential equations by integral transforms

The results established in §§8.6, 8.7 which are required in the application of integral transforms to the solution of some of the partial differential equations considered in the last chapter can be summarized as follows.

(a) Fourier sine transform

$$\bar{f}_s(p) = \int_0^\infty f(x)\sin px\,dx, \quad (1)$$

$$f(x) = \frac{2}{\pi}\int_0^\infty \bar{f}_s(p)\sin xp\,dp, \quad (2)$$

$$\int_0^\infty \frac{\partial^2 f}{\partial x^2}\sin px\,dx = pf(0)-p^2\bar{f}_s(p), \quad (3)$$

where $f(0)$ is the value of f when $x = 0$.

(b) Fourier cosine transform

$$\bar{f}_c(p) = \int_0^\infty f(x)\cos px\,dx, \quad (4)$$

$$f(x) = \frac{2}{\pi}\int_0^\infty \bar{f}_c(p)\cos xp\,dp, \quad (5)$$

$$\int_0^\infty \frac{\partial^2 f}{\partial x^2}\cos px\,dx = -f'(0)-p^2\bar{f}_c(p), \quad (6)$$

where $f'(0)$ is the value of $\partial f/\partial x$ when $x = 0$.

(*c*) *Hankel transform*

$$\bar{f}_H(p) = \int_0^\infty \rho f(\rho) J_n(p\rho)\, d\rho, \tag{7}$$

$$f(\rho) = \int_0^\infty p\bar{f}_H(p) J_n(\rho p)\, dp, \tag{8}$$

$$\int_0^\infty \rho\left(\frac{\partial^2 f}{\partial \rho^2} + \frac{1}{\rho}\frac{\partial f}{\partial \rho} - \frac{n^2}{\rho^2} f\right) J_n(p\rho)\, d\rho = -p^2 \bar{f}_H(p). \tag{9}$$

The method of using these transforms in the solution of differential equations is similar to that of the Laplace transform and is briefly as follows. By forming the transform of a given partial differential equation in n independent variables, the equation is reduced to one in $(n-1)$ variables. In other words, a transform is employed to *exclude* an independent variable from a differential equation. Repeated use of transforms can therefore ultimately reduce the partial equation to an *ordinary* differential equation with one independent variable—it can, in fact, reduce it to an algebraic equation, but this is seldom worth while as it is usually fairly simple to write down the solution of the ordinary differential equation obtained when all but one of the variables have been excluded. Once the solution of the ordinary differential equation has been found, the solution of the original partial equation is given by the use of the appropriate inversion formulae.

Some guidance on the suitability of a particular transform for the above purpose may be helpful. The transforms considered in this section should only be used when the range of the variable selected for exclusion is from 0 to ∞. When a finite range is involved, use should be made of the finite transforms considered in §8.9. A term $(\partial^2 f/\partial x^2)$ or, of course, $(\partial^2 f/\partial y^2)$, $(\partial^2 f/\partial z^2)$ with appropriate change of letters, can be removed from a differential equation by a sine or cosine transform. A glance at equations (3) and (6) then shows that the sine transform should be used when the value of f at $x = 0$ is known and the cosine transform when $(\partial f/\partial x)$ is given at $x = 0$. A group of terms like

$$\frac{\partial^2 f}{\partial \rho^2} + \frac{1}{\rho}\frac{\partial f}{\partial \rho} - \frac{n^2}{\rho^2} f$$

can be excluded by the use of the Hankel transform of order n and a particularly important special case arises when $n = 0$, in which case the terms excluded are

$$\frac{\partial^2 f}{\partial \rho^2} + \frac{1}{\rho}\frac{\partial f}{\partial \rho}.$$

The following worked examples should make clear the details of the process.

Example 8. *A semi-infinite transmission line $x > 0$ with negligible inductance and leakance is initially at zero potential and at time $t = 0$ a constant potential V_0 is applied at the end $x = 0$. Find an expression for the potential at time t at a point in the line at distance x from the end at which the potential is applied.*

The potential V is given by equation (18) of §7.3 with $L = G = 0$, so that

$$\frac{\partial^2 V}{\partial x^2} = RC\frac{\partial V}{\partial t}, \tag{10}$$

where R and C are respectively the resistance and capacitance per unit length of the line. The initial and boundary conditions are

$$V = 0 \quad \text{when } t = 0, \tag{11}$$

and

$$V = V_0 \quad \text{when } x = 0. \tag{12}$$

Here the term $(\partial^2 V/\partial x^2)$ is excluded from equation (10) by using a sine transform, this being appropriate since the value of V when $x = 0$ is known. Multiplying equations (10) and (11) by $\sin px$ and integrating with respect to x between 0 and ∞,

$$\int_0^\infty \frac{\partial^2 V}{\partial x^2} \sin px\, dx = RC\int_0^\infty \frac{\partial V}{\partial t} \sin px\, dx,$$

$$\int_0^\infty V \sin px\, dx = 0 \quad \text{when } t = 0.$$

Writing $\bar{V}_s = \int_0^\infty V \sin px\, dx$ and using equation (3), these two equations can be written, since $V = V_0$ when $x = 0$,

$$pV_0 - p^2\bar{V}_s = RC\frac{d\bar{V}_s}{dt}, \tag{13}$$

$$\bar{V}_s = 0 \quad \text{when } t = 0. \tag{14}$$

In deriving equation (13) it has been assumed that

$$\int_0^\infty \frac{\partial V}{\partial t} \sin px\, dx = \frac{\partial}{\partial t}\int_0^\infty V \sin px\, dx,$$

and similar assumptions are made in subsequent examples. For the functions appearing in physical applications, such a procedure is usually justified. The solution of the ordinary differential equation (13) is easily found to be

$$\bar{V}_s = A \exp(-p^2 t/RC) + V_0/p$$

where A is an arbitrary constant. Substitution in (14) shows that $A = -V_0/p$, and hence

$$\bar{V}_s = \frac{V_0}{p}\{1 - \exp(-p^2 t/RC)\}.$$

The required value of V is now given by the inversion formula (2) for the sine transform; it is

$$V = \frac{2V_0}{\pi}\int_0^\infty \{1 - \exp(-p^2 t/RC)\}\frac{\sin xp}{p}\, dp.$$

Since $\int_0^\infty p^{-1} \sin xp\, dp = \frac{1}{2}\pi\ (x > 0)$, this can be written

$$V = V_0 - \frac{2V_0}{\pi}\int_0^\infty \exp(-p^2 t/RC)\frac{\sin xp}{p}\, dp.$$

Example 9. *The magnetic potential Ω in the field of a circular disc of radius a and strength ω, magnetized parallel to its axis, satisfies Laplace's equation, is equal to $2\pi\omega$ on the disc itself and vanishes at exterior points in the plane of the disc. Using cylindrical polar coordinates (ρ, ϕ, z) and taking the disc as $\rho < a, z = 0$, show that at a point for which $z > 0$*

$$\Omega = 2\pi a\omega \int_0^\infty e^{-pz} J_1(ap) J_0(\rho p)\, dp.$$

From symmetry, the potential Ω is clearly independent of the coordinate ϕ and Laplace's equation in cylindrical polar coordinates therefore takes the form

$$\frac{\partial^2 \Omega}{\partial \rho^2} + \frac{1}{\rho}\frac{\partial \Omega}{\partial \rho} + \frac{\partial^2 \Omega}{\partial z^2} = 0, \tag{15}$$

the boundary conditions being

$$\Omega = \begin{rcases} 2\pi\omega, & (\rho < a) \\ 0, & (\rho > a) \end{rcases}, \quad \text{when } z = 0, \tag{16}$$

and

$$\Omega \to 0 \quad \text{as} \quad z \to \infty. \tag{17}$$

The first two terms can be removed from equation (15) by using the Hankel transform of order zero given by

$$\overline{\Omega}_H = \int_0^\infty \rho \Omega J_0(p\rho)\, d\rho. \tag{18}$$

Hence, multiplying equations (15), (16) and (17) by $\rho J_0(p\rho)$ and integrating with respect to ρ between 0 and ∞, we obtain with the help of (9),

$$-p^2 \overline{\Omega}_H + \frac{d^2 \overline{\Omega}_H}{dz^2} = 0, \tag{19}$$

with

$$\overline{\Omega}_H = \int_0^a 2\pi\omega\rho J_0(p\rho)\, d\rho = \frac{2\pi\omega a}{p} J_1(ap) \quad \text{when } z = 0, \tag{20}$$

$$\overline{\Omega}_H \to 0 \quad \text{as} \quad z \to \infty. \tag{21}$$

The solution of the ordinary differential equation (19) is

$$\overline{\Omega}_H = A\, e^{pz} + B\, e^{-pz},$$

and use of equations (20), (21) gives the values of the arbitrary constants A, B as

$$A = 0, \quad B = \frac{2\pi\omega a}{p} J_1(ap).$$

Hence

$$\overline{\Omega}_H = \frac{2\pi\omega a}{p} J_1(ap)\, e^{-pz},$$

and the inversion formula (9) with $n = 0$ immediately leads to

$$\Omega = 2\pi\omega a \int_0^\infty e^{-pz} J_1(ap) J_0(\rho p)\, dp. \tag{22}$$

Example 10. *The material of the semi-infinite solid $z > 0$ is of diffusivity k and the initial temperature of the solid is zero. From time $t = 0$, the surface $z = 0$ is kept at unit temperature over the strip $|x| < a$ and at zero temperature outside the strip. Show that the temperature V at time t at the point (x, y, z) is given by*

$$V = \frac{4}{\pi^2} \int_0^\infty \int_0^\infty [1 - \exp\{-(p^2+q^2)kt\}] \frac{q}{p(p^2+q^2)} \sin zq \sin ap \cos xp\, dq\, dp.$$

The temperature V is clearly independent of the coordinate y and hence satisfies the differential equation

$$\frac{\partial^2 V}{\partial x^2}+\frac{\partial^2 V}{\partial z^2}=\frac{1}{k}\frac{\partial V}{\partial t}, \tag{23}$$

with the initial and boundary conditions

$$V=0 \quad \text{when } t=0, \tag{24}$$

$$V=\left.\begin{matrix}1, & |x|<a\\ 0, & |x|>a\end{matrix}\right\} \quad \text{when } z=0. \tag{25}$$

The symmetry of the problem shows that it is sufficient to consider only points for which $0<x<\infty$ together with the condition

$$\frac{\partial V}{\partial x}=0 \quad \text{when } x=0. \tag{26}$$

The partial differential equation (23) contains three independent variables x, z, t and two transforms used successively are now necessary to reduce it to an ordinary differential equation. Since the value of $(\partial V/\partial x)$ at $x=0$ is given by (26), the cosine transform

$$\bar{V}_c=\int_0^\infty V\cos px\,dx$$

is suitable for excluding the term $(\partial^2 V/\partial x^2)$. This transform, used in the usual way, gives

$$-p^2\bar{V}_c+\frac{\partial^2 \bar{V}_c}{\partial z^2}=\frac{1}{k}\frac{\partial \bar{V}_c}{\partial t}, \tag{27}$$

$$\bar{V}_c=0 \quad \text{when } t=0, \tag{28}$$

$$\bar{V}_c=\int_0^a \cos px\,dx=\frac{\sin pa}{p} \quad \text{when } z=0. \tag{29}$$

To remove the term $(\partial^2\bar{V}_c/\partial z^2)$, a sine transform is suitable since, by equation (29), the value of $\bar{V}_c$ is given when $z=0$. Hence writing

$$\bar{V}_s'=\int_0^\infty \bar{V}_c\sin qz\,dz, \tag{30}$$

equations (27), (28) become, when use is made of (29),

$$-p^2\bar{V}_s'+\frac{q\sin pa}{p}-q^2\bar{V}_s'=\frac{1}{k}\frac{d\bar{V}_s'}{dt}, \tag{31}$$

with

$$\bar{V}_s'=0 \quad \text{when } t=0. \tag{32}$$

The solution of the ordinary differential equation (31) is

$$\bar{V}_s'=A\exp\{-(p^2+q^2)kt\}+\frac{q\sin pa}{p(p^2+q^2)}.$$

The arbitrary constant A is given by equation (32) as

$$A=-\frac{q\sin pa}{p(p^2+q^2)},$$

and hence

$$\bar{V}_s'=\frac{q\sin pa}{p(p^2+q^2)}[1-\exp\{-(p^2+q^2)kt\}].$$

The expression for $\bar{V}_c$ is now given by the inversion formula appropriate to the sine transform introduced in equation (30), so that

$$\bar{V}_c = \frac{2}{\pi}\frac{\sin pa}{p}\int_0^\infty [1-\exp\{-(p^2+q^2)kt\}]\frac{q}{p^2+q^2}\sin zq\,dq.$$

The final expression for V is given by the inversion formula for the cosine transform: hence

$$V = \frac{4}{\pi^2}\int_0^\infty\int_0^\infty [1-\exp\{-(p^2+q^2)kt\}]\frac{q}{p(p^2+q^2)}\sin zq \sin ap \cos xp\,dq\,dp.$$

Exercises 8(*b*)

1. The material of the semi-infinite solid $z > 0$ is of diffusivity k, the initial temperature of the solid is zero and, from time $t = 0$, a constant flow of heat is supplied to the plane surface. If V is the temperature in the solid at time t at depth z below the surface, the boundary condition is $(\partial V/\partial z) = -C$ (constant) when $z = 0$. Show that the temperature at the surface at time t is given by

$$\frac{2C}{\pi}\int_0^\infty \frac{(1-e^{-ktp^2})}{p^2}\,dp.$$

2. Use the sine transform to show that the solution of the partial differential equation

$$\frac{\partial^2 V}{\partial x^2} = \frac{\partial V}{\partial t}, \qquad (x > 0, \qquad t > 0),$$

with the conditions $V = \cos t$ when $x = 0$, $V = 0$ when $t = 0$ is given by

$$V = \frac{2}{\pi}\int_0^\infty (\sin t + p^2\cos t - p^2 e^{-tp^2})\frac{p \sin xp}{p^4+1}\,dp.$$

3. The displacement y at time t of a point at distance x from a fixed end $x = 0$ of a semi-infinite string satisfies the wave equation

$$\frac{\partial^2 y}{\partial x^2} = \frac{1}{c^2}\frac{\partial^2 y}{\partial t^2}.$$

The initial shape of the string is given by $y = f(x)$ and it is initially at rest. Use the sine transform to show that

$$y = \frac{1}{(2\pi)^{1/2}}\int_0^\infty \bar{f}_s(p)\{\sin (x+ct)p + \sin (x-ct)p\}\,dp,$$

where $\bar{f}_s(p)$ is the sine transform of $f(x)$. Deduce that

$$2y = f(x+ct)+f(x-ct).$$

4. Show that the solution of the two-dimensional form of Laplace's equation for V inside the semi-infinite strip $x > 0$, $0 < y < b$ such that $V = f(x)$ when $y = 0$, $V = 0$ when $y = b$ and $V = 0$ when $x = 0$, is given by

$$V = \frac{2}{\pi}\int_0^\infty f(u)\,du\int_0^\infty \frac{\sinh (b-y)p}{\sinh bp}\sin xp \sin up\,dp.$$

5. Use the cosine transform to show that the steady temperature in the semi-infinite solid $z > 0$ when the temperature at the surface $z = 0$ is kept at unity over the strip $|x| < a$ and at zero outside this strip, is

$$\frac{1}{\pi}\left\{\tan^{-1}\left(\frac{a+x}{z}\right)+\tan^{-1}\left(\frac{a-x}{z}\right)\right\}.$$

[The result $\int_0^\infty e^{-sp}p^{-1}\sin rp\,dp = \tan^{-1}(r/s),\ r > 0,\ s > 0,$ may be assumed].

6. Use cylindrical polar coordinates and a sine transform to show that the steady temperature in the semi-infinite annulus bounded by the plane $z = 0$ and the cylindrical surfaces $\rho = a$, $\rho = b$ $(a > b)$, when the surfaces $z = 0$, $\rho = a$ are maintained at zero and the surface $\rho = b$ is kept at temperature $f(z)$ is

$$\frac{2}{\pi}\int_0^\infty\int_0^\infty F(a, b, \rho, p)f(u)\sin zp\sin up\,dp\,du,$$

where

$$F(a, b, \rho, p) = \frac{I_0(p\rho)K_0(pa)-I_0(pa)K_0(p\rho)}{I_0(pb)K_0(pa)-I_0(pa)K_0(pb)}.$$

7. Show that the solution of the differential equation

$$\frac{\partial\theta}{\partial t} = a\frac{\partial^2\theta}{\partial x^2}-b\theta, \qquad (x > 0,\quad t > 0),$$

(in which a and b are constants) subject to the conditions

$$\theta = 0 \text{ when } t = 0, \qquad \theta = 1 \text{ when } x = 0,$$

is

$$\theta = \frac{2a}{\pi}\int_0^\infty\left\{1-e^{-(b+ap^2)t}\right\}\frac{p}{b+ap^2}\sin xp\,dp.$$

8. Heat is supplied at a constant rate Q per unit area per unit time over a circular area of radius a in the plane $z = 0$ to an infinite solid of thermal conductivity K. Show that the steady temperature at a point distant ρ from the axis of the circular area and distant z from its plane is

$$\frac{Qa}{2K}\int_0^\infty\frac{e^{-|z|p}}{p}J_0(\rho p)J_1(ap)\,dp.$$

9. The circular disc $z = 0$, $\rho < 1$ is electrically charged to unit potential. Show that the potential V at the point (ρ, ϕ, z) is given by

$$V = \int_0^\infty A(p)e^{-|z|p}J_0(\rho p)\,dp,$$

where $A(p)$ is given by the "dual" integral equations

$$\left.\begin{aligned}\int_0^\infty pA(p)J_0(\rho p)\,dp &= 1, \qquad (0 \leq \rho < 1),\\ \int_0^\infty p^2A(p)J_0(\rho p)\,dp &= 0, \qquad (\rho > 1).\end{aligned}\right\}$$

10. By applying the cosine transform $\bar{V}_c = \int_0^\infty V \cos px\, dx$ followed by the sine transform $\bar{V}_s' = \int_0^\infty \bar{V}_c \sin qz\, dz$, reduce the problem specified by

$$\frac{\partial^2 V}{\partial x^2} + \frac{\partial^2 V}{\partial z^2} = 0, \qquad (x > 0,\ \ z > 0),$$

with the boundary conditions

$$\frac{\partial V}{\partial x} = 0 \text{ when } x = 0 \quad \text{and} \quad V = \left.\begin{matrix} 1, & x < a, \\ 0, & x > a, \end{matrix}\right\} \text{ when } z = 0,$$

to the solution of the algebraic equation

$$(p^2+q^2)\bar{V}_s' = \frac{q \sin pa}{p}.$$

Given that, when r and s are positive,

$$\int_0^\infty \frac{q \sin sq}{p^2+q^2}\, dq = \frac{\pi}{2} e^{-sp}, \qquad \int_0^\infty \frac{e^{-sp}}{p} \sin rp\, dp = \tan^{-1}\left(\frac{r}{s}\right),$$

show that

$$V = \frac{1}{\pi}\left\{\tan^{-1}\left(\frac{a+x}{z}\right) + \tan^{-1}\left(\frac{a-x}{z}\right)\right\}.$$

8.9 Finite transforms

So far we have used only transforms defined by definite integrals in which the upper limit has been infinite. When this limit is equal to a finite quantity a (say), the transforms containing trigonometrical and Bessel functions as kernels are known respectively as *finite* Fourier and Hankel transforms. These are used in the solution of differential equations in a way similar to that already described, but the variable now excluded ranges from 0 to a and both the inversion formulae and those giving the transforms of derivatives require modification.

Commencing with the *finite sine transform* of $f(x)$ defined by

$$_s(p) = \int_0^a f(x) \sin \frac{p\pi x}{a}\, dx, \tag{1}$$

where p is now a positive integer, we obtain the required inversion formula by making use of the theory of Fourier series. Thus, if $f(x)$ can be expanded in a sine series in the range $0 < x < a$, the coefficient a_p of $\sin(p\pi x/a)$ in that series is given by

$$a_p = \frac{2}{a}\int_0^a f(x) \sin \frac{p\pi x}{a}\, dx = \frac{2}{a}\bar{f}_s(p).$$

Hence the required inversion formula is

$$f(x) = \frac{2}{a}\sum_{p=1}^{\infty} \bar{f}_s(p) \sin\frac{p\pi x}{a}. \tag{2}$$

Similarly, for the *finite cosine transform* defined by

$$\bar{f}_c(p) = \int_0^a f(x) \cos\frac{p\pi x}{a}\, dx, \qquad (p = 0, 1, 2, 3, \ldots), \tag{3}$$

the inversion formula is

$$f(x) = \frac{1}{a}\bar{f}_c(0) + \frac{2}{a}\sum_{p=1}^{\infty} \bar{f}_c(p) \cos\frac{p\pi x}{a}. \tag{4}$$

We define the *finite Hankel transform of $f(\rho)$ of order n* by

$$\bar{f}_H(p_i) = \int_0^a \rho f(\rho) J_n(p_i\rho)\, d\rho, \tag{5}$$

where p_i is the ith positive root of the equation $J_n(pa) = 0$. Working as above but now making use of the theory of Fourier-Bessel series [§6.9 (*e*)], the inversion formula is

$$(x) = \frac{2}{a^2}\sum_{i=1}^{\infty} \bar{f}_H(p_i)\frac{J_n(p_i\rho)}{J_{n+1}^2(p_i a)}. \tag{6}$$

Other finite Hankel transforms can be defined by equation (5) by placing different interpretations on p_i. For example, p_i might be taken as the ith root of the equation $hJ_n(pa) + kJ_n'(pa) = 0$ where h, k are constants. Such transforms do, in fact, have their uses in the solution of physical problems but lack of space forbids a detailed discussion in this book.

The finite sine transform of $(\partial^2 f/\partial x^2)$ can be found by integrating by parts; thus

$$\int_0^a \frac{\partial^2 f}{\partial x^2} \sin\frac{p\pi x}{a}\, dx = \left[\frac{\partial f}{\partial x} \sin\frac{p\pi x}{a}\right]_0^a - \frac{p\pi}{a}\int_0^a \frac{\partial f}{\partial x} \cos\frac{p\pi x}{a}\, dx,$$

and, since the first term on the right vanishes at both limits through the term $\sin(p\pi x/a)$, a second integration gives

$$\int_0^a \frac{\partial^2 f}{\partial x^2} \sin\frac{p\pi x}{a}\, dx = -\frac{p\pi}{a}\left[f \cos\frac{p\pi x}{a}\right]_0^a - \frac{p^2\pi^2}{a^2}\int_0^a f \sin\frac{p\pi x}{a}\, dx$$

$$= \frac{p\pi}{a}\{(-1)^{p+1} f(a) + f(0)\} - \frac{p^2\pi^2}{a^2}\bar{f}_s(p). \tag{7}$$

Similarly,

$$\int_0^a \frac{\partial^2 f}{\partial x^2} \cos\frac{p\pi x}{a}\,dx = (-1)^p f'(a) - f'(0) - \frac{p^2\pi^2}{a^2}\bar{f}_c(p), \tag{8}$$

where $f'(a)$ and $f'(0)$ are the values of $(\partial f/\partial x)$ when $x = a$ and $x = 0$ respectively.

To obtain the corresponding result for the Hankel transform defined in (5), we have

$$\int_0^a \rho\left(\frac{\partial^2 f}{\partial \rho^2} + \frac{1}{\rho}\frac{\partial f}{\partial \rho}\right) J_n(p_i\rho)\,d\rho = \int_0^a \frac{\partial}{\partial \rho}\left(\rho\frac{\partial f}{\partial \rho}\right) J_n(p_i\rho)\,d\rho$$

$$= \left[\rho\frac{\partial f}{\partial \rho} J_n(p_i\rho)\right]_0^a - p_i\int_0^a \rho\frac{\partial f}{\partial \rho} J_n'(p_i\rho)\,d\rho.$$

The first term on the right vanishes at the upper limit since p_i satisfies the equation $J_n(pa) = 0$, and it vanishes also at the lower limit. Hence, after a second integration by parts

$$\int_0^a \rho\left(\frac{\partial^2 f}{\partial \rho^2} + \frac{1}{\rho}\frac{\partial f}{\partial \rho}\right) J_n(p_i\rho)\,d\rho = -p_i\Big[\rho f J_n'(p_i\rho)\Big]_0^a$$

$$+ p_i\int_0^a f\{p_i\rho J_n''(p_i\rho) + J_n'(p_i\rho)\}\,d\rho. \tag{9}$$

The first term on the right gives $ap_i f(a)J_n'(p_i a)$ and, since $J_n(p_i\rho)$ satisfies Bessel's equation,

$$p_i\rho J_n''(p_i\rho) + J_n'(p_i\rho) = \left(\frac{n^2}{p_i\rho} - p_i\rho\right) J_n(p_i\rho).$$

Hence, we obtain after a slight rearrangement,

$$\int_0^a \rho\left(\frac{\partial^2 f}{\partial \rho^2} + \frac{1}{\rho}\frac{\partial f}{\partial \rho} - \frac{n^2}{\rho^2} f\right) J_n(p_i\rho)\,d\rho = ap_i f(a)J_n'(p_i a) - p_i^2\int_0^a \rho f J_n(p_i\rho)\,d\rho$$

$$= ap_i f(a)J_n'(p_i a) - p_i^2 \bar{f}_H(p_i). \tag{10}$$

8.10 The solution of partial differential equations by finite transforms

In the application of finite transforms to the solution of partial differential equations the results of the previous section are required. For easy reference, these are summarized below.

(*a*) *Finite Fourier sine transform*

$$_s(p) = \int_0^a f(x)\sin\frac{p\pi x}{a}\,dx, \qquad (p = 1, 2, 3, \ldots), \tag{1}$$

$$f(x) = \frac{2}{a}\sum_{p=1}^{\infty} \bar{f}_s(p)\sin\frac{p\pi x}{a}, \tag{2}$$

$$\int_0^a \frac{\partial^2 f}{\partial x^2}\sin\frac{p\pi x}{a}\,dx = \frac{p\pi}{a}\{(-1)^{p+1}f(a)+f(0)\}-\frac{p^2\pi^2}{a^2}\bar{f}_s(p), \tag{3}$$

where $f(a)$, $f(0)$ are respectively the values of $f(x)$ when $x = a$ and $x = 0$.

(b) Finite Fourier cosine transform

$$\bar{f}_c(p) = \int_0^a f(x)\cos\frac{p\pi x}{a}\,dx, \qquad (p = 0, 1, 2, 3, \ldots), \tag{4}$$

$$f(x) = \frac{1}{a}\bar{f}_c(0)+\frac{2}{a}\sum_{p=1}^{\infty}\bar{f}_c(p)\cos\frac{p\pi x}{a}, \tag{5}$$

$$\int_0^a \frac{\partial^2 f}{\partial x^2}\cos\frac{p\pi x}{a}\,dx = (-1)^p f'(a)-f'(0)-\frac{p^2\pi^2}{a^2}\bar{f}_c(p), \tag{6}$$

where $f'(a), f'(0)$ are respectively the values of $(\partial f/\partial x)$ when $x = a$ and $x = 0$.

(c) Finite Hankel transform

$$\bar{f}_H(p_i) = \int_0^a \rho f(\rho)J_n(p_i\rho)\,d\rho, \quad \text{where } J_n(p_i a) = 0, \tag{7}$$

$$f(x) = \frac{2}{a^2}\sum_{i=1}^{\infty}\bar{f}_H(p_i)\frac{J_n(p_i\rho)}{J_{n+1}^2(p_i a)}, \tag{8}$$

$$\int_0^a \rho\left(\frac{\partial^2 f}{\partial \rho^2}+\frac{1}{\rho}\frac{\partial f}{\partial \rho}-\frac{n^2}{\rho^2}f\right)J_n(p_i\rho)\,d\rho = ap_i f(a)J_n'(p_i a)-p_i^2\ {}_H(p_i), \tag{9}$$

where $f(a)$ is the value of $f(\rho)$ when $\rho = a$.

Finite transforms are used to exclude independent variables with finite ranges from a partial differential equation in a similar way to that in which the transforms considered in §8.8 were employed to exclude variables with infinite ranges. Similar remarks to those given in the previous section also apply when considering the suitability of a particular transform. Thus the finite sine transform is used to exclude a term like $(\partial^2 f/\partial x^2)$ when the values of f are given at $x = 0$ and $x = a$ (say), while the cosine transform is appropriate if the values of $(\partial f/\partial x)$ are given for these values of x. The finite Hankel transform is used to exclude the variable ρ in problems involving cylindrical polar co-ordinates when the range of ρ is finite.

Some examples of partial differential equations which can be solved by using finite transforms are given below. The use of these transforms does not solve problems which cannot be solved by the classical methods of Fourier and Fourier-Bessel series described in the last chapter. It does, however, facilitate their solution in that the same "drill" can be used whatever limits appear in the definite integral defining the transform.

Example 11. *A "lossless" transmission line of length a is initially at zero potential, and at time $t = 0$ a constant voltage V_0 is applied at the end $x = a$, the end $x = 0$ being earthed throughout. Show that at time t, the voltage V at a point at distance x from the earthed end is given by*

$$V = \frac{V_0 x}{a} + \frac{2V_0}{\pi}\sum_{p=1}^{\infty}\frac{(-1)^p}{p}\sin\frac{p\pi x}{a}\cos\frac{p\pi ct}{a},$$

where $c = (LC)^{-1/2}$ and L, C are respectively the inductance and capacitance per unit length of line.

In a "lossless" line the resistance R and leakage conductance G can be taken to be zero, and by equation (18) of §7.3, V satisfies the equation

$$\frac{\partial^2 V}{\partial x^2} = \frac{1}{c^2}\frac{\partial^2 V}{\partial t^2}, \tag{10}$$

where $c = (LC)^{-1/2}$. The boundary conditions are

$$V = 0 \quad \text{when } x = 0, \qquad V = V_0 \quad \text{when } x = a, \qquad t > 0, \tag{11}$$

while, since the initial voltage and current are zero,

$$V = \frac{\partial V}{\partial t} = 0 \quad \text{when } t = 0. \tag{12}$$

The term $(\partial^2 V/\partial x^2)$ can be excluded from equation (10) by using the finite sine transform

$$\bar{V}_s = \int_0^a V \sin\frac{p\pi x}{a}\,dx, \quad (p = 1, 2, 3, \ldots), \tag{13}$$

since, by equation (11), the values of V are known when $x = 0$ and $x = a$. Multiplying equations (10) and (12) by $\sin(p\pi x/a)$ and integrating with respect to x between 0 and a, we have

$$\int_0^a \frac{\partial^2 V}{\partial x^2}\sin\frac{p\pi x}{a}\,dx = \frac{1}{c^2}\int_0^a \frac{\partial^2 V}{\partial t^2}\sin\frac{p\pi x}{a}\,dx,$$

$$\int_0^a V\sin\frac{p\pi x}{a}\,dx = \int_0^a \frac{\partial V}{\partial t}\sin\frac{p\pi x}{a}\,dx = 0 \quad \text{when } t = 0.$$

Using (13), (3) and (11) these give

$$\frac{p\pi}{a}\{(-1)^{p+1}V_0\} - \frac{p^2\pi^2}{a^2}\bar{V}_s = \frac{1}{c^2}\frac{d^2\bar{V}_s}{dt^2} \tag{14}$$

with

$$\bar{V}_s = \frac{d\bar{V}_s}{dt} = 0 \quad \text{when } t = 0. \tag{15}$$

The solution of the ordinary differential equation (14) is

$$\bar{V}_s = A\cos\frac{p\pi ct}{a} + B\sin\frac{p\pi ct}{a} - (-1)^p\frac{aV_0}{p\pi},$$

where A and B are arbitrary constants. Substitution in (15) leads to

$$A = (-1)^p \frac{aV_0}{p\pi}, \quad B = 0,$$

so that

$$\bar{V}_s = (-1)^p \frac{aV_0}{p\pi}\left(\cos\frac{p\pi ct}{a} - 1\right).$$

The inversion formula (2) for the finite sine transform now gives

$$V = \frac{2V_0}{\pi}\sum_{p=1}^{\infty}\frac{(-1)^p}{p}\left(\cos\frac{p\pi ct}{a} - 1\right)\sin\frac{p\pi x}{a}. \tag{16}$$

Now it is known that for $0 < x < a$,

$$x = \frac{2a}{\pi}\sum_{p=1}^{\infty}\frac{(-1)^{p+1}}{p}\sin\frac{p\pi x}{a},$$

and the required result is obtained when use is made of this in (16).

Example 12. *The material of a long circular cylinder of radius a is of diffusivity k. The cylinder is initially at constant temperature V_0 and its curved surface is maintained at zero temperature. Find an expression for the temperature V at time t of a point in the cylinder at distance ρ from its axis.*

This is Example 8 of §7.7 and the equations for solution are

$$\frac{\partial^2 V}{\partial \rho^2} + \frac{1}{\rho}\frac{\partial V}{\partial \rho} = \frac{1}{k}\frac{\partial V}{\partial t}, \tag{17}$$

with $V = V_0$ when $t = 0$ and $V = 0$ when $\rho = a$. Using the finite Hankel transform of order zero given by

$$\bar{V}_H = \int_0^a \rho V J_0(p_i \rho)\, d\rho,$$

where p_i is a positive root of $J_0(pa) = 0$, equation (17) transforms into

$$-p_i^2 \bar{V}_H = \frac{1}{k}\frac{d\bar{V}_H}{dt} \tag{18}$$

when use is made of equation (9) and the fact that $V = 0$ when $\rho = a$. The initial condition $V = V_0$ when $t = 0$ transforms into

$$\bar{V}_H = \int_0^a \rho V_0 J_0(p_i \rho)\, d\rho = \frac{V_0 a}{p_i} J_1(p_i a) \quad \text{when } t = 0, \tag{19}$$

using the result (12) of §6.9. The solution of equation (18) is

$$\bar{V}_H = A \exp(-k p_i^2 t),$$

and the arbitrary constant A is given by (19). Hence

$$\bar{V}_H = \frac{V_0 a}{p_i} J_1(p_i a) \exp(-k p_i^2 t),$$

and substitution in the inversion formula (8) leads to

$$V = \frac{2V_0}{a}\sum_{i=1}^{\infty}\exp(-k p_i^2 t)\frac{J_0(p_i \rho)}{p_i J_1(p_i a)}.$$

8.11 Other transforms

Fourier and Hankel transforms are not the only ones which have been successfully used in the solution of partial differential equations with assigned boundary and initial conditions. The Laplace transform has been so used but it is not considered here, as its inversion formula involves a contour integral and a consideration of such integrals is outside the scope of the present work. In the Laplace transform solution of ordinary differential equations it was possible to avoid employing the inversion formula by using instead the short table given in §8.2, but for problems involving partial differential equations, either a very much extended table would have to be available or the inversion formula used.

Similar remarks apply to the Mellin transform with kernel x^{p-1}. Use has also been made of finite transforms with Legendre and Jacobi polynomials as kernels. These latter are of limited application but they have proved useful in special problems.

Exercises 8(*c*)

1. Use a finite sine transform to find the solution of the equation

$$\frac{\partial^2 V}{\partial x^2}+\frac{\partial^2 V}{\partial y^2} = 0, \qquad (0 < x < \pi, \quad y > 0),$$

with the conditions $V = 0$ when $x = 0$ and $x = \pi$, $V = 1$ when $y = 0$ and $V \to 0$ when $y \to \infty$.

2. Find a function V which satisfies the two-dimensional form of Laplace's equation inside the square $0 < x < a$, $0 < y < a$, and is such that $V = C$ (constant) on the side $y = a$ and that V vanishes on the other three sides of the square.

3. Show that the temperature at time t in a large slab of material of diffusivity k whose faces $x = 0$, $x = \pi$ are thermally insulated and whose initial temperature is $f(x)$ is

$$\frac{1}{\pi}\int_0^\pi f(u)\,du+\frac{2}{\pi}\sum_{p=1}^{\infty} e^{-kp^2t}\cos px\int_0^\pi f(u)\cos pu\,du.$$

4. Find a solution of the wave equation

$$c^2\frac{\partial^2 V}{\partial x^2} = \frac{\partial^2 V}{\partial t^2}, \qquad (0 < x < \pi, \quad t > 0),$$

such that $V = \sin t$ when $x = 0$, $V = 0$ when $x = \pi$, $V = (\partial V/\partial t) = 0$ when $t = 0$.

5. Solve the wave equation

$$c^2\frac{\partial^2 V}{\partial x^2} = \frac{\partial^2 V}{\partial t^2}, \qquad (0 < x < \pi, \quad t > 0)$$

subject to the conditions $(\partial V/\partial x) = 0$ when $x = 0$, $(\partial V/\partial x) = K$ (constant) when $x = \pi$, $V = (\partial V/\partial t) = 0$ when $t = 0$.

6. The cross-section of a long bar of diffusivity k is the square $0 < x < \pi$, $0 < y < \pi$. If the four faces of the bar are maintained at zero temperature, and if the initial temperature is unity, use repeated finite sine transforms to show that the temperature at time t is $\phi(x).\phi(y)$ where

$$\phi(x) = (4/\pi) \sum_{p=1}^{\infty} (2p-1)^{-1} \exp\{-k(2p-1)^2 t\} \sin(2p-1)x.$$

7. The material of the sphere $r < a$ is of diffusivity k and it is initially at zero temperature. From time $t > 0$ its surface is kept at temperature unity. Show that the temperature V at time t is given by

$$V = \frac{U}{r} \quad \text{where} \quad \frac{\partial U}{\partial t} = k\frac{\partial^2 U}{\partial r^2}.$$

Deduce, by using a finite sine transform, that

$$V = 1 + \frac{2a}{\pi r} \sum_{p=1}^{\infty} \frac{(-1)^p}{p} \exp\left(-k\frac{p^2\pi^2}{a^2}t\right) \sin\frac{p\pi r}{a}.$$

8. A thin circular membrane of unit radius and uniform surface density σ is fixed round its edge and stretched by a tension T. It is displaced symmetrically from its equilibrium position with velocity $f(\rho)$ and allowed to vibrate. Use a finite Hankel transform to show that the displacement z at time t and radial distance ρ is given by

$$z = \frac{2}{c} \sum_{i=1}^{\infty} \frac{\sin cp_i t}{p_i} \frac{J_0(p_i \rho)}{J_1^2(p_i)} \int_0^1 uf(u) J_0(p_i u)\, du$$

where $c^2 = Tg/\sigma$ and p_i satisfies the equation $J_0(p) = 0$.

9. The membrane of Exercise 8 above is set in motion from rest and is subject to a uniform constant pressure P acting over its whole surface for time $t > 0$. Show that

$$z = \frac{P}{T}\left\{\frac{1}{4}(1-\rho^2) - 2\sum_{i=1}^{\infty} \frac{J_0(p_i\rho)}{p_i^3 J_1(p_i)} \cos p_i ct\right\}.$$

10. ϕ satisfies the differential equation

$$\frac{\partial^2 \phi}{\partial \rho^2} + \frac{1}{\rho}\frac{\partial \phi}{\partial \rho} - \frac{\phi}{\rho^2} + \frac{\partial^2 \phi}{\partial z^2} = 0, \qquad (0 \leqq \rho < 1, \quad -h < z < h),$$

and the boundary conditions

$$\phi = 0 \quad \text{when} \quad \rho = 1,$$

$$\phi = \rho \quad \text{when} \quad z = h, \qquad \phi = 0 \quad \text{when} \quad z = -h.$$

Show that

$$\phi = 2\sum_{i=1}^{\infty} \frac{1}{p_i J_2(p_i)} \frac{\sinh(h+z)p_i}{\sinh 2hp_i} J_1(p_i\rho)$$

where $J_1(p_i) = 0$.

CHAPTER 9

GRAPHICAL AND NUMERICAL METHODS

9.1 Introductory

Comparatively few of the differential equations that arise in physical problems have solutions which can be expressed in terms of known functions. Solutions of linear equations can usually be expressed as infinite series, but, even then, the work involved in computing the numerical value of the solution for a particular value of the independent variable can be very great and the limited range of convergence of such series often further complicates matters. For these reasons graphical and numerical methods of finding solutions are of paramount importance. Graphical methods yield approximate values of the solutions and often give useful information as to their nature. These approximate values can be refined to any desired degree of accuracy by the powerful numerical methods considered in this and the following chapter.

The numerical integration of differential equations can be carried out without any mechanical aid to computing, but some kind of desk calculating machine is a considerable help. The methods given can also be used on electronic computors which usually have programmes prepared for carrying out some or all of these methods.

We begin this chapter by considering the graphical method of isoclinals for drawing the integral curves of first order differential equations and for reducing second order equations. We then consider some of the standard methods of initiating and continuing the numerical solutions of differential equations. In the final section we indicate briefly how such methods can be used with electronic computors.

9.2 The method of isoclinals

If the first order equation

$$y' = f(x, y) \tag{1}$$

has a solution $F(x, y) = 0$, the function $F(x, y)$ contains an arbitrary constant and the solution $F(x, y) = 0$ represents a family of integral curves of the equation (1).

If $y' = c$, a constant, in equation (1) a curve can be drawn whose equation is

$$f(x, y) = c, \tag{2}$$

which is the locus of all points at which the slope of integral curves has the value c; this curve is called an *isoclinal* of the differential equation

(1). We shall assume that $f(x, y)$ is a one-valued function of x and y so that the slope of the integral curves is uniquely determined at each point (x, y). It follows from this that, in general, no two integral curves will cross.

Integral curves of the family of solutions may be sketched approximately by first drawing a number of isoclinals $f(x, y) = c_1$, $f(x, y) = c_2, \ldots$. The isoclinal $f(x, y) = c_1$ is crossed by a number of short lines of slope c_1 and these short lines will indicate the slope of an integral curve where it crosses the isoclinal. Similarly the isoclinal $f(x, y) = c_2$ is crossed by short lines of slope c_2 and so on.

A picture of the family of integral curves then begins to appear. If

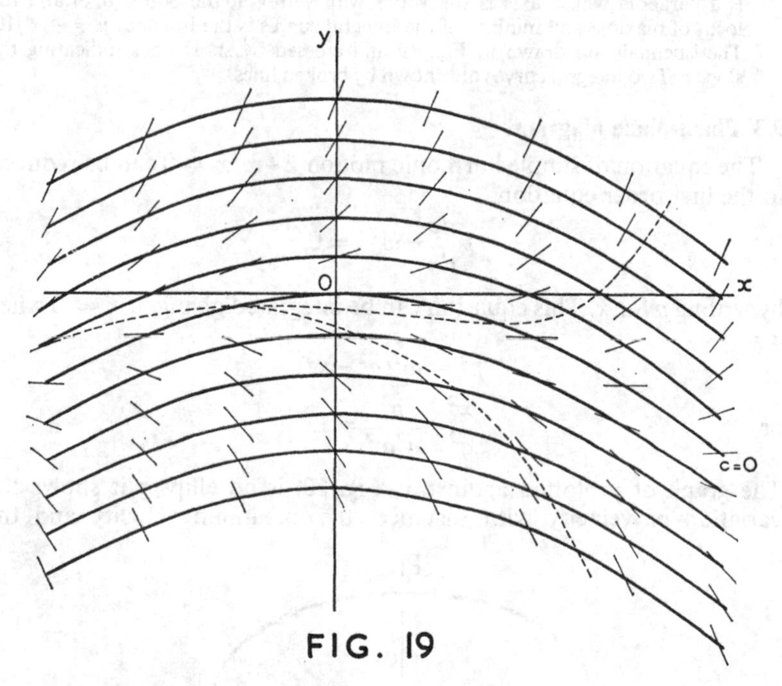

FIG. 19

the isoclinals are drawn fairly close together and the short crossing lines on neighbouring isoclinals are made to meet about halfway between them a form of polygon emerges which can be smoothed off to give an integral curve. It should be noted that the isoclinal $f(x, y) = 0$ is of special importance since it gives the locus of the maxima and minima or points of inflexion of the integral curves.

The locus of points where $y'' = 0$ is also of interest since (if $y''' \neq 0$) it is the locus of points of inflexion of the integral curves. The slope of an isoclinal $f(x, y) = c$ at a point (x, y) is $-\dfrac{\partial f}{\partial x}\Big/\dfrac{\partial_,}{\partial y}$, and on an integral curve if $y'' = 0$ we have $y' = -\dfrac{\partial f}{\partial x}\Big/\dfrac{\partial f}{\partial y}$. Hence, at a point of inflexion, the slope of an integral curve is that of the isoclinal through the point, that is, the short crossing line is a tangent to the isoclinal.

For reasonable accuracy the solution of differential equations by the method of isoclinals should be carried out under drawing-office conditions.

Example 1. *Sketch the isoclinals and portions of the integral curves of the differential equation* $y' = y+x^2/10$.

The isoclinals corresponding to $y' = c$ are the curves $y-c = -x^2/10$. This is a parabola whose axis is the y-axis with vertex at the point $(0, c)$ and the locus of maxima and minima of the integral curves is the isoclinal $y = -x^2/10$. The isoclinals are drawn in Fig. 19 and crossed by short lines indicating the slope c. Two integral curves are shown by broken lines.

9.3 Phase-plane diagrams

The equation of simple harmonic motion $\ddot{x}+\omega^2 x = 0$ can be reduced to the first order equation

$$p\frac{dp}{dx}+\omega^2 x = 0,$$

by writing p for $\dot{x}$. This equation can be integrated giving, if $\dot{x} = 0$ when $x = a$,

$$p^2 = \omega^2(a^2-x^2),$$

or

$$\frac{x^2}{a^2}+\frac{p^2}{a^2\omega^2} = 1.$$

The graph of p plotted against x, Fig. 20, is an ellipse; it shows the variation of velocity with distance, the maximum velocity and the

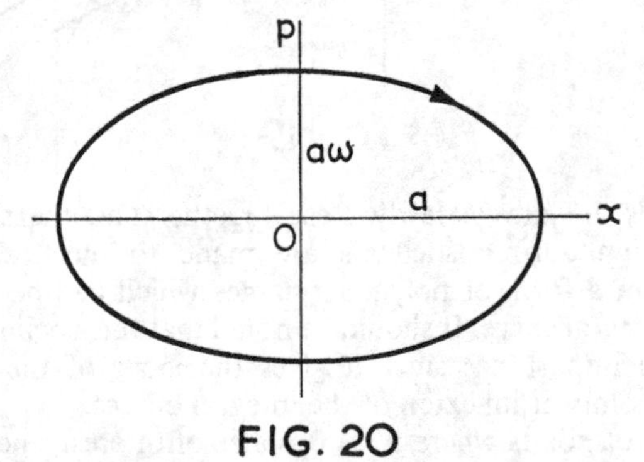

FIG. 20

amplitude. It also shows the periodicity of the motion. As the time increases, the point (p, x) goes round the ellipse in a clockwise direction, each circuit corresponding to one complete oscillation.

Such a diagram of velocity against distance is called a *phase-plane diagram*. These diagrams are of considerable value in discussing motion governed by non-linear equations of the form

$$\ddot{x} = f(\dot{x}, x),$$

which, by writing $\dot{x} = p$, becomes

$$\frac{dp}{dx} = \frac{1}{p} f(p, x).$$

If this equation can be solved, the phase-plane diagram can be constructed and the motion discussed. The method of isoclinals is often used to find the form of the integral curves of equations of this type which cannot be solved directly.

Example 2. *Construct the phase-plane diagram for the solution of the differential equation* $\ddot{x}+|\dot{x}|\dot{x}+x = 0$ *with* $x = 1$ *and* $\dot{x} = 0$ *when* $t = 0$.

The equation of motion of a particle moving in a straight line under an attraction proportional to distance from a fixed point in a medium offering a resistance proportional to the square of its velocity is of the form

$$\frac{d^2X}{dT^2}+k\left|\frac{dX}{dT}\right|\frac{dX}{dT}+\omega^2X = 0.$$

The modulus of dX/dT must be inserted to ensure that the resistance of the medium opposes the motion. The substitution $x = kX$, $t = \omega T$ reduces this equation to the given equation which may be taken as the canonical form of

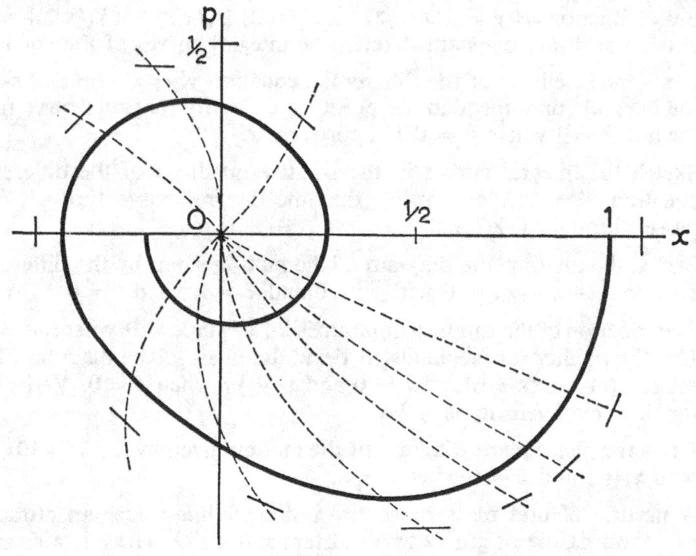

FIG. 21

the equation of simple harmonic motion with damping proportional to the square of the velocity. The substitution $p = \dot{x}$ leads to the equations

$$p < 0, \quad p\frac{dp}{dx} = +p^2 - x, \tag{1}$$

$$p > 0, \quad p\frac{dp}{dx} = -p^2 - x. \tag{2}$$

The isoclinals of these equations are

$$p < 0, \quad p - x/p = c,$$

$$p > 0, \quad -p - x/p = c.$$

These are the parabolas $(p-c/2)^2 = x+c^2/4$ and $(p+c/2)^2 = -(x-c^2/4)$. The isoclinals are shown by broken lines in Fig. 21 and the integral curve is easily sketched in. It will be seen that the amplitude of the oscillation diminishes rapidly.

Exercises 9(*a*)

1. Draw isoclinals of the differential equation $y' = x-y$ and sketch in an integral curve through the origin. Solve the equation and verify that the integral curve is correct.
2. Use the method of isoclinals to sketch the integral curves of the differential equation $y' = x^2+y^2$, and verify that, if $y = 0$ when $x = 0$, $y = 0{\cdot}35$ approximately when $x = 1$.
3. Sketch the integral curves of the differential equation $y' = x-y^2/10$. If $y = 1$ when $x = 0$ show that when $x = 1$, $y = 1{\cdot}38$ approximately.
4. By means of the substitution $y = -2xz'/z$ reduce the equation $2xy' = x^2+y^2$ to an equation in z and show that its solution in terms of Bessel functions is $y = x\{J_1(x/2)+AY_1(x/2)\}/\{J_0(x/2)+AY_0(x/2)\}$ where A is an arbitrary constant. Sketch the integral curves of the equation.
5. Draw the isoclinals of the differential equation $y' = x^2-y^2$ and sketch the integral curve through the point (0, 1). Verify that this curve meets the line $x = 1$ where $y = 0{\cdot}75$ approximately.
6. Sketch the integral curves in the positive quadrant of the differential equation $y' = 2xy$, and verify that the integral curve through (0, 1) passes through (1, 2·72).
7. Sketch the phase-plane diagram of the motion given by the differential equation $\ddot{x}+\dot{x}^2-x\dot{x} = 0$ with $x = 0$ and $\dot{x} = 1$ when $t = 0$.
8. The equation of the simple pendulum is $\ddot{x}+\omega^2 \sin x = 0$, where $\omega^2 = g/l$. Use the method of isoclinals to draw the phase-plane diagram of the motion for the case where $\dot{x} = 0$ and $x = \frac{1}{2}\pi$ when $t = 0$. Verify that the maximum velocity is $\sqrt{2\omega}$.
9. Draw the phase-plane diagram of the motion given by $\ddot{x}+2\dot{x}+10x = 0$ with $x = 1$ and $\dot{x} = 0$ when $t = 0$.
10. A particle of unit mass moves in a straight line under an attraction $\omega^2 x$ towards the origin O when distant x from O. There is a constant frictional force μg opposing motion. Show that $\ddot{x}+\omega^2(x\pm k) = 0$,

where $k = \mu g/\omega^2$, and the positive sign is taken if $\dot{x}$ is positive, the negative sign if $\dot{x}$ is negative. Draw the phase-plane diagram for the motion if initially $x = 10k$ and $\dot{x} = 0$ and $\omega = 1$. Show that the particle first comes to rest when $x = -8k$ and finally comes to rest after time 5π.

9.4 Numerical solutions of first order equations

The numerical solution of a differential equation comprises a set of values of the dependent variable y corresponding to values of the independent variable x over a given range. The intervals between the values of x must be sufficiently small to permit interpolation so that the value of y corresponding to any value of x in the range can be deduced.

The solution of the first order differential equation

$$\frac{dy}{dx} = f(x, y) \tag{1}$$

for the initial condition $y = y_0$ when $x = x_0$ is

$$y - y_0 = \int_{x_0}^{x} f(x, y)\, dx,$$

but this integral cannot be evaluated since the value of y under the integral sign is unknown. We can, however, find approximately the value y_1 corresponding to $x_0 + h$ when h is small, and the approximation can be made as close as is desired by taking h sufficiently small. When this has been done the process is repeated to find the value y_2 corresponding to $x_0 + 2h$, and so on until a table of values of y has been built up.

This is called a *step by step method* of obtaining a solution. Such methods have the disadvantage that the error in approximating to values of y is always to a certain extent cumulative and may become large unless some form of checking process is included. For this reason methods of solution involving finite differences are commonly used, most of these being variations of the Adams-Bashforth process considered in §9.10.

Finite difference methods involve two formulae; the first is a *prediction* formula for estimating the value of y at the end of an interval, the estimate being based on values of y at the ends of preceding intervals; the second is a *correction* formula in which the estimated value is checked against previous values of y, the estimate being recalculated if a discrepancy is found.

In order to make use of finite difference formulae it is necessary initially to know the values of y at the ends of several successive intervals. These starting values are found by using methods such as those of Taylor, Picard and Euler, or the formulae of Runge-Kutta.

Such processes are generally too cumbersome for use over a long range of integration and lack the simple self-checking facilities of finite difference methods. Runge-Kutta formulae are, however, sometimes used for large numbers of steps on electronic computors.

In the following sections we shall consider some of the methods of starting a solution and then show how the solution can be continued by finite difference and other methods.

9.5 Initiation of a solution by Taylor's series

Let y be the solution of the differential equation

$$\frac{dy}{dx} = f(x, y), \tag{1}$$

for the initial condition $y = y_0$ when $x = x_0$, and suppose that the solution can be expanded as a Taylor series in the neighbourhood of x_0. Then, if $y = y_1$ when $x = x_0+h$, we have

$$y_1 = y_0 + h\left(\frac{dy}{dx}\right)_0 + \frac{h^2}{2!}\left(\frac{d^2y}{dx^2}\right)_0 + \frac{h^3}{3!}\left(\frac{d^3y}{dx^3}\right)_0 + \ldots, \tag{2}$$

for sufficiently small values of h. Now

$$\frac{dy}{dx} = f(x, y),$$

$$\frac{d^2y}{dx^2} = \frac{\partial f}{\partial x} + \frac{dy}{dx}\frac{\partial f}{\partial y} = \frac{\partial f}{\partial x} + f\frac{\partial f}{\partial y},$$

$$\frac{d^3y}{dx^3} = \left(\frac{\partial}{\partial x} + f\frac{\partial}{\partial y}\right)\left(\frac{\partial f}{\partial x} + f\frac{\partial f}{\partial y}\right),$$

and further successive derivatives of y with respect to x can be calculated.

A good approximation can be obtained for y_1 by summing a number of terms of the series (2). The remainder after n terms of the series is

$$R_n = \frac{h^n}{n!}\left(\frac{d^ny}{dx^n}\right)_\theta, \tag{3}$$

where θ indicates the value of the nth derivative when $x = x_0 + \theta h$, and $0 < \theta < 1$. Putting $\theta = 0$, this gives an estimate of the *truncation error* in taking the solution as the sum of n terms of the series. If a required degree of accuracy is required for y_1 the size of interval that must be used can be deduced or, alternatively, the number of terms of the series that must be summed.

If the value y_1 corresponding to x_0+h is being computed, the value y_{-1} corresponding to x_0-h can be obtained with little extra labour, thus giving in all three initial values of y.

Example 3. *Find correct to four places of decimals values of y corresponding to $x = 0{\cdot}2$ and $x = -0{\cdot}2$ for the solution of the differential equation*

$$(dy/dx) = x - y^2/10,$$

with the initial condition $y = 1$ when $x = 0$.

Since the solution is required to be correct to four places of decimals, we shall in this example work to five places and round off the answers to four places. We have

$$\frac{dy}{dx} = x - \frac{y^2}{10}, \qquad \left(\frac{dy}{dx}\right)_0 = -0{\cdot}1,$$

$$\frac{d^2y}{dx^2} = 1 - \frac{y}{5}\frac{dy}{dx}, \qquad \left(\frac{d^2y}{dx^2}\right)_0 = 1{\cdot}02,$$

$$\frac{d^3y}{dx^3} = -\frac{1}{5}\left\{\left(\frac{dy}{dx}\right)^2 + y\frac{d^2y}{dx^2}\right\}, \qquad \left(\frac{d^3y}{dx^3}\right)_0 = -0{\cdot}206,$$

$$\frac{d^4y}{dx^4} = -\frac{1}{5}\left(3\frac{dy}{dx}\frac{d^2y}{dx^2} + y\frac{d^3y}{dx^3}\right), \qquad \left(\frac{d^4y}{dx^4}\right)_0 = 0{\cdot}1024.$$

Thus we have the Taylor series

$$y = 1 - 0{\cdot}1h + 0{\cdot}51h^2 - 0{\cdot}0343h^3 + 0{\cdot}0043h^4 + \ldots.$$

If $h = 0{\cdot}2$ the term in h^4 is $0{\cdot}000007$; this is a permissible truncation error, so that

$$y_1 = 1 - 0{\cdot}02 + 0{\cdot}0204 - 0{\cdot}00027 = 1{\cdot}0001,$$

and, for $h = -0{\cdot}2$,

$$y_{-1} = 1 + 0{\cdot}02 + 0{\cdot}0204 + 0{\cdot}00027 = 1{\cdot}0407.$$

Taylor's theorem can be used to initiate the solution of an equation of any order provided that successive derivatives can be calculated from the differential equation.

Example 4. *The Lane-Emden function of index n is the solution of the equation*

$$\frac{d^2y}{dx^2} + \frac{2}{x}\frac{dy}{dx} + y^n = 0,$$

satisfying the initial conditions $y = 1$ and $y' = 0$ when $x = 0$. Calculate three values of the function of index 2 to four places of decimals.

We have $xy'' + 2y' + xy^2 = 0$. Differentiation gives

$$xy''' + 3y'' + y^2 + 2xyy' = 0, \qquad xy^{\text{iv}} + 4y''' + 4yy' + 2x(y')^2 + 2xyy'' = 0.$$

Putting $(y)_0 = 1$ and $(y')_0 = 0$, we have

$$(y'')_0 = -\tfrac{1}{3}, \qquad (y''')_0 = 0.$$

Differentiating again,

$$xy^{\text{v}} + 5y^{\text{iv}} + 6(y')^2 + 6yy'' + 2xyy''' + 6xy'y'' = 0,$$

$$xy^{\text{vi}} + 6y^{\text{v}} + 24y'y'' + 8yy''' + 8xy'y''' + 2xyy^{\text{iv}} + 6x(y'')^2 = 0,$$

giving

$$(y^{\text{iv}})_0 = \tfrac{2}{5}, \qquad (y^{\text{v}})_0 = 0.$$

Differentiating once more and putting $x = 0$,

$$7(y^{\text{vi}})_0+40(y')_0(y''')_0+30(y'')_0^2+10(y)_0(y^{\text{iv}})_0 = 0,$$

and hence

$$(y^{\text{vi}})_0 = -\tfrac{22}{21}.$$

Thus we obtain the series

$$y_1 = 1-\frac{h^2}{6}+\frac{h^4}{60}-\frac{11h^6}{7560}+\ldots.$$

If the first three terms are to give a solution correct to four places of decimals we must have

$$\frac{11h^6}{7560} < 0{\cdot}00005,$$

that is

$$(10h)^6 < 34{,}364, \qquad 10h < 5 \quad \text{roughly.}$$

We may therefore use three terms of the series to compute values of y for $h = \pm 0{\cdot}2$ and $0{\cdot}4$. Thus

$$y(-0{\cdot}2) = 1+0{\cdot}00667+0{\cdot}00003 = 1{\cdot}0067,$$
$$y(0{\cdot}2) = 1-0{\cdot}00667+0{\cdot}00003 = 0{\cdot}9934,$$
$$y(0{\cdot}4) = 1-0{\cdot}02667+0{\cdot}00043 = 0{\cdot}9738.$$

Exercises 9(*b*)

1. Show that a Taylor series expansion of the solution of the equation $y' = 1-xy$ with $y = 0$ when $x = 0$ is $y = x-x^3/3+x^5/15-x^7/105+\ldots$. Calculate the value of y when $x = \frac{1}{2}$ to four places of decimals.

2. Show that a Taylor series expansion of the solution of the equation $y'+y^2 = 1$ with $y = 0$ when $x = 0$ is

$$y = x-x^3/3+2x^5/15-17x^7/315+\ldots.$$

Use the series to calculate to three places of decimals tanh (0·2), tanh (0·4) and tanh (0·6).

3. Show that the solution of the equation $y'+y^2 = x^2$ with $y = 0$ when $x = 0$ can be expressed as the series $y = x^3/3-x^7/63+2x^{11}/2079-\ldots$. Calculate values of y for $x = \pm 0{\cdot}2$ and $x = \pm 0{\cdot}4$ to four places of decimals.

4. Show that a solution of the equation $y'(y^2+1) = x^2$ with $y = 0$ when $x = 0$ is $y = x^3/3-x^9/81+x^{15}/729-\ldots$. Calculate the value of y to four places of decimals when $x = \pm 0{\cdot}5$ and $x = \pm 1$.

5. Show that the Lane-Emden function of index 3, which is the solution of the equation $y''+2y'/x+y^3 = 0$ with $y = 1$ and $y' = 0$ when $x = 0$, has a Taylor series $1-x^2/6+x^4/40\ldots$ with truncation error $19x^6/5040$. Find values of the function for $x = \pm 0{\cdot}2$ and $x = \pm 0{\cdot}4$ to four places of decimals.

6. Show that a solution of the equation $(y^2+1)y'' = (2y-1)(y')^2$ with $y = 0$ and $y' = 1$ when $x = 0$ can be expressed as the Taylor series $x-x^2/2+2x^3/3-3x^4/4+11x^5/12-\ldots$. Calculate y when $x = -0{\cdot}2$ and $x = +0{\cdot}2$.

9.6 Picard's method

Picard's method of finding a solution of the differential equation $y' = f(x, y)$ with $y = y_0$ when $x = x_0$ is to take the formal solution

$$y = y_0 + \int_{x_0}^{x} f(x, y)\, dx,$$

and substitute an approximate value of y, possibly $y = y_0$, under the integral sign. When the integration has been performed this gives a second approximation to y. This second approximation is then substituted under the integral sign and a third approximation obtained. The process is continued until the difference between two successive approximations is less than the allowable error in the value of y.

The process can be shown* to be convergent if $f(x, y)$ is a one-valued continuous function of x and y in the neighbourhood of x_0 and y_0. When this is so it provides a solution of the differential equation.

If the successive integrations can be carried out the method gives a formal solution of the differential equation, possibly as an infinite series. If the integrations can not be performed an interval h can be chosen and numerical methods of integration used over the range $x = x_0$ to $x = x_0+h$ to obtain successive approximations to the value y_1 corresponding to x_0+h. It is not essential to take y_0 as the first approximation to y; other approximations such as $y_0(1+x-x_0)$ will sometimes make the integration easier and give more rapid convergence of the process.

Example 5. *Find, correct to four places of decimals, solutions of the differential equation $y' = x^2+y^2$, with $y = 0$ when $x = 0$, for $x = 0{\cdot}4$ and $x = 0{\cdot}8$.*

Taking $y_0 = 0$ as the first approximation, we have for a second approximation

$$y_1 = \int_0^x x^2 dx = \frac{x^3}{3},$$

and for a third approximation

$$y_2 = \int_0^x \left(x^2+\frac{x^6}{9}\right) dx = \frac{x^3}{3}+\frac{x^7}{63}.$$

Repeating the procedure we have

$$y_3 = \int_0^x \left(x^2+\frac{x^6}{9}+\frac{2x^{10}}{189}+\frac{x^{14}}{63\times 63}\right) dx$$

$$= \frac{x^3}{3}+\frac{x^7}{63}+\frac{2x^{11}}{2079}+\frac{x^{15}}{15\times 63\times 63},$$

and similarly

$$y_4 = \frac{x^3}{3}+\frac{x^7}{63}+\frac{2x^{11}}{2079}+\frac{39x^{15}}{63\times 99\times 105}+\ldots.$$

* See, for example, E. L. Ince, *Ordinary Differential Equations*, Longmans Green and Co., London, 1927, reprinted by Dover Publications Inc., New York, 1956, §3.2.

The term in x^{15} can be neglected when $x = 0{\cdot}8$ and the term in x^{11} when $x = 0{\cdot}4$, and we have

$$\text{when } x = 0{\cdot}4, \quad y = 0{\cdot}02133+0{\cdot}00003 = 0{\cdot}0214,$$
$$\text{when } x = 0{\cdot}8, \quad y = 0{\cdot}17067+0{\cdot}00333+0{\cdot}00008 = 0{\cdot}1741.$$

9.7 Euler's method

Euler's method identifies the integral curve of the equation $y' = f(x, y)$ with its tangent at the point (x_0, y_0) over a short range x_0 to x_0+h. Thus, since the slope of the tangent is $f(x_0, y_0)$, we have

$$y_1 - y_0 = hf(x_0, y_0). \tag{1}$$

This rough approximation can be improved upon by taking the gradient of the integral curve as the mean of the slopes at x_0 and x_0+h, that is, using the approximate value obtained for y_1, we obtain an improved value $(y_1)_1$ given by

$$(y_1)_1 - y_0 = \tfrac{1}{2}h\{f(x_0, y_0)+f(x_0+h, y_1)\}. \tag{2}$$

This process can be repeated until there is agreement to a required degree of accuracy between successive approximations.

Example 6. *Find the value of y when $x = 0{\cdot}2$ on the integral curve of the equation $y' = x^2-2y$ through the point $x = 0$, $y = 1$.*

We have $(y')_0 = -2$, giving $y_1-y_0 = -0{\cdot}4$ and $y_1 = 0{\cdot}6$. At the point $(0{\cdot}2, 0{\cdot}6)$, $y' = -1{\cdot}16$, the mean gradient is $-1{\cdot}58$, giving $y_1 = 1-0{\cdot}316 = 0{\cdot}684$. At the point $(0{\cdot}2, 0{\cdot}684)$, $y' = -1{\cdot}328$, the mean gradient is $-1{\cdot}664$, giving $y_1 = 1-0{\cdot}3328 = 0{\cdot}6672$. Hence, we may take the value of y_1 to two places of decimals as $0{\cdot}67$.

Exercises 9(c)

1. Use Picard's method to approximate to the solution of the equation $y'+2xy^2 = 0$ with $y = 1$ when $x = 0$, and hence show that
$$y = 1/(1+x^2).$$

2. Obtain by Picard's method successive approximations to the solution of the equation $y' = x^2/(1+y^2)$ with $y = 0$ when $x = 0$. Calculate to three places of decimals the numerical values of three successive approximations when $x = 1$.

3. Apply Picard's method to find the solution of the equation $y' = x-y^2/10$ with $y = 1$ when $x = 0$, in the form
$$y = 1-x/10+51x^2/100-103x^3/3000+\ldots.$$

4. Use Picard's method to approximate to the solution of the equation $y' = x^2+xy+y^2$ with $y = 0$ when $x = 0$, and calculate the value of y to four places of decimals when $x = 0{\cdot}5$.

5. Find the value of the solution of the equation $y' = 2x - y/x$, with $y = 1$ when $x = 1$, for $x = 1{\cdot}2$.

6. Show that repeated applications of Euler's method to find the solution of the equation $y' + 2xy^2 = 0$, with $y = 1$ when $x = 0$, for $x = h$ leads to the solution $y = 1/(1+h^2)$.

7. Use Euler's method to find the value of y when $x = 0{\cdot}2$ in the solution of the equation $y' = x - y^2$, with $y = 0$ when $x = 0$.

8. Use Euler's method to find the value of y when $x = 0{\cdot}3$ in the solution of the equation $y' = x - y^2/10$, with $y = 1$ when $x = 0$, and compare the result with that obtained in Example 7, §9.8.

9.8 Runge-Kutta formulae

Two approximations, due to Runge and Kutta, are frequently used to obtain starting values for the solution of the equation $y' = f(x, y)$ with $y = y_0$ when $x = x_0$. The formulae give an approximate value y_1 corresponding to the value $x_0 + h$, for small values of h.

(i) *Third order approximation*

$$y_1 - y_0 = \tfrac{1}{6}(k_1 + 4k_2 + k_3), \tag{1}$$

where

$$k_1 = hf(x_0, y_0), \tag{2}$$

$$k_2 = hf(x_0 + h/2, y_0 + k_1/2), \tag{3}$$

$$k_3 = hf(x_0 + h, y_0 + 2k_2 - k_1). \tag{4}$$

This approximation is equivalent to Simpson's rule for the approximate integration of $f(x, y)$, and the error in y_1 is of order h^4.

(ii) *Fourth order approximation*

$$y_1 - y_0 = \tfrac{1}{6}(k_1 + 2k_2 + 2k_3 + k_4), \tag{5}$$

where

$$k_1 = hf(x_0, y_0), \tag{6}$$

$$k_2 = hf(x_0 + h/2, y_0 + k_1/2), \tag{7}$$

$$k_3 = hf(x_0 + h/2, y_0 + k_2/2), \tag{8}$$

$$k_4 = hf(x_0 + h, y_0 + k_3). \tag{9}$$

With this formula the error in y_1 is of order h^5. The analogy with Simpson's rule is evident.

These formulae are established by considering the Taylor series expansion of y in powers of h, the successive derivatives of y being obtained as in §9.5. We have

$$y-y_0 = hf_0+\tfrac{1}{2}h^2(p_0+f_0q_0)+\tfrac{1}{6}h^3(r_0+2f_0s_0+f_0{}^2t_0+p_0q_0+f_0q_0{}^2) +O(h^4), \qquad (10)$$

where

$$f=f(x,y), \quad p=\frac{\partial f}{\partial x}, \quad q=\frac{\partial f}{\partial y}, \quad r=\frac{\partial^2 f}{\partial x^2}, \quad s=\frac{\partial^2 f}{\partial x\,\partial y}, \quad t=\frac{\partial^2 f}{\partial y^2},$$

and the suffix 0 denotes the values of these quantities when $h = 0$, that is when $x = x_0$ and $y = y_0$.

In the third order approximation (1) we can expand each of the quantities k_1, k_2 and k_3 in powers of h in a similar manner. Thus

$$k_1 = hf(x_0, y_0) = hf_0,$$

$$k_2 = hf(x_0+\tfrac{1}{2}h, y_0+\tfrac{1}{2}k_1)$$

$$= hf_0+\tfrac{1}{2}h^2(p_0+f_0q_0)+\tfrac{1}{8}h^3(r_0+2f_0s_0+f_0{}^2t_0)+O(h^4).$$

Now

$$2k_2-k_1 = hf_0+h^2(p_0+f_0q_0)+\ldots,$$

so that

$$\left[\frac{d}{dh}(2k_2-k_1)\right]_{h=0} = f_0, \quad \left[\frac{d^2}{dh^2}(2k_2-k_1)\right]_{h=0} = 2(p_0+f_0q_0).$$

Hence

$$k_3 = hf(x_0+h, y_0+2k_2-k_1)$$

$$= hf_0+h^2(p_0+f_0q_0)+\tfrac{1}{2}h^3\{r_0+2f_0s_0+f_0{}^2t_0+2q_0(p_0+f_0q_0)\}+O(h^4).$$

Therefore

$$\tfrac{1}{6}(k_1+4k_2+k_3)$$

$$= hf_0+\tfrac{1}{2}h^2(p_0+f_0q_0)+\tfrac{1}{6}h^3(r_0+2f_0s_0+f_0{}^2t_0+p_0q_0+f_0q_0{}^2)+O(h^4).$$

This agrees with the Taylor series expansion (10) as far as the term in h^3 and hence the formula is established. The fourth order approximation is established in a similar manner by taking another term of the Taylor series.

It is evident from the above that it would be wrong to expect the Runge-Kutta formulae to yield more accurate information than the Taylor series method of §9.5. They are, however, occasionally more simple to apply, but become rather complicated if pursued for more than two or three steps.

Example 7. *Find, correct to three places of decimals, values of y corresponding to* $x = 0{\cdot}1,\ 0{\cdot}2,\ 0{\cdot}3$ *for the solution of the differential equation* $y' = x - y^2/10$, *with the boundary condition* $y = 1$ *when* $x = 0$.

With $h = 0{\cdot}1$, $h^4 = 0{\cdot}0001$ and we may use the Runge-Kutta third order approximation.

First step:

$$\begin{aligned} x_0 &= 0 \\ y_0 &= 1 \\ f_0 &= -0{\cdot}1 \\ k_1 &= -0{\cdot}01 \\ y_0 + \tfrac{1}{2}k_1 &= 0{\cdot}995 \\ k_2 &= 0{\cdot}1\{0{\cdot}05 - 0{\cdot}1(0{\cdot}995)^2\} = -0{\cdot}0049 \\ 2k_2 - k_1 &= 0{\cdot}0002 \\ k_3 &= 0{\cdot}1\{0{\cdot}1 - 0{\cdot}1(1{\cdot}0002)^2\} = 0 \\ \tfrac{1}{6}(k_1 + 4k_2 + k_3) &= -0{\cdot}0049 \\ y_1 &= 0{\cdot}9951. \end{aligned}$$

Second step:

$$\begin{aligned} x_0 &= 0{\cdot}1 \\ y_0 &= 0{\cdot}9951 \\ f_0 &= 0{\cdot}0010 \\ k_1 &= 0{\cdot}0001 \\ y_0 + \tfrac{1}{2}k_1 &= 0{\cdot}9952 \\ k_2 &= 0{\cdot}1\{0{\cdot}15 - 0{\cdot}1(0{\cdot}9952)^2\} = 0{\cdot}0051 \\ 2k_2 - k_1 &= 0{\cdot}0101 \\ k_3 &= 0{\cdot}1\{0{\cdot}20 - 0{\cdot}1(1{\cdot}0052)^2\} = 0{\cdot}0099 \\ \tfrac{1}{6}(k_1 + 4k_2 + k_3) &= 0{\cdot}0051 \\ y_1 &= 1{\cdot}0002. \end{aligned}$$

Third step:

$$\begin{aligned} x_0 &= 0{\cdot}2 \\ y_0 &= 1{\cdot}0002 \\ f_0 &= 0{\cdot}1000 \\ k_1 &= 0{\cdot}01 \\ y_0 + \tfrac{1}{2}k_1 &= 1{\cdot}0052 \\ k_2 &= 0{\cdot}1\{0{\cdot}25 - 0{\cdot}1(1{\cdot}0052)^2\} = 0{\cdot}0149 \\ 2k_2 - k_1 &= 0{\cdot}0198 \\ k_3 &= 0{\cdot}1\{0{\cdot}30 - 0{\cdot}1(1{\cdot}02)^2\} = 0{\cdot}0196 \\ \tfrac{1}{6}(k_1 + 4k_2 + k_3) &= 0{\cdot}0149 \\ y_1 &= 1{\cdot}0151. \end{aligned}$$

Hence, to three places of decimals, the values of y are 0·995, 1·000 and 1·015. The reader should compare this example with Example 3 of §9.5 and note the small rounding errors which make the fourth place of decimals inaccurate in the Runge-Kutta method. The values of y obtained from the Taylor series to four places of decimals are 0·9951, 1·0001, 1·0150.

Exercises 9(*d*)

Use the Runge-Kutta formulae to calculate three steps in the solutions of the following equations with the interval h that is given, to the requisite number of places of decimals.

1. $y' = x-y$ with $y = 0$ when $x = 0$, $h = 0{\cdot}2$, four places of decimals.
2. $y' = x-y^2$ with $y = 1$ when $x = 0$, $h = 0{\cdot}25$, three places of decimals.
3. $y' = x^2+y^2$ with $y = 0$ when $x = 0$, $h = 0{\cdot}2$, four places of decimals.
4. $y' = (x-y)/(x+y)$ with $y = 1$ when $x = 0$, $h = 0{\cdot}2$, four places of decimals.
5. $y' = x^2+xy+y^2$ with $y = 0$ when $x = 0$, $h = 0{\cdot}25$, four places of decimals.
6. $y' = 1+\dfrac{y}{x}$ with $y = 1$ when $x = 1$, $h = 0{\cdot}2$, four places of decimals.
7. $y' = (x^2+y^2)/(x+y)$, $y = 1$ when $x = 0$, $h = 0{\cdot}1$, three places of decimals.
8. $y'+y^2+y/x+1 = 0$, $y = 0$ when $x = 0$, $h = 0{\cdot}2$, four places of decimals. Verify that the solution is $y = -J_1(x)/J_0(x)$.

9.9 Finite differences

We now consider a few points from the theory of finite differences which are necessary in order to understand the formulae used in the integration of differential equations.

Consider the function $f(x) = x^3+x$. We can tabulate values of the function for $x = 0, 1, 2, 3, \ldots$, and the differences between successive values. We can also tabulate the differences of these differences, and so on. Thus

x	$f(x)$	$\nabla f(x)$	$\nabla^2 f(x)$	$\nabla^3 f(x)$	$\nabla^4 f(x)$
0	0				
		2			
1	2		6		
		8		6	
2	10		12		0
		20		6	
3	30		18		0
		38		6	
4	68		24		0
		62		6	
	130		30		
		92			
6	222				

In this table each column $\nabla f, \nabla^2 f, \ldots$, represents the difference between the numbers in the preceding column immediately above and below its own level. It will be noticed that the third differences here are all equal and the fourth differences zero; this is because $f(x)$ is a polynomial of the third degree. It will also be noticed that each value of the function can be obtained by summing the function and the differences in a diagonal line starting from the preceding function. Thus $130+62+24+6 = 222$, and the value of the function when $x = 7$ is $222+92+30+6 = 350$.

We shall use the notation of *backward differences*, so that if $f_0, f_1, f_2, \ldots$, are successive values of the function,

$$\nabla f_1 = f_1 - f_0, \qquad \nabla f_2 = f_2 - f_1, \qquad \text{and in general } \nabla f_n = f_n - f_{n-1}.$$

Similarly,

$$\nabla^2 f_2 = \nabla f_2 - \nabla f_1 = f_2 - 2f_1 + f_0,$$

and generally

$$\nabla^r f_n = \nabla^{r-1} f_n - \nabla^{r-1} f_{n-1}.$$

Thus the last diagonal of the table for $f(x) = x^3+x$ gives the values of

$$f_6, \nabla f_6, \nabla^2 f_6, \nabla^3 f_6, \nabla^4 f_6,$$

and we have seen that

$$f_7 = (1+\nabla+\nabla^2+\nabla^3+\ldots)f_6.$$

The following theorems on finite differences are quoted without proof; reference may be made to one of the standard texts on the subject.*

(i) $$\nabla^n f_r = f_r - nf_{r-1} + \frac{n(n-1)}{2!} f_{r-2} - \frac{n(n-1)(n-2)}{3!} f_{r-3} + \ldots. \qquad (1)$$

The coefficients here are those of the binomial expansion. This is illustrated in the table of x^3+x where $\nabla^4 f = 0$, giving

$$f_r - 4f_{r-1} + 6f_{r-2} - 4f_{r-3} + f_{r-4} = 0$$

for all values of r.

(ii) $$\nabla^n f_0 = h^n f_0^{(n)} + \text{higher powers of } h, \qquad (2)$$

where h is the interval in the values of x and $f_0^{(n)} = \left[\frac{d^n}{dx^n} f(x)\right]_{x=0}$. This

* For example, L. M. Milne-Thompson, *The Calculus of Finite Differences*, Macmillan, 1951.

theorem is proved by assuming a Taylor expansion of $f_r = f(rh)$ in powers of h. It follows that the $(n+1)$th difference of a polynomial of degree n is zero.

(iii) $$f_{n+1} = (1+\nabla+\nabla^2+\nabla^3+\dots)f_n. \quad (3)$$

This is the theorem in which we have already noticed that each value of the function in the difference table is the sum of the preceding diagonal terms.

(iv) $$f\{(n+x)h\} = \left\{1+x\nabla+\frac{x(x+1)}{2!}\nabla^2+\frac{x(x+1)(x+2)}{3!}\nabla^3+\dots\right\}f_n. \quad (4)$$

This is the *Newton-Gregory interpolation formula* expressed in terms of backward differences. This formula is based on the assumption that $f(x)$ is a polynomial in x, but it is used to approximate to interpolated values of other functions.

9.10 Adams-Bashforth formulae

In the solution of the differential equation $y' = f(x, y)$ a certain number of values of y must be found by one of the methods described in the previous sections. At least three values are necessary but five or more are often desirable in order to initiate a finite difference process. We shall suppose these starting values are tabulated with the corresponding values of $f(x, y)$ at an interval h in x. Assuming that we have five starting values we shall denote these quantities by the symbols x_r, y_r and f_r, where $r = -4, -3, -2, -1, 0$ and form a difference table of the values of f. Thus

x	y	$f(x, y)$	∇f	$\nabla^2 f$	$\nabla^3 f$	$\nabla^4 f$
x_{-4}	y_{-4}	f_{-4}				
			∇f_{-3}			
x_{-3}	y_{-3}	f_{-3}		$\nabla^2 f_{-2}$		
			∇f_{-2}		$\nabla^3 f_{-1}$	
x_{-2}	y_{-2}	f_{-2}		$\nabla^2 f_{-1}$		$\nabla^4 f_0$
			∇f_{-1}		$\nabla^3 f_0$	
x_{-1}	y_{-1}	f_{-1}		$\nabla^2 f_0$		
			∇f_0			
x_0	y_0	f_0				

The next value y_1 is obtained by using the Newton-Gregory interpolation formula, §9.9 (4), and integrating it for values of x between x_0 and x_1. Thus

$$y_1 - y_0 = h\int_0^1 f(x_0 + xh)\,dx$$
$$= h\int_0^1 \left\{1 + x\nabla + \frac{x(x+1)}{2!}\nabla^2 + \frac{x(x+1)(x+2)}{3!}\nabla^3 + \ldots\right\} f_0\,dx.$$

Now

$$\int_0^1 x\,dx = \frac{1}{2}, \quad \int_0^1 \frac{x(x+1)}{2!}\,dx = \frac{5}{12},$$
$$\int_0^1 \frac{x(x+1)(x+2)}{3!}\,dx = \frac{3}{8}, \quad \int_0^1 \frac{x(x+1)(x+2)(x+3)}{4!}\,dx = \frac{251}{720}, \ldots,$$

and hence

$$y_1 - y_0 = h\{1 + \tfrac{1}{2}\nabla + \tfrac{5}{12}\nabla^2 + \tfrac{3}{8}\nabla^3 + \tfrac{251}{720}\nabla^4 + \ldots\}f_0. \qquad (1)$$

This is the Adams-Bashforth *prediction formula* by which the value y_1 is estimated using as many of the differences of y_0 as are available or are necessary to give a required degree of accuracy. The term in ∇^5 in the formula (1) is $h(95/288)\nabla^5 f_0$, and this can be used as an estimate of the truncation error in stopping at the fourth difference. From equation (2) §9.9, this is approximately $\frac{1}{3}h^6 f_0^{(5)}$.

The next stage consists of calculating the value of $f(x_1, y_1)$ and inserting f_1 and its differences $\nabla f_1, \nabla^2 f_1, \ldots$ in the difference table.

We then turn to the *correction formula*

$$y_1 - y_0 = h\{1 - \tfrac{1}{2}\nabla - \tfrac{1}{12}\nabla^2 - \tfrac{1}{24}\nabla^3 - \tfrac{19}{720}\nabla^4 - \ldots\}f_1. \qquad (2)$$

This formula is obtained by integrating the Newton-Gregory interpolation formula for $f(x_1 - xh)$ between x_0 and x_1 and, since the coefficients are smaller than in equation (1), this formula gives a smaller truncation error. Thus the term in ∇^5 is $-h(3/160)\nabla^5 f_1$ compared with $h(95/288)\nabla^5 f_0$ in (1).

The values of f_1 and its differences are substituted in the correction formula (2) and a new value of y_1 is obtained which we may call $(y_1)_1$. If the two values of y_1 do not differ significantly we may accept the value $(y_1)_1$ as correct and proceed to the next value y_2. If however there is a difference of 5 units or more in the last decimal place, f_1 and its differences should be recalculated using the value $(y_1)_1$ and the correction formula applied again. This process should be continued until consecutive values of y_1 are in agreement.

Example 8. *Calculate to four places of decimals three further steps in the solution of the equation $y' = x - y^2/10$, with $y = 1$ when $x = 0$, given the values*

$x =$	$-0{\cdot}2$	$-0{\cdot}1$	$0{\cdot}1$	$0{\cdot}2$
$y =$	$1{\cdot}04068$	$1{\cdot}01513$	$0{\cdot}99507$	$1{\cdot}00013.$

We first calculate the corresponding values of $f(x) = x - y^2/10$ and form the difference table as far as the broken line.

	x	y	$f = x - y^2/10$	∇f	$\nabla^2 f$	$\nabla^3 f$	$\nabla^4 f$
x_{-4}	−0·2	1·04068	−0·30830				
				10525			
x_{-3}	−0·1	1·01513	−0·20305		−220		
				10305		13	
x_{-2}	0	1·00000	−0·10000		−207		−5
				10098		8	
x_{-1}	0·1	0·99507	+0·00098		−199		−7
				09899		1	
x_0	0·2	1·00013	0·09997		−198		−7
				09701		−6	
x_1	0·3	1·01499	0·19698		−204		−2
				09497		−8	
x_2	0·4	1·03945	0·29195		−212		
				09285			
x_3	0·5	1·07331	0·38480				

Using formula (1)

$$y_1 - y_0 = 0{\cdot}1 \left\{\begin{array}{lr} 09998 - 83 & \\ 04950 & 2 \\ 00003 & \\ \hline 14951 & 85 \end{array}\right. \qquad \begin{array}{r} = 0{\cdot}014866 \\ 1{\cdot}00013 \\ \hline y_1 = 1{\cdot}01500. \end{array}$$

Hence $f_1 = 0{\cdot}19698$ and the differences are entered below the broken line. Applying formula (2)

$$(y_1)_1 - y_0 = 0{\cdot}1 \left\{\begin{array}{r} 19698 - 04851 \\ 17 \\ \hline 19715 \end{array}\right. \qquad \begin{array}{r} = 0{\cdot}014864 \\ 1{\cdot}00013 \\ \hline (y_1)_1 = 1{\cdot}01499. \end{array}$$

Accepting this value for $(y_1)_1$ we use formula (1) to find y_2

$$y_2 - y_1 = 0{\cdot}1 \left\{\begin{array}{lr} 19698 - 83 & \\ 04850 & 3 \\ \hline 24548 & 86 \end{array}\right. \qquad \begin{array}{r} = 0{\cdot}024462 \\ 1{\cdot}01499 \\ \hline y_2 = 1{\cdot}03945. \end{array}$$

Hence $f_2 = 0{\cdot}29195$ and the differences are entered. Then formula (2) gives

$$(y_2)_1 - y_1 = 0{\cdot}1 \left\{\begin{array}{r} 29195 - 04748 \\ 17 \\ \hline 29212 \end{array}\right. \qquad \begin{array}{r} = 0{\cdot}024464 \\ 1{\cdot}01499 \\ \hline (y_2)_1 = 1{\cdot}03945. \end{array}$$

Using formula (1) again we have

$$y_3 - y_2 = 0{\cdot}1 \left\{\begin{array}{lr} 29195 - 85 & \\ 04748 & 2 \\ \hline 33943 & 87 \end{array}\right. \qquad \begin{array}{r} = 0{\cdot}033856 \\ 1{\cdot}03945 \\ \hline y_3 = 1{\cdot}07331. \end{array}$$

Hence $f_3 = 0{\cdot}38480$ and the differences are entered in the table. Using formula (2) we have

$$(y_3)_1 - y_2 = 0{\cdot}1 \left\{\begin{array}{r} 38480 - 04642 \\ 18 \\ \hline 38498 \end{array}\right. \qquad \begin{array}{r} = 0{\cdot}033856 \\ 1{\cdot}03945 \\ \hline (y_3)_1 = 1{\cdot}07331. \end{array}$$

We thus have the three values of y, 1·0150, 1·0395 and 1·0733.

9.11 Use of Simpson's rule

Simpson's rule may be used for forward integration of the equation $y' = f(x, y)$ without bringing in finite differences. The appropriate formula is

$$y_1 - y_{-1} = \tfrac{1}{3}h(f_{-1} + 4f_0 + f_1). \tag{1}$$

To estimate the error in using Simpson's rule we write the Adams-Bashforth correction formula §9.10 (2) for $y_0 - y_{-1}$, namely

$$y_0 - y_{-1} = h(1 - \tfrac{1}{2}\nabla - \tfrac{1}{12}\nabla^2 - \tfrac{1}{24}\nabla^3 - \tfrac{19}{720}\nabla^4 - \ldots)f_0.$$

Adding this to the prediction formula §9.10 (1) we have

$$y_1 - y_{-1} = h(2 + \tfrac{1}{3}\nabla^2 + \tfrac{1}{3}\nabla^3 + \tfrac{29}{90}\nabla^4 + \ldots)f_0. \tag{2}$$

Now $f_{-1} = (1 - \nabla)f_0$ and $f_1 = (1 + \nabla + \nabla^2 + \nabla^3 + \ldots)f_0$, so that

$$\tfrac{1}{3}h(f_{-1} + 4f_0 + f_1) = h(2 + \tfrac{1}{3}\nabla^2 + \tfrac{1}{3}\nabla^3 + \tfrac{1}{3}\nabla^4 + \ldots)f_0.$$

Hence, using equation (2)

$$y_1 - y_{-1} = \tfrac{1}{3}h(f_{-1} + 4f_0 + f_1) - h(\tfrac{1}{90}\nabla^4 + \ldots)f_0.$$

Thus the truncation error in using the formula (1) is $(h/90)\nabla^4 f_0$, that is $(h^5/90)\, f_0^{\text{iv}}$.

Simpson's rule is used in the solution of the differential equation by taking the fifth differences in the value of f as zero, that is

$$\nabla^5 f_1 = f_1 - 5f_0 + 10f_{-1} - 10f_{-2} + 5f_{-3} - f_{-4} = 0. \tag{3}$$

Now if the five values y_0 to y_{-4} are known, the corresponding values of $f(x, y)$ can be found and f_1 calculated from equation (3). This value of f_1 is then used in equation (1) to give $y_1 - y_{-1}$.

The correction process consists in substituting the value obtained for y_1 in $f(x, y)$ to give a new value of f_1. This new value of f_1 is substituted in Simpson's rule to give a second estimate of y_1 and, if there is agreement, the integration is continued using the corrected value of f_1. If there is disagreement the process is repeated.

Example 9. *Use Simpson's rule to calculate to four places of decimals three further steps in the solution of the equation* $y' = x - y^2/10$, *with* $y = 1$ *when* $x = 0$, *given the values*

$$x = \quad -0{\cdot}2 \qquad -0{\cdot}1 \qquad 0{\cdot}1 \qquad 0{\cdot}2$$

$$y = \quad 1{\cdot}04068 \qquad 1{\cdot}01513 \qquad 0{\cdot}99507 \qquad 1{\cdot}00013.$$

The five initial values of $f(x, y)$ are, taking f_0 when $x = 0{\cdot}2$, f_{-1} when $x = 0{\cdot}1$, etc.,

$$\begin{aligned} f_0 &= 0{\cdot}09997 \\ f_{-1} &= 0{\cdot}00098 \\ f_{-2} &= -0{\cdot}10000 \\ f_{-3} &= -0{\cdot}20305 \\ f_{-4} &= -0{\cdot}30830. \end{aligned}$$

Hence using equation (3) we have

$$\begin{array}{rl} f_1 = 0{\cdot}49985 & -0{\cdot}00980 = 0{\cdot}19700. \\ 1{\cdot}01525 & 1{\cdot}00000 \\ & 0{\cdot}30830 \\ \hline 1{\cdot}51510 & 1{\cdot}31810 \end{array}$$

Using Simpson's rule (1),

$$y_1 - y_{-1} = \tfrac{1}{3}\times 0{\cdot}1 \begin{pmatrix} 0{\cdot}19700 \\ 0{\cdot}39988 \\ 0{\cdot}00098 \end{pmatrix} = \begin{array}{r} 0{\cdot}01993 \\ 0{\cdot}99507 \\ \hline \end{array}$$

$$0{\cdot}59786 \quad y_1 = 1{\cdot}01500.$$

With this value of y_1 we find $f_1 = 0{\cdot}19698$ and this does not alter the value of y_1. Again using equation (3),

$$\begin{array}{rl} f_2 = 0{\cdot}98490 & -0{\cdot}99970 = 0{\cdot}29195, \\ 0{\cdot}00980 & 0{\cdot}20305 \\ 0{\cdot}50000 & \\ \hline 1{\cdot}49470 & 1{\cdot}20275 \end{array}$$

and using Simpson's rule

$$y_2 - y_0 = \tfrac{1}{3}\times 0{\cdot}1 \begin{pmatrix} 0{\cdot}29195 \\ 0{\cdot}78792 \\ 0{\cdot}09997 \end{pmatrix} = \begin{array}{r} 0{\cdot}039328 \\ 1{\cdot}00013 \\ \hline \end{array}$$

$$1{\cdot}17984 \quad y_2 = 1{\cdot}03946.$$

With this value of y_2 we find $f_2 = 0{\cdot}29195$ and this does not alter the value of y_2. Again using equation (2) we have

$$\begin{array}{rl} f_3 = 1{\cdot}45975 & -1{\cdot}96980 = 0{\cdot}38475, \\ 0{\cdot}99970 & 0{\cdot}00490 \\ & 0{\cdot}10000 \\ \hline 2{\cdot}45945 & 2{\cdot}07470 \end{array}$$

and using Simpson's rule

$$y_3 - y_1 = \tfrac{1}{3}\times 0{\cdot}1 \begin{pmatrix} 0{\cdot}38475 \\ 1{\cdot}16780 \\ 0{\cdot}19698 \end{pmatrix} = \begin{array}{r} 0{\cdot}058318 \\ 1{\cdot}01500 \\ \hline \end{array}$$

$$1{\cdot}74953 \quad y_3 = 1{\cdot}07332.$$

With this value of y_3 we find $f_3 = 0{\cdot}38480$ and y_3 is unaltered. Hence, to four places of decimals, $y_1 = 1{\cdot}0150$, $y_2 = 1{\cdot}0395$ and $y_3 = 1{\cdot}0733$.

9.12 Milne's method

A useful method of forward integration, due to Milne, uses the prediction formula

$$y_1 - y_{-3} = \tfrac{4}{3}h(2f_0 - f_{-1} + 2f_{-2}), \tag{1}$$

for which the truncation error is $(14/45)h\nabla^4 f_0$. This can be used with Simpson's rule

$$y_1 - y_{-1} = \tfrac{1}{3}h(f_1 + 4f_0 + f_{-1}) \tag{2}$$

as a correction formula, with truncation error $(1/90)h\nabla^4 f_0$.

Milne's formula is derived from the Adams-Bashforth formula §9.10 (1) for the four intervals between x_1 and x_{-3}, giving

$$y_1 - y_{-3} = h(1+\tfrac{1}{2}\nabla+\tfrac{5}{12}\nabla^2+\tfrac{3}{8}\nabla^3+\tfrac{251}{720}\nabla^4+\dots)(f_0+f_{-1}+f_{-2}+f_{-3}).$$

Now

$$f_{-1} = (1-\nabla)f_0,\quad f_{-2} = (1-2\nabla+\nabla^2)f_0,\quad f_{-3} = (1-3\nabla+3\nabla^2-\nabla^3)f_0,$$

so that

$$f_0+f_{-1}+f_{-2}+f_{-3} = (4-6\nabla+4\nabla^2-\nabla^3)f_0.$$

Thus

$$\begin{aligned} y_1 - y_{-3} &= h(1+\tfrac{1}{2}\nabla+\tfrac{5}{12}\nabla^2+\tfrac{3}{8}\nabla^3+\dots)(4-6\nabla+4\nabla^2-\nabla^3)f_0 \\ &= 4h(1-\nabla+\tfrac{2}{3}\nabla^2+\tfrac{7}{90}\nabla^4+\dots)f_0, \end{aligned}$$

and, since $(1-\nabla+\frac{2}{3}\nabla^2)f_0=\frac{1}{3}(2f_0-f_{-1}+2f_{-2})$, the formula is established.

Example 10. *Use Milne's method to calculate three steps in the solution of the equation $y' = x-y^2/10$, using the starting values given in Example 9.*

The formula (1) gives

$$y_1-y_{-3} = \frac{0{\cdot}4}{3}(0{\cdot}19994-0{\cdot}20098) = -0{\cdot}000139$$

$$\underline{1{\cdot}01513}$$

$$y_1 = 1{\cdot}01499.$$

Hence $f_1 = 0{\cdot}19698$, and formula (2) gives

$$y_1-y_{-1} = \frac{0{\cdot}1}{3}\begin{pmatrix}0{\cdot}19698\\0{\cdot}39988\\0{\cdot}00098\end{pmatrix} = 0{\cdot}019928 \qquad \underline{0{\cdot}99507}$$

$$\underline{0{\cdot}59784} \qquad y_1 = 1{\cdot}01500.$$

Using formula (1) again

$$y_2-y_{-2} = \frac{0{\cdot}4}{3}\begin{pmatrix}0{\cdot}39396-0{\cdot}09997\\0{\cdot}00196\end{pmatrix} = 0{\cdot}039455 \qquad \underline{1{\cdot}00000}$$

$$y_2 = 1{\cdot}03946.$$

Hence $f_2 = 0{\cdot}29195$, and formula (2) gives

$$y_2-y_0 = \frac{0{\cdot}1}{3}\begin{pmatrix}0{\cdot}29195\\0{\cdot}78792\\0{\cdot}09997\end{pmatrix} = 0{\cdot}039326 \qquad \underline{1{\cdot}00013}$$

$$\underline{1{\cdot}17984} \qquad y_2 = 1{\cdot}03946.$$

Using formula (1) again

$$y_3-y_{-1} = \frac{0{\cdot}4}{3}\begin{pmatrix}0{\cdot}58390-0{\cdot}19698\\0{\cdot}19994\end{pmatrix} = 0{\cdot}078248 \qquad \underline{0{\cdot}99507}$$

$$\underline{0{\cdot}78384} \qquad y_3 = 1{\cdot}07332.$$

Hence $f_3 = 0{\cdot}38480$, and formula (2) gives

$$y_3-y_1 = \frac{0{\cdot}1}{3}\begin{pmatrix}0{\cdot}38480\\1{\cdot}16780\\0{\cdot}19698\end{pmatrix} = 0{\cdot}058318 \qquad \underline{1{\cdot}01500}$$

$$\underline{1{\cdot}74958} \qquad y_3 = 1{\cdot}07332.$$

Thus the values of y_1, y_2 and y_3 to four places of decimals are as found in Examples 8 and 9.

9.13 Deferred approach to a limit

A process due to L. F. Richardson called the *deferred approach to a limit* can be used in step by step integration over a range. In this process a simple integration formula can be used and an error tolerated at each step, the cumulative error being allowed for when the integration is completed. The method can be used in straightforward integration or in obtaining the solution of a differential equation.

Thus to find the value of $y = \int_0^a f(x)\,dx$, we divide the range of integration into intervals x_0 to x_1, x_1 to x_2, . . ., each of length h, and using the trapezoidal rule we have

$$y_1 - y_0 = \tfrac{1}{2}h(f_0 + f_1),$$

$$y_2 - y_1 = \tfrac{1}{2}h(f_1 + f_2),$$

and so on. Now the error in using the trapezoidal rule for integration over an interval of length h is of order h^3 and, since the number of intervals in the range is inversely proportional to h, the cumulative error at the end of the range is of order h^2. If the integration is carried out with two different intervals, say h and $2h$, giving results $y(a, h)$ and $y(a, 2h)$ respectively, the errors of the results will be proportional to h^2 and $4h^2$, and hence an estimate of the error in using the interval h is $\tfrac{1}{3}\{y(a, 2h) - y(a, h)\}$. Thus the final estimate is

$$y(a,0) = y(a,h) - \tfrac{1}{3}\{y(a,2h) - y(a,h)\}$$

$$= \tfrac{1}{3}\{4y(a,h) - y(a,2h)\}.$$

In the same way, if we seek to integrate the equation $y' = f(x, y)$ over the range 0 to a we can simplify the computation by using only two terms of the Adams-Bashforth prediction and correction formulae. These are respectively

$$y_1 - y_0 = \tfrac{1}{2}h(3f_0 - f_{-1}),$$

and

$$y_1 - y_0 = \tfrac{1}{2}h(f_0 + f_1),$$

and the error in each step will be of order h^3. There is also an error of order h^2 in the value of $f(x, y)$ at each step, giving a further error of order h^3 in each interval. Thus the cumulative error over the a/h intervals will be of order h^2 and this may be eliminated by taking intervals of h and $2h$ so as to estimate the error. The advantage of the method lies in the fact that, since the cumulative error can be reduced, a simple integration formula can be used.

Exercises 9(*e*)

1. Continue the integration of the differential equation $y' = x-y^2/10$, with $y = 1$ when $x = 0$, up to the value $x = 1$, using the values up to $x = 0{\cdot}5$ obtained in Example 8.

2. Continue the integration of the differential equation $y'+y^2 = x^2$ with $y = 0$ when $x = 0$ up to $x = 1$, given that when $x = \pm 0{\cdot}2$, $y = \pm 0{\cdot}00267$ and when $x = \pm 0{\cdot}4$, $y = \pm 0{\cdot}02131$.

3. The differential equation $y'(y^2+1) = x^2$, with $y = 0$ when $x = 0$ has a solution which is $\pm 0{\cdot}0416$ when $x = \pm\frac{1}{2}$ and $\pm 0{\cdot}3222$ when $x = \pm 1$. Find the value of the solution when $x = 4$.

4. The differential equation $y'+y^2+y/x+1 = 0$ with $y = 0$ when $x = 0$ has $y = \mp 0{\cdot}10050$ when $x = \pm 0{\cdot}2$ and $y = \mp 0{\cdot}20410$ when $x = \pm 0{\cdot}4$. Find the values of y when $x = 1$ and when $x = 2$.

5. Show that the differential equation $y'+y(1-y^2)^{1/2} = 0$ with $y = 1$ when $x = 0$, has solution $y = 0{\cdot}99502$ when $x = \pm 0{\cdot}1$ and $y = 0{\cdot}98033$ when $x = \pm 0{\cdot}2$. Find correct to four decimal places the value of the solution when $x = 1$.

6. Find to three decimal places the values of the solution of the equation $y' = (x^2+y^2)/(x+y)$, with $y = 1$ when $x = 0$, for $x = 0{\cdot}5$ and $x = 1$.

7. Obtain an approximate formula for step by step numerical integration of the differential equation $y' = 1-y^2$ where $y = 0$ when $x = 0$. Use your formula to tabulate the values of y from $x = 0$ to $x = 0{\cdot}6$ in steps of $0{\cdot}1$, giving values correct to three decimal places. Integrate the differential equation and obtain the true value of y to three decimal places when $x = 0{\cdot}5$. (L.U.)

8. Obtain a formula for step by step numerical integration of the equation $y' = f(x, y)$. If $y' = x+y^2/10$ and $y = 1$ when $x = 0$, (*a*) draw rough graphs of the solutions using the method of isoclines, (*b*) show by any method that the values of y at $x = -0{\cdot}1$ and $0{\cdot}1$ are $0{\cdot}99507$ and $1{\cdot}01513$ respectively, (*c*) extend the solution to find the values of y at $x = 0{\cdot}2$, $0{\cdot}3$, to four decimal places. (L.U.)

9.14 Simultaneous differential equations

Numerical solutions of simultaneous differential equations are obtained in much the same way as those of single equations. Consider the first order equations

$$\frac{dy}{dx} = f(x, y, z), \qquad \frac{dz}{dx} = g(x, y, z), \tag{1}$$

with the boundary conditions $y = y_0$ and $z = z_0$ when $x = x_0$. Starting values may be obtained by any of the methods used for single equations, each step in the process of integration being carried out independently for each equation except for the substitution in the equations. It is often convenient to use Taylor series, calculating successive derivatives from the equations (1). Thus

$$\frac{d^2y}{dx^2} = \frac{\partial f}{\partial x} + \frac{\partial f}{\partial y}\frac{dy}{dx} + \frac{\partial f}{\partial z}\frac{dz}{dx} = \frac{\partial f}{\partial x} + f\frac{\partial f}{\partial y} + g\frac{\partial f}{\partial z},$$

$$\frac{d^2z}{dx^2} = \frac{\partial g}{\partial x} + \frac{\partial g}{\partial y}\frac{dy}{dx} + \frac{\partial g}{\partial z}\frac{dz}{dx} = \frac{\partial g}{\partial x} + f\frac{\partial g}{\partial y} + g\frac{\partial g}{\partial z},$$

and higher derivatives are obtained in a similar way.

When starting values have been obtained, forward integration may be carried out by constructing difference tables for f and g and using the Adams-Bashforth prediction and correction formulae of §9.10. Alternatively formulae such as those of §9.11 and §9.12 may be used as for single equations.

Example 11. *Find to four places of decimals the solutions between $x = 0$ and $x = 0{\cdot}5$ of the equations $y' = \frac{1}{2}(y+z)$, $z' = \frac{1}{2}(y^2-z^2)$, with $y = z = 1$ when $x = 0$.*

Using Taylor series to calculate starting values we have $y_0' = 1$, $z_0' = 0$, and differentiating

$$2y'' = y'+z', \quad 2y''' = y''+z'', \quad 2y^{iv} = y'''+z''',$$

$$z'' = yy'-zz', \quad z''' = yy''-zz''+(y')^2-(z')^2,$$

$$z^{iv} = yy'''-zz'''+3y'y''-3z'z''.$$

Hence

$$(y'')_0 = \tfrac{1}{2}, \quad (y''')_0 = \tfrac{3}{4}, \quad (y^{iv})_0 = \tfrac{5}{8},$$

$$(z'')_0 = 1, \quad (z''')_0 = \tfrac{1}{2}, \quad (z^{iv})_0 = \tfrac{7}{4},$$

and we have the series for y and z when $x = h$

$$y = 1+h+\tfrac{1}{4}h^2+\tfrac{1}{8}h^3+\tfrac{5}{192}h^4+\ldots,$$

$$z = 1+\tfrac{1}{2}h^2+\tfrac{1}{12}h^3+\tfrac{7}{96}h^4+\ldots.$$

These terms give five-figure accuracy when $h = \pm 0{\cdot}1$ and $\pm 0{\cdot}2$ and we have the following table of values of y, z, $f = \frac{1}{2}(y+z)$, $g = \frac{1}{2}(y^2-z^2)$:

	x	y	z	f	g
x_{-4}	−0·2	0·80904	1·01945	0·91424	−0·19237
x_{-3}	−0·1	0·90238	1·00492	0·95365	−0·09779
x_{-2}	0	1·00000	1·00000	1·00000	0
x_{-1}	0·1	1·10263	1·00509	1·05386	0·10279
x_0	0·2	1·21104	1·02078	1·11591	0·21231

Since $\nabla^4 f_0$ and $\nabla^4 g_0$ are both approximately 10^{-4} and the truncation error of Milne's method is $(14/45)10^{-5}$ with $h = 0{\cdot}1$ we shall use this method. We have

$$y_1 - y_{-3} = \frac{0{\cdot}4}{3} \times 3{\cdot}17796 \qquad z_1 - z_{-3} = \frac{0{\cdot}4}{3} \times 0{\cdot}32183$$

$$= 0{\cdot}42373 \qquad = 0{\cdot}04291$$

$$0{\cdot}90238 \qquad 1{\cdot}00492$$

$$y_1 = 1{\cdot}32611. \qquad z_1 = 1{\cdot}04783.$$

Hence

$$f_1 = 1{\cdot}18697, \qquad g_1 = 0{\cdot}33031.$$

Using these values of f_1 and g_1 with Simpson's rule as a correction formula we get the same values of y_1 and z_1.

Proceeding we have

$$\begin{aligned} y_2 - y_{-2} &= \frac{0{\cdot}4}{3} \times 3{\cdot}36575 & z_2 - z_{-2} &= \frac{0{\cdot}4}{3} \times 0{\cdot}65389 \\ &= 0{\cdot}44877 & &= 0{\cdot}08719 \\ &\ 1{\cdot}00000 & &\ 1{\cdot}00000 \\ y_2 &= 1{\cdot}44877. & z_2 &= 1{\cdot}08719. \end{aligned}$$

Hence

$$f_2 = 1{\cdot}26798, \qquad g_2 = 0{\cdot}45848.$$

With these values of f_2 and g_2 Simpson's rule gives the same values of y_2 and z_2.

Proceeding

$$\begin{aligned} y_3 - y_{-1} &= \frac{0{\cdot}4}{3} \times 3{\cdot}58081 & z_3 - z_{-1} &= \frac{0{\cdot}4}{3} \times 1{\cdot}01127 \\ &= 0{\cdot}47744 & &= 0{\cdot}13484 \\ &\ 1{\cdot}10263 & &\ 1{\cdot}00509 \\ y_3 &= 1{\cdot}58007. & z_3 &= 1{\cdot}13993. \end{aligned}$$

Hence $f_3 = 1{\cdot}36000$, $g_3 = 0{\cdot}59859$ and Simpson's rule confirms the values of y_3 and z_3.

We have finally

x	0	0·1	0·2	0·3	0·4	0·5
y	1	1·1026	1·2110	1·3261	1·4488	1·5801
z	1	1·0051	1·0208	1·0478	1·0872	1·1399

9.15 Second order differential equations

Numerical solutions of equations of the second and higher orders can be obtained by adaptation of the methods of the previous sections. The general form of a second order equation is

$$y'' = f(x, y, y'), \tag{1}$$

and a solution is required to satisfy certain boundary conditions. If these conditions relate to a single value of x, so that $y = y_0$ and $y' = y_0'$ when $x = x_0$, we have what is called a *one-point boundary problem* or a *marching problem*, and a solution can be built up by step by step integration starting from the point x_0. If the conditions relate to two values of x, that is $y = y_0$ when $x = x_0$ and $y = y_1$ when $x = x_1$ or $y' = y_1'$ when $x = x_1$, we have a *two-point boundary problem* or a *jury problem* and in this case the solution, if one exists, can be much more complicated. If the differential equation is linear it is usually sufficient to calculate two solutions relative to the point $x = x_0$ and to choose a combination of these solutions to satisfy the conditions at $x = x_0$ and $x = x_1$. If, however, the equation is non-linear it may be necessary to calculate a number of solutions starting from x_0 before one is found which is satisfactory at $x = x_1$.

A differential equation of any order can be reduced to a system of simultaneous equations of the first order by introducing new variables. Thus for the second order equation $y'' = f(x, y, y')$ writing $z = y'$, we have the simultaneous equations

$$\frac{dy}{dx} = z,$$

$$\frac{dz}{dx} = f(x, y, z),$$

and the boundary conditions $y = y_0$ and $y' = y_0'$ when $x = x_0$ become $y = y_0$ and $z = z_0 = y_0'$ when $x = x_0$. Numerical solutions of these equations can be found by the methods of §9.14.

In the same way a differential equation of the nth order can be reduced to a system of n first order simultaneous equations by writing $y' = z$, $z' = w$, etc.

Example 12. *Find to three places of decimals a solution of the equation*

$$y''+2xy'-4y = 0 \quad \textit{with} \quad y = y' = 1 \quad \textit{when} \quad x = 0.$$

The corresponding simultaneous equations are

$$y' = z,$$

$$z' = 4y-2xz,$$

with $y = z = 1$ when $x = 0$.
By successive differentiations we easily obtain the Taylor series

$$y = 1+h+2h^2+\frac{h^3}{3}-\frac{h^5}{30}+\frac{h^7}{210}-\ldots,$$

$$z = 1+4h+h^2-\frac{h^4}{6}+\frac{h^6}{30}-\ldots.$$

From these series we obtain starting values for $x = -0{\cdot}2$, $0{\cdot}2$, $0{\cdot}4$ and the corresponding values of $f(=z)$ and $g(=4y-2xz)$.

$$\begin{array}{llll} x = -0{\cdot}2, & y_{-3} = 0{\cdot}8773, & z_{-3} = 0{\cdot}2397 = f_{-3}, & g_{-3} = 3{\cdot}6052 \\ \quad\; 0, & y_{-2} = 1{\cdot}0000, & z_{-2} = 1{\cdot}0000 = f_{-2}, & g_{-2} = 4{\cdot}0000 \\ \quad\; 0{\cdot}2, & y_{-1} = 1{\cdot}2827, & z_{-1} = 1{\cdot}8397 = f_{-1}, & g_{-1} = 4{\cdot}3949 \\ \quad\; 0{\cdot}4, & y_0 = 1{\cdot}7410, & z_0 = 2{\cdot}7559 = f_0, & g_0 = 4{\cdot}7593. \end{array}$$

Using the method of §9.12, we have

$$\begin{array}{ll} y_1-y_{-3} = \frac{0{\cdot}8}{3}\times 5{\cdot}6721 & z_1-z_{-3} = \frac{0{\cdot}8}{3}\times 13{\cdot}1237 \\ \qquad\quad = 1{\cdot}5126 & \qquad\quad = 3{\cdot}4997 \\ \qquad\qquad 0{\cdot}8773 & \qquad\qquad 0{\cdot}2397 \\ \hline \qquad y_1 = 2{\cdot}3899. & \qquad z_1 = 3{\cdot}7394. \end{array}$$

Hence

$$f_1 = 3{\cdot}7394, \qquad g_1 = 5{\cdot}0723.$$

Simpson's rule then gives

$$\begin{aligned} y_1 - y_{-1} &= \frac{0\cdot 2}{3} \times 16\cdot 6027 & z_1 - z_{-1} &= \frac{0\cdot 2}{3} \times 28\cdot 5044 \\ &= 1\cdot 1068 & &= 1\cdot 9003 \\ & \underline{1\cdot 2827} & & \underline{1\cdot 8397} \\ y_1 &= 2\cdot 3895. & z_1 &= 3\cdot 7400. \end{aligned}$$

The corrected values of y_1 and z_1 may be accepted and the recalculated values $f_1 = 3\cdot 7400$, $g_1 = 5\cdot 0700$ used in the next step; the integration then proceeds normally.

9.16 Other methods for second order equations

The second order differential equation

$$y'' = f(x, y, y'), \tag{1}$$

with one-point boundary conditions, can be solved by direct integration using one or other of the integration formulae of the previous sections and integrating twice. Starting values

$$y_0, y_{-1}, y_{-2}, \ldots \quad \text{and} \quad y'_0, y'_{-1}, y'_{-2}, \ldots$$

may be obtained from a Taylor series or by other methods.

The following procedure is recommended for forward integration when starting values have been obtained.

(1) Use a prediction formula to find y'_1, e.g.

$$y'_1 - y'_{-3} = \frac{4h}{3}(2y''_0 - y''_{-1} + 2y''_{-2}). \tag{2}$$

(2) With the value obtained for y'_1 use a correction formula to find y_1, e.g.,

$$y_1 - y_{-1} = \frac{h}{3}(y'_1 + 4y'_0 + y'_{-1}). \tag{3}$$

(3) With the values obtained for y_1 and y_1' calculate y_1'' from the differential equation (1).

(4) With this value of y_1'' use a correction formula to recalculate y'_1, e.g.

$$y'_1 - y'_{-1} = \frac{h}{3}(y''_1 + 4y''_0 + y''_{-1}). \tag{4}$$

(5) If the two values obtained for y'_1 differ significantly, y_1 must be recalculated from equation (3) and hence y''_1, to give another value of y'_1.

Instead of the above formulae it is often convenient to use the Adams-Bashforth prediction and correction formulae or those of §9.11.

Example 13. *Find to three places of decimals a solution of the equation*

$$y'' = 4y - 2xy' \quad \textit{with} \quad y = y' = 1 \quad \textit{when} \quad x = 0.$$

Accepting the starting values from Example 12 we have

x	y	y'	y''
$-0{\cdot}2$	$y_{-3} = 0{\cdot}8773$	$y_{-3}' = 0{\cdot}2397$	$y_{-3}'' = 3{\cdot}6052$
0	$y_{-2} = 1{\cdot}0000$	$y_{-2}' = 1{\cdot}0000$	$y_{-2}'' = 4{\cdot}0000$
$0{\cdot}2$	$y_{-1} = 1{\cdot}2827$	$y_{-1}' = 1{\cdot}8397$	$y_{-1}'' = 4{\cdot}3949$
$0{\cdot}4$	$y_0 = 1{\cdot}7410$	$y_0' = 2{\cdot}7559$	$y_0'' = 4{\cdot}7593.$

Using equation (2)

$$y_1' - y_{-3}' = 3{\cdot}4997,$$
$$y_1' = 3{\cdot}7394.$$

Using equation (3)

$$y_1 - y_{-1} = 1{\cdot}1068,$$
$$y_1 = 2{\cdot}3895,$$

and hence

$$y_1'' = 5{\cdot}0707.$$

Recalculating y_1' from equation (4)

$$y_1' - y_{-1}' = 1{\cdot}9002,$$
$$y_1' = 3{\cdot}7399.$$

Recalculating y_1 from equation (3)

$$y_1 - y_{-1} = 1{\cdot}1069,$$
$$y_1 = 2{\cdot}3896.$$

Hence $y_1'' = 5{\cdot}0705$ and, accepting the recalculated values of y_1 and y_1', we proceed to the next step.

9.17 High-speed electronic computors

All the processes for finding numerical solutions of differential equations described in this chapter can be carried out on electronic computors. Advice on the methods most suitable for use with computors is given in the National Physical Laboratory pamphlet referred to in §7.2, page 192, and a brief account of the working of such computors may be of interest to the reader.

Suppose then that it is desired to find a solution of the equation $y' = f(x, y)$ with certain initial conditions and that it has been decided to use Milne's method of solution given in §9.12. We may assume that starting values $y_0, y_{-1}, \ldots,$ corresponding to $x_0, x_{-1}, \ldots,$ have been found and that values of $f_0, f_{-1}, \ldots,$ have been calculated. These starting values can in fact be found by the computor but they are sometimes found separately.

Starting values are stored in the machine, each in a separate store, that is, each value is imprinted on a tape which enters the computor and is automatically converted into a number in the binary scale. Thus the number 25 becomes 11001 and the store can be visualized as a row of lamps some of which are illuminated so as to give the sequence 11001. The machine is capable of adding, subtracting, multiplying or dividing the numbers in any two stores and placing the result of the operation in another store.

Another tape is prepared which will enter the computor with coded instructions for the process of solving the equation. Thus the first instruction will be to double the number f_0 in store 8 (say) and put the result in store 10; this may be imprinted on the tape as 8 $J\phi$ 10. The next instruction will be to subtract f_{-1} in store 7 from store 10 placing the result $2f_0 - f_{-1}$ in store 11, then to place $2f_{-2}$ in store 12 and the sum of stores 11 and 12 in store 13. Store 13 now contains $2f_0 - f_{-1} + 2f_{-2}$, this is to be multiplied by $\frac{4}{3}$ and by h, the interval in x, and y_{-3} added, leading to y_1 in store 16.

The next set of instructions on the tape will be to compute f_1 and recalculate y_1, using the correction formula. Let us suppose that this value $(y_1)_1$ arrives in store 28. The difference $(y_1)_1 - y_1$ is then to be calculated and placed in store 29 and instructions are given that, if this difference is greater than a certain small error margin ε, f_1 must be recalculated and the process repeated or repeated with the interval halved. The final values obtained for y_1 and f_1 may, if desired, be printed (reconverted to the decimal scale) but in any case they are removed to the stores formerly occupied by y_0 and f_0 while each of the starting values is moved back or deleted. The computor is then ready for the next step in which the routine is repeated.

The solution of a first order differential equation involving 100 steps of which the results of 10 or so are to be printed may be completed in a few seconds. Of course the preparation of the routine of instructions to the computor is a lengthy business, but this routine may be prepared for general values of the function $f(x, y)$ and a subroutine tape prepared giving instructions for the calculation of the particular function $f(x, y)$ which appears in the equation that is to be solved. The calculation of $f(x, y)$ may indeed be very complex in certain cases involving the computing of such functions as $\sin x$, $\log_e x$, etc.

Exercises 9(f)

Solve the following differential equations for five steps with the interval given and to the accuracy shown:

1. $y' = z$, $z' = y - xz$, with $y = z = 1$ when $x = 0$; $h = 0{\cdot}2$, four places of decimals.

2. $y' = (1+z)y$, $z' = (1+y)z$, with $z = 0{\cdot}2$ and $y = 1{\cdot}2$ when $x = 0$; $h = 0{\cdot}1$, three places of decimals.

3. $y' = (z-x)y$, $z' = (z+y)x$, with $y = 1$ and $z = 0$ when $x = 0$; $h = 0{\cdot}1$, three places of decimals.

4. $y' = z$, $z' = 2xz + (1-x^2)y$, $y = z = 1$ when $x = 0$; $h = 0{\cdot}2$, three places of decimals.

5. $y'' + y'/x + y = 0$, $y = 1$ and $y' = 0$ when $x = 0$; $h = 0{\cdot}4$, four places of decimals.

6. $y'' = xy$, $y = 0$ and $y' = 1$ when $x = 0$; $h = 0{\cdot}5$, four places of decimals.
7. $y''+2yy' = 0$, $y = 0$ and $y' = 1$ when $x = 0$; $h = 0{\cdot}2$, four places of decimals.
8. $y''-y^2 = 0$, $y = y' = 1$ when $x = 0$; $h = 0{\cdot}2$, three places of decimals.
9. $y''+y^3 = 0$, $y = 0$ and $y' = 1$ when $x = 0$; $h = 0{\cdot}2$, four places of decimals.
10. $\ddot{y}+y+y^2 = 0$, $y = 1$ and $\dot{y} = 0$ when $t = 0$; $h = 0{\cdot}2$, four places of decimals.

CHAPTER 10

THE RELAXATION METHOD

10.1 Introduction

The numerical solution of ordinary differential equations with one-point boundary conditions was discussed in the previous chapter. Here we describe a method which is useful in solving ordinary differential equations with two-point boundary conditions and show how the method can be extended to deal also with the problems involving *partial* differential equations in two independent variables. This method was developed by Sir Richard Southwell and his co-workers and, for a reason which will be explained later, is known as the *relaxation method.*

The basis of the method lies in the replacement of derivatives in the differential equation and boundary conditions by their finite-difference approximations, and in the subsequent approximate solution of the resulting simultaneous algebraic equations. It is in the second of these processes that the relaxation method makes its special contribution: a very approximate solution of the simultaneous equations is guessed and the resulting errors are then systematically removed by what is called the "relaxation process". In carrying out this process, the solver is the master rather than the servant of the method, and he can often shorten his work considerably by using the experience and judgment he gains with practice at this type of work.

It is not proposed to discuss the more mathematical aspects of the relaxation method in this chapter. Such topics as the convergence of the relaxation process, the errors caused by the replacement of derivatives by their finite-difference approximations, etc., have been investigated, and the results of such investigations can be found in the original papers and in treatises devoted to the subject. It is sufficient to say here that the method outlined in this chapter works in practice and that its great power has been clearly demonstrated by the many important engineering and physical problems which have been recently solved by its use.

10.2 The solution of ordinary differential equations

To fix ideas, we consider first the simple differential equation

$$\frac{d^2y}{dx^2}+F(x) = 0, \tag{1}$$

with the two-point boundary conditions

$$y = y_a \text{ when } x = a, \qquad y = y_b \text{ when } x = b, \qquad (a < b).$$

Here it is supposed that $F(x)$ is a function of x whose numerical values are known at a set of values of x equally spaced in the interval (a, b) and the solution aimed at is a set of numerical values of y corresponding to these values of x.

The first step in the solution is to replace the term d^2y/dx^2 in equation (1) by its finite-difference approximation. To do this, subdivide the interval (a, b) by a series of points at equal intervals h and consider any three* consecutive points of subdivision, x_3, x_0, x_1 (Fig. 22). If suffix 0

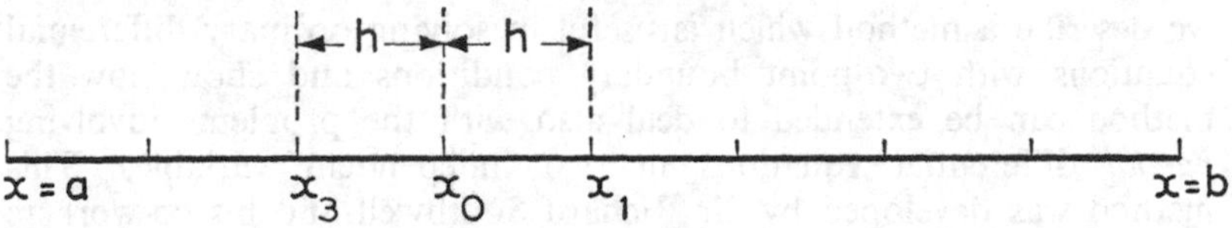

FIG. 22

denotes values at x_0, we have in the neighbourhood of this point,

$$y = y_0+\left(\frac{dy}{dx}\right)_0(x-x_0)+\frac{1}{2!}\left(\frac{d^2y}{dx^2}\right)_0(x-x_0)^2+\frac{1}{3!}\left(\frac{d^3y}{dx^3}\right)_0(x-x_0)^3+\ldots.$$

Writing $x_1 = x_0+h$, $x_3 = x_0-h$ and using y_1, y_3 to denote the values of y corresponding to these values of x,

$$\left.\begin{aligned} y_1 &= y_0+h\left(\frac{dy}{dx}\right)_0+\frac{h^2}{2}\left(\frac{d^2y}{dx^2}\right)_0+\frac{h^3}{6}\left(\frac{d^3y}{dx^3}\right)_0+\frac{h^4}{24}\left(\frac{d^4y}{dx^4}\right)_0+\ldots, \\ y_3 &= y_0-h\left(\frac{dy}{dx}\right)_0+\frac{h^2}{2}\left(\frac{d^2y}{dx^2}\right)_0-\frac{h^3}{6}\left(\frac{d^3y}{dx^3}\right)_0+\frac{h^4}{24}\left(\frac{d^4y}{dx^4}\right)_0-\ldots. \end{aligned}\right\} \quad (2)$$

By addition,

$$y_1+y_3-2y_0 = h^2\left(\frac{d^2y}{dx^2}\right)_0 + \text{terms of order } h^4 \text{ and above.} \quad (3)$$

Hence, neglecting terms of order h^4 and above, the differential equation (1) can, in the neighbourhood of $x = x_0$, be replaced by the algebraic equation

$$y_1+y_3-2y_0+h^2F(x_0) = 0. \quad (4)$$

There is a relation of this type at each point of subdivision of the interval (a, b) and we therefore have a set of simultaneous algebraic equations for solution.

This set of simultaneous algebraic equations, of which equation (4)

* The reason for denoting the left-hand point by x_3 rather than by x_{-1} is to enable the notation to be preserved when the method is extended to partial differential equations.

is typical, is solved approximately by the following method. We write

$$R_0 \equiv y_1 + y_3 - 2y_0 + h^2 F(x_0), \tag{5}$$

and R_0 is called the *residual* at $x = x_0$. When correct values of y at each point of subdivision have been found, the residuals at each point should, from equation (4), be zero. The solution is started by calculating the *initial residuals* at each point of subdivision from equation (5), using the given value of $F(x_0)$ and *guessed* values of y_1, y_3 and y_0. These initial residuals now have to be reduced to zero (or liquidated) by a "relaxation process".

The basis of this process is the *relaxation pattern*, which shows the effect on the residuals of a unit increment in the value of y at a given point of subdivision. Because of the term $-2y_0$ in equation (5), the effect of such an increment is to decrease the value of the residual R_0 at $x = x_0$ by two units. This, however, is not the only effect of the

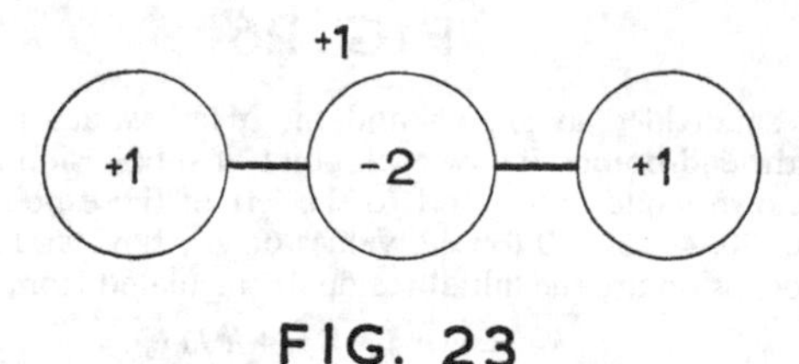

FIG. 23

increment, for the formulae for the residuals R_1, R_3 at $x = x_1$ and $x = x_3$ each contain a term y_0 and each of these residuals is therefore increased by one unit. The total effect is conveniently shown by the pattern of Fig. 23, which shows that when the value of y is increased by unity at a chosen point, the residual at the point in question is altered by -2 and those at the two neighbouring points by $+1$. This relaxation pattern holds at every point of subdivision of the range (a, b) except

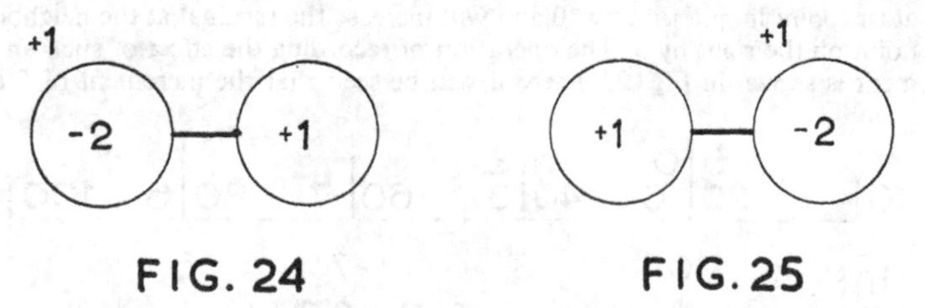

FIG. 24 FIG. 25

the first and last. Since the value of y is given at $x = a$, this is the *correct* value, and thus there is no residual for which any change has to be recorded to the left of the first point of subdivision. The relaxation pattern for this point is therefore that shown in Fig. 24, and, for similar reasons, that for the last point of subdivision is as depicted in Fig. 25.

The details of the actual relaxation process are best shown by describing the work for a particular case; this is done in Example 1 below.

Example 1. *Obtain a numerical solution of the differential equation $y''+F(x) = 0$ given that $y = 0$ when $x = 0$, $y = 100$ when $x = 5$ and values of $F(x)$ are given by the table*:

x	1	2	3	4
$F(x)$	10	3	−7	6.

Here we take the interval $h = 1$ and, as our initial guesses for the values of y we assume 20, 40, 60, 80, these being linearly interpolated between the end-values of 0 and 100. The relaxation process is carried out on a *relaxation diagram*, and Fig. 26 shows the first stages of such a diagram. The range $x = 0$

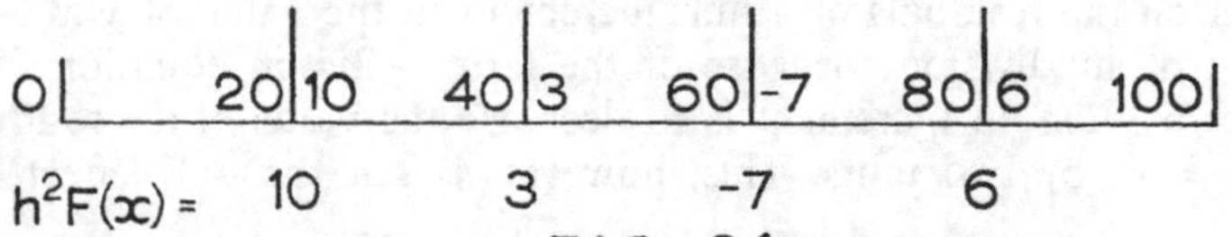

0 20 | 10 40 | 3 60 | -7 80 | 6 100

$h^2F(x)$ = 10 3 -7 6

FIG. 26

to $x = 5$ is subdivided as shown and the given values $y = 0$, $y = 100$ are recorded at the end-points. Below each point of subdivision the given values of $h^2F(x)$ are shown while above and to the left of these points are written the initial guesses 20, 40, 60, 80 for the values of y. Above and to the right of the points of subdivision are the initial residuals calculated from equation (5),

$$R_0 = y_1+y_3-2y_0+h^2F(x_0).$$

Thus the residual at the first point of subdivision is

$$40+0-2(20)+10 = 10,$$

that at the second is

$$60+20-2(40)+3 = 3,$$

and so on.

Adjustments are now made to the guessed values of y until these residuals are liquidated. Starting with the biggest residual 10 (at the first point of subdivision), this can be removed by giving an increment of 5 to the value of y at this point. Using the relaxation pattern of Fig. 24, such an increment will reduce the residual at the point in question by 10 and will increase the residual at the neighbouring point on the right by 5. The operation of recording the effect of such an increment is shown in Fig. 27 where it will be seen that the increment of 5 in y is

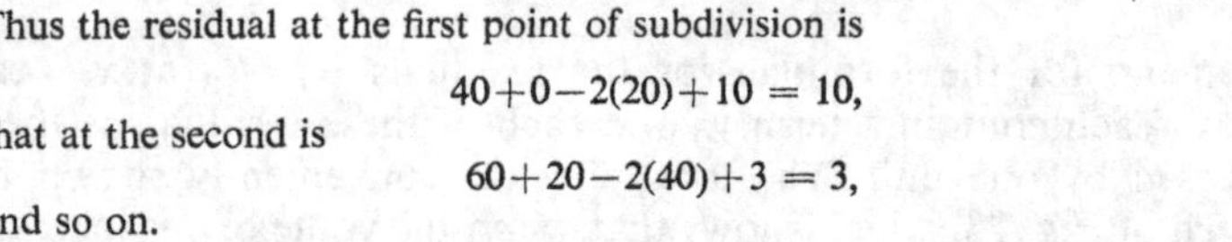

5 | 0 | 8

0 20 | 10 40 | 3 60 | -7 80 | 6 100

$h^2F(x)$ = 10 3 -7 6

FIG. 27

shown and the adjusted residuals of 0 (10−10) and 8 (3+5) are recorded. The largest residual is now that at the second point of subdivision, and this is removed by giving an increment of 4 to the value of y at this point. The effect of this is given by the relaxation pattern of Fig. 23 (Figs. 24 and 25 only being used at the first and last point of subdivision). The results at this stage of the work are shown in Fig. 28. The residual 6 at the last point of subdivision is now removed and the work proceeds in this way until the stage shown in Fig. 29 is reached. The liquidation process has now been carried through until the absolute value of each individual residual is less than 2 and the algebraic sum of all the residuals

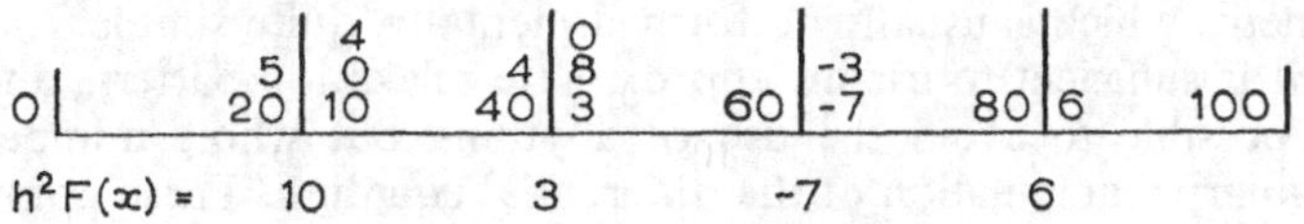

0 | 5 20 | 4 0 10 | 4 40 | 0 8 3 | 60 | −3 −7 | 80 | 6 | 100

$h^2F(x) =$ 10 3 −7 6

FIG. 28

is approximately zero. The final values of y, obtained by adding the various increments made during the process to the initial guessed values, are recorded to the left of the points of subdivision and above the upper line. As a check, the final residuals are recalculated from equation (5) and these are shown to the right of the points of subdivision above the upper line. In Fig. 29 these re-

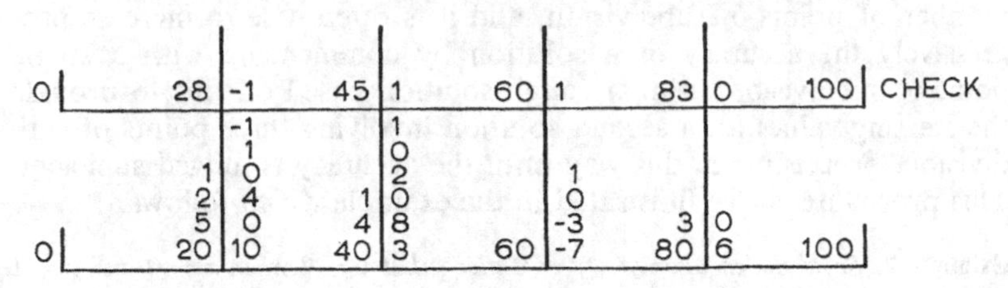

FIG. 29

calculated residuals agree with those previously recorded but, if slips had been made during the liquidation process, the values of y and the recalculated residuals recorded above the upper line would be used as the starting point of a fresh liquidation process. The required approximate solution to this example is therefore given by

$$x = 1 \quad 2 \quad 3 \quad 4$$
$$y = 28 \quad 45 \quad 60 \quad 83.$$

10.3 Some practical hints

Now that one example of the numerical solution of a two-point boundary value problem has been worked by the relaxation method, it will be profitable to give a few practical hints on the process.

Firstly, in all but the simplest examples such as that just considered, the columns of figures in the relaxation diagram tend to become inconveniently long. The only quantities, however, that really matter at a given stage of the calculation are the *current* residuals and the *total* of the assumed values of the wanted function y and the increments made up to the end of the stage in question. Thus any unduly lengthy column can be erased provided that the total of the y column and its current residual are re-recorded. It is therefore wise to work in pencil and to use tracing cloth for the relaxation diagram. Such cloth is very resistant to repeated erasures and, by drawing the framework of the diagram on the reverse side, it is possible to obviate its continual re-drawing.

Secondly, great accuracy is not essential at every stage of the liquidation process, and it is often convenient to sacrifice it in order to keep the arithmetic, which is usually performed mentally, quite simple. In some cases it is sufficient to use an approximate relaxation pattern, and it is often possible to avoid the use of anything but whole numbers by preliminary manipulation of the differential equation. These points are illustrated in Example 2.

The accuracy of the solution obtained increases, of course, as the interval h between successive points of subdivision of the range decreases, since the finite-difference approximation to the differential equation neglects terms in h^4 and above. On the other hand, the amount of work involved in the relaxation process is very dependent on the number of points of subdivision, and it is often wise to increase progressively the accuracy of a solution by commencing with a single point of subdivision, using a rough solution based on this to provide the starting values for a second solution involving three points of subdivision, proceeding in this way until the accuracy is judged sufficient. This procedure is also illustrated in the example which follows.

Example 2. *Solve the equation $y''+y = 0$ given that $y = 0$ when $x = 0$ and $y = 1$ when $x = \pi/2$.*

If we are aiming at three-figure accuracy, decimals can be avoided by writing $y = 10^{-3}Y$ and solving the boundary-value problem $Y''+Y = 0$ with $Y = 0$ when $x = 0$, $Y = 1000$ when $x = \pi/2$. Using equation (3) of §10.2, the finite-difference approximation to this differential equation is $Y_1+Y_3-2Y_0+h^2Y_0 = 0$ and the residual at $x = x_0$ is given by

$$R_0 \equiv Y_1+Y_3-(2-h^2)Y_0. \tag{1}$$

First taking $h = \pi/4$ and using one point of subdivision, the residual at $x = x_0$ will be zero if we take

$$\left(2-\frac{\pi^2}{16}\right)Y_0 = Y_1+Y_3 = 0+1000,$$

so that a rough starting value at the mid-point of the range is given by

$$Y_0 = \frac{1000}{2-\dfrac{\pi^2}{16}} = 723.$$

Interpolating linearly between 0, 723 and 723, 1000, rough starting values for a second approximate solution with three points of subdivision ($h = \pi/8$) are 362, 723, 862 and the residuals for this solution are, by equation (1), given by

$$R_0 \equiv Y_1+Y_3-\left(2-\frac{\pi^2}{64}\right)Y_0 = Y_1+Y_3-1{\cdot}846\ Y_0.$$

Thus the initial residual at the first point of subdivision in Fig. 30 is given by $R_0 = 723+0-1{\cdot}846\times 362 = 55$, and the other two residuals are calculated in a similar way. As only a rough solution is required at this stage, we have here approximated to 1·846 by 2 and liquidated the residuals in Fig. 30 by the relaxation patterns of Figs. 23, 24, 25, and the results of this diagram are used as starting values for the next approximation. Linearly interpolated values are used as starting values of Y at $x = \pi/16$, $x = 3\pi/16$, etc., and the next approximate solution is shown in Fig. 31. In this part of the work $h = \pi/16$ and the formula

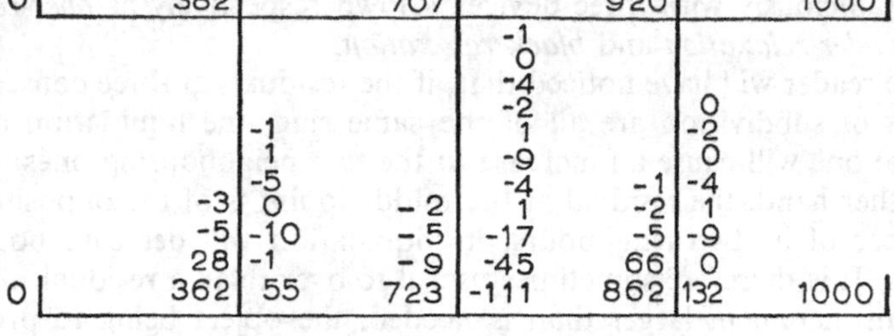

0	382		707		920		1000
	-3 -5 28 362	-1 1 -5 0 -10 -1 55	-2 -5 -9 723	-1 0 -4 -2 1 -9 -4 1 -17 -45 -111	-1 -2 -5 66 862	0 -2 0 -4 1 -9 0 132	
0							1000

FIG. 30

for the residuals is

$$R_0 \equiv Y_1+Y_3-\left(2-\frac{\pi^2}{256}\right)Y_0 = Y_1+Y_3-1{\cdot}9615\ Y_0. \tag{2}$$

Although this formula is used for calculating the initial residuals and checking at the end, it is sufficient to approximate to 1·9615 by 2 in the actual relaxation process. On the upper line of Fig. 31 are given the values of Y obtained as the

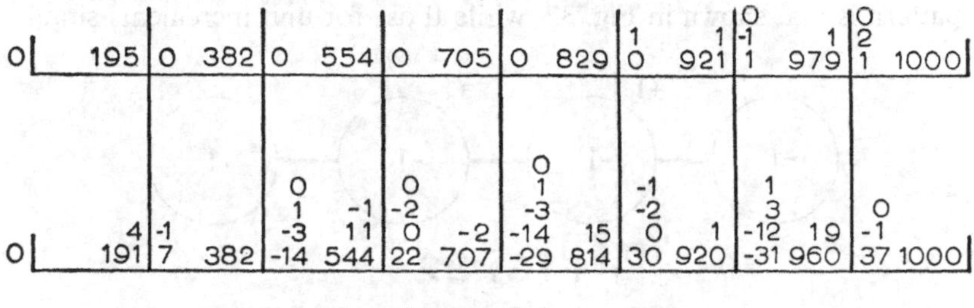

0	195	0	382	0	554	0	705	0	829	1 0	1 921	0 -1 1	1 979	0 2 1	1000
0	4 191	-1 7	382	0 1 -3 -14	-1 11 544	0 -2 0 22	-2 707	0 1 -3 -14 -29	15 814	-1 -2 0 30	1 920	1 3 -12 -31	19 960	0 -1 37	1000

FIG. 31

sums of the starting values and the increments made during the process. The residuals calculated from formula (2) are also shown on this line. As these differ very little from those arrived at in the main part of the work (when 1·9615 was approximated to by 2), little further liquidation of residuals is here necessary. In general a further approximate solution (made by again halving the interval h) should be obtained to establish the accuracy of the results found, but it is not proposed to do this here. For this example, an exact solution ($y = \sin x$) is easily found and the comparison between this solution and the values derived from Fig. 31 is given below.

x	0	$\pi/16$	$\pi/8$	$3\pi/16$	$\pi/4$	$5\pi/16$	$3\pi/8$	$7\pi/16$	$\pi/2$
y (exact)	0	0·195	0·383	0·555	0·707	0·831	0·924	0·981	1·000
$y = 10^{-3}Y$ (from Fig. 31)	0	0·195	0·382	0·554	0·705	0·829	0·922	0·980	1·000

10.4 Devices for speeding up the relaxation process

So far the relaxation process has been carried out by systematically liquidating the currently largest residual. This procedure is known as *point relaxation*, and although this is the basic operation it should be used judiciously with three devices known respectively as *over-relaxation*, *under-relaxation* and *block relaxation.*

The reader will have noticed that, if the residuals at three consecutive points of subdivision are all of the same sign, the liquidation of the middle one will cause an increase in the two neighbouring ones. If, on the other hand, the residual at the middle point is of the opposite sign to those of its two neighbours, its liquidation will decrease both the others. It is therefore sometimes useful to over-relax a residual, that is, give an increment larger than is needed, the object being to produce residuals of opposite signs at neighbouring points and then to liquidate these residuals at a later stage by point relaxation. Similarly, when adjacent residuals are fairly large and of opposite signs, under-relaxation is often useful.

The liquidation process can often be greatly accelerated by simultaneously applying increments at two or more adjacent points of subdivision. Unit *block relaxation patterns* are easily constructed by combining two or more of the point patterns of Fig. 23. Thus, for unit increments simultaneously applied at two adjacent points, the block pattern is that shown in Fig. 32, while those for unit increments simul-

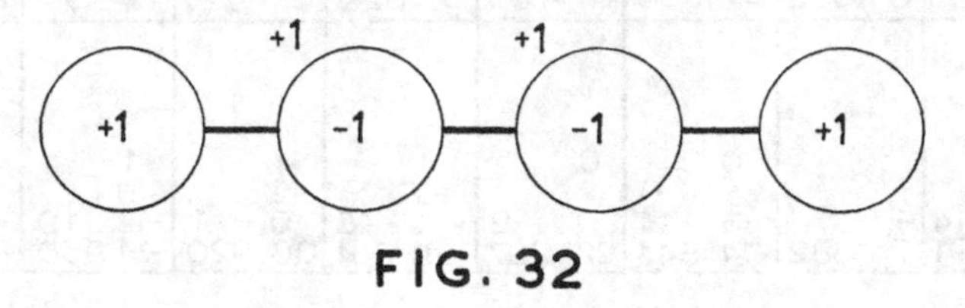

FIG. 32

taneously applied at three and four adjacent points are shown respectively in Figs. 33 and 34.

Block relaxation is particularly useful when the residuals at a given stage of the work are all of one sign. It is then useful to apply block relaxation to a group of centrally situated points of subdivision and to follow this by more extensive block operations. The details are shown in

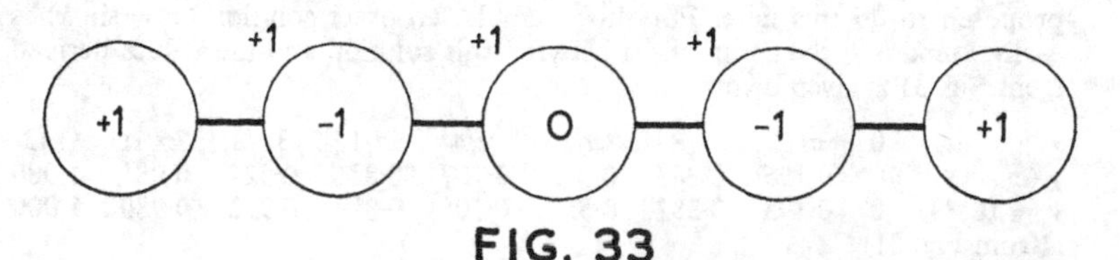

FIG. 33

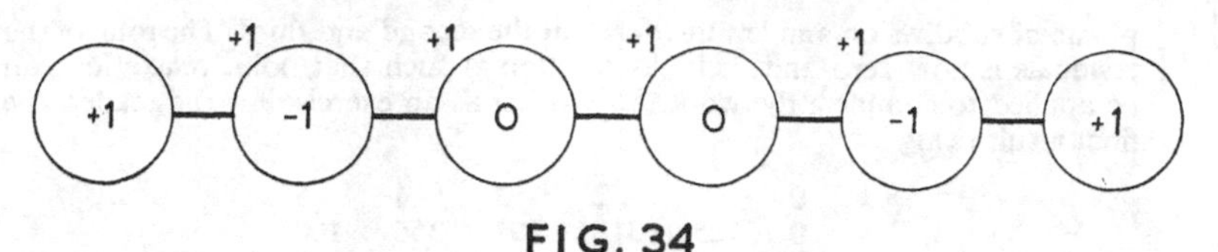

FIG. 34

Example 3 below and it will be seen that, after such operations, the total of the residuals is drastically reduced and the distribution of the individual residuals is then much more favourable to the application of point relaxation.

Example 3. *Solve the differential equation $y''+F(x) = 0$ given that $y = 0$ when $x = 0$, $y = 100$ when $x = 5$ and $F(x)$ is given by*:

x	1	2	3	4
$F(x)$	18	60	126	216.

Here $h = 1$ and the residuals are given by

$$R_0 \equiv y_1+y_3-2y_0+F(x_0).$$

Using the linearly interpolated values 20, 40, 60, 80 as initial values of y, the initial residuals are easily found from the above formula to be 18, 60, 126 and 216. These are all of one sign and their total is 420. The total residual at the two central points is 186, and we first apply an increment of (186/2) at each of these two points. The effect of this block is, by Fig. 32, to reduce the residuals at each of the two central points by 93 and to increase those at the other two points by a like amount. The position at the end of this operation is as shown in Fig. 35.

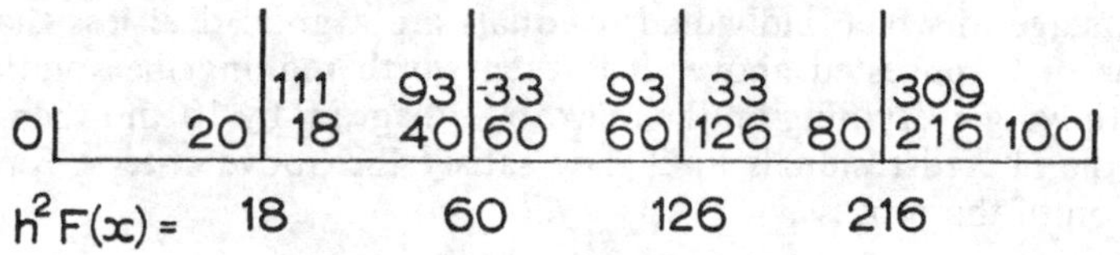

111 93 -33 93 33 309
0 20 18 40 60 60 126 80 216 100

$h^2F(x) =$ 18 60 126 216

FIG. 35

The total of the residuals is still 420, and we next apply a block of half this amount at each of the four points of subdivision. Using Fig. 34 (slightly modified because the end-points of the range are now involved), we obtain the result shown in Fig. 36, the residuals being reduced by 210 at the first and fourth

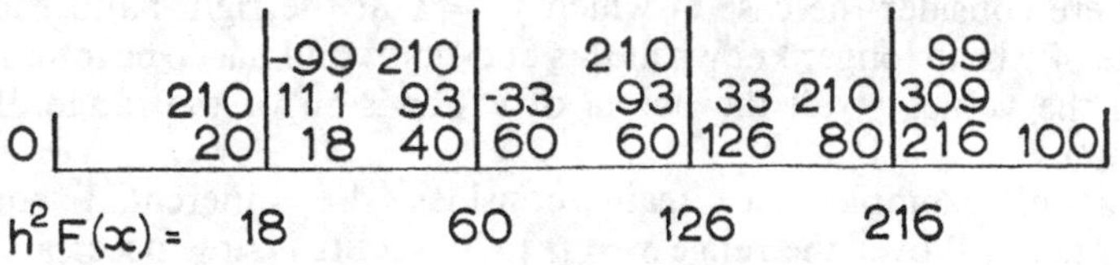

-99 210 210 99
210 111 93 -33 93 33 210 309
0 20 18 40 60 60 126 80 216 100

$h^2F(x) =$ 18 60 126 216

FIG. 36

points of subdivision and left unaltered at the second and third. The total of the residuals is now zero and their distribution is such that point relaxation can be applied to complete the work. This is left as an exercise for the reader, the final result being :

x	0	1	2	3	4	5
y	0	165	311	397	356	100.

10.5 The completion of the liquidation process

It has already been suggested (see Example 1) that the liquidation process is complete when each individual residual (irrespective of its sign) is approximately zero. Block relaxation is often useful in effecting this last requirement.

One additional criterion is really necessary to ensure that the process has been taken as far as possible. This is to ensure that those individual residuals which are not zero should be well distributed along the range of integration. It would be wrong, for example, to consider the process to have been completed if the residuals remaining were

$$1, 1, 0, 1, 0, -1, -1, -1, 0,$$

as, in this case, the positive residuals are bunched together on the left of the range and the negative ones on the right. In such cases, block relaxation is again often helpful in improving the distribution of the residuals.

Sometimes the finite-difference approximation to the differential equation is sufficiently accurate to justify taking the liquidation process to a stage in which individual residuals are a good deal less than the value of 2 suggested above. It is then worth refining the solution by multiplying everything on the relaxation diagram by 10 and then relaxing the altered residuals until they satisfy the above criteria for completion of the process.

10.6 Cases in which the derivative is specified at an end of the range

So far we have only considered the solution of second order differential equations in which the values of the wanted function y have been given at each end of the range of integration. In some problems the value of dy/dx is specified instead at one or both ends and, to fix ideas, we here consider the case in which $y' = k$ at the right-hand end. The value of y is no longer known at this end-point and has to be found along with the values at all the points of subdivision of the range of integration.

As an example, we again consider the differential equation $y''+F(x) = 0$ over the range $x = a$ to $x = b$. It is now necessary to be able to define the residual at the end-point $x = b$ but this cannot be done by equation (5) of §10.2 as it stands, for when the point $x = x_n$

is made to coincide with $x = b$, there is no point x_1 and no value y_1. However, we can say that $x = x_1$ is a "fictitious" point (Fig. 37) and, by subtracting equations (2) of §10.2, we find that

x_3 x_0 $(x = b)$ x_1

FIG. 37

$$\left(\frac{dy}{dx}\right)_0 = \frac{1}{2h}(y_1 - y_3), \tag{1}$$

if terms of order h^3 and above are neglected. Hence

$$y_1 = y_3 + 2hk, \tag{2}$$

for $(dy/dx)_0 = k$ when the point $x = x_0$ is taken at $x = b$. Substituting for y_1 from (2) in equation (5) of §10.2, the residual at the end-point $x = b$ is given by

$$R_0 \equiv 2y_3 - 2y_0 + 2hk + h^2F(b). \tag{3}$$

Modifications have also to be made in the relaxation patterns to be used at the end-point $x = b$ and at the neighbouring point to its left. These are easily found to be those given in Fig. 38 (*a*) and (*b*) respectively.

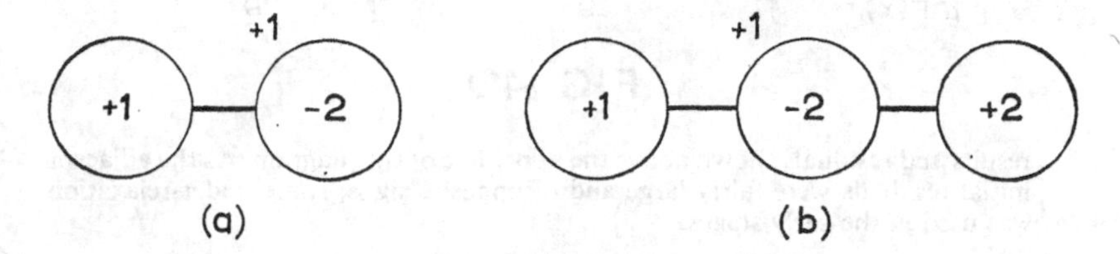

FIG. 38

Example 4. *Find the values of y at integer values of x satisfying the differential equation $y'' + F(x) = 0$ given that $y = 0$ when $x = 0$, $y' = 0$ when $x = 4$ and $F(x)$ is given by:*

x	0	1	2	3	4
$F(x)$	0	17	29	37	39.

A very rough first estimate of the value of y when $x = 4$ can be found by equating to zero the residual given by equation (3) above and taking $y_3 = 0$, $k = 0$, $h = 4$, $F = 39$. This gives $-2y_0 + 4^2 \times 39 = 0$, leading to $y_0 = 312$. A rough estimate of the value of y when $x = 2$ can now be found from equation (4) of §10.2 taking $y_3 = 0$, $y_1 = 312$, $h = 2$, $F = 29$. This gives

$$312 - 2y_0 + 2^2 \times 29 = 0,$$

leading to $y_0 = 214$. Starting values of 0, 214, 312 are now taken for a first relaxation solution with $h = 2$. The residual at $x = 2$ is calculated from equation (5) of §10.2 and that at $x = 4$ from equation (3) above, these residuals being respectively 0 and -40 as shown in Fig. 39. Over-relaxing at the right-

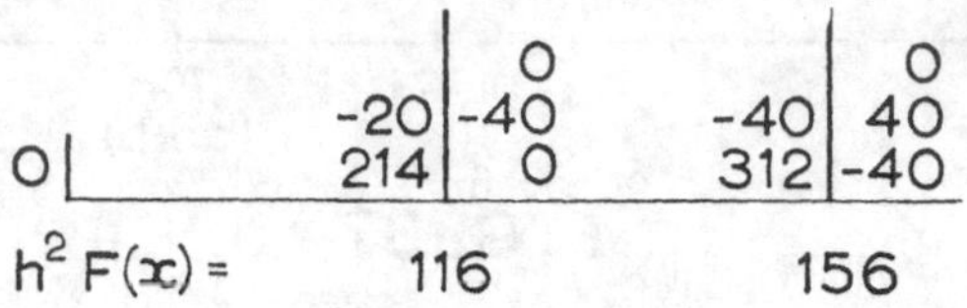

0
-20 -40
214 0
-40 40
312 -40
0
0
$h^2 F(x) =$
116
156

FIG. 39

hand point by applying an increment of -40 using the pattern of Fig. 38(*a*) and following this with an increment of -20 at the middle point using the pattern of Fig. 38(*b*), the values of 0, 194 and 272 are obtained as starting values for the next stage of the work.

Using linearly interpolated values and $h = 1$, starting values and the initial residuals are shown in Fig. 40. Also shown are the details leading to the final

0
101 1
186 -1
241 1
260 1
0
97 17
194 -29
233 37
272 -39
$h^2 F(x) =$
17
29
37
39

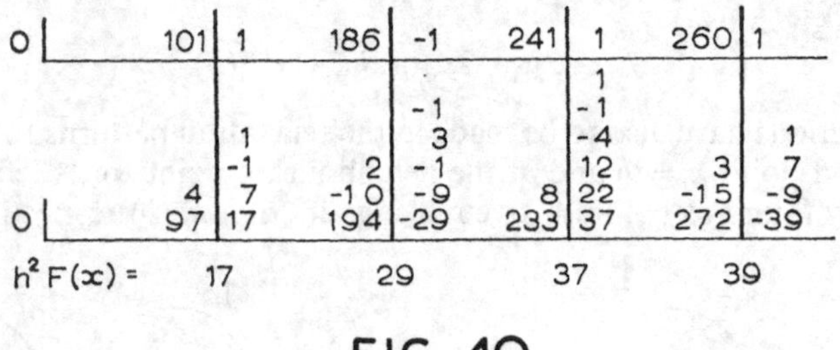

FIG. 40

results and residuals shown above the upper line of the diagram. As the adjacent initial residuals were fairly large and of opposite signs, some under-relaxation was used in the early stages.

The amount of work involved in a relaxation solution increases greatly with the number of points of subdivision used and it is sometimes possible to limit the number of such points. For example, if the boundary conditions are $y = C$ at $x = 0$ and $x = a$ and the differential equation is of a form giving symmetry about the mid-point of the range, the problem is best treated as one over half the range with the boundary conditions $y = C$ at $x = 0$ and $y' = 0$ at $x = \frac{1}{2}a$. This type of device for shortening the work is even more important when partial differential equations are under consideration.

10.7 Extension to other differential equations

To enable the principles of the relaxation process to be shown as clearly as possible, the examples chosen have involved only very simple differential equations such as $y''+F(x) = 0$. The method can, however, be extended to other differential equations at the expense of some loss of simplicity in the mental arithmetic required during the liquidation process.

Suppose, for example, that the differential equation under discussion is $y''+ay'+by+F(x) = 0$ where a, b are constants and $F(x)$ is a prescribed function of x. Using the finite-difference approximations for y'' and y' given respectively by equation (3) of §10.2 and equation (1) of §10.6, the formula for the residuals is, neglecting h^3 and higher powers,

$$\begin{aligned} R_0 &\equiv y_1+y_3-2y_0+\tfrac{1}{2}ah(y_1-y_3)+bh^2y_0+h^2F(x_0) \\ &= (1+\tfrac{1}{2}ah)y_1+(1-\tfrac{1}{2}ah)y_3-(2-bh^2)y_0+h^2F(x_0), \end{aligned}$$

and the relaxation pattern is that shown in Fig. 41. Thus, in performing

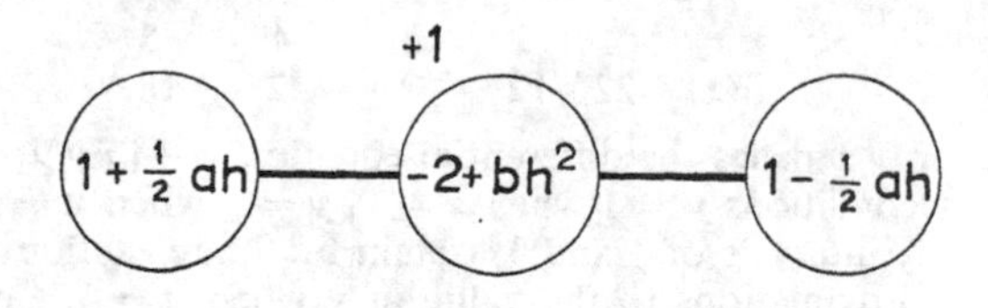

FIG. 41

the liquidation process, multipliers as shown in the diagram have to be used instead of the simpler multipliers 1, -2, 1 so far employed. By using suitable approximations to these multipliers, the bulk of the arithmetic can still be performed mentally but the correct multipliers must, of course, be used in checking the final residuals.

If the coefficients in the differential equation are functions of x, instead of constants, the relaxation pattern will change from point to point. Although, with practice, the relaxation method can still be used, a good deal of additional complexity is introduced. Again, for equations of higher order than the second, the relaxation pattern will involve more than three neighbouring points and this adds yet another complexity.

As this chapter is intended only as an introduction to the method, it is not proposed to go into the details of the solution of these more complicated equations here. A consideration of the equation

$$y''+F(x) = 0,$$

of which some detailed solutions have been given, is sufficient for the extension of the method to some of the more important partial differential equations of mathematical physics and it is in the solution of such equations that the relaxation method has had its most striking successes.

Exercises 10(*a*)

1. Using an interval $h = 1$, solve the differential equation $y''+F(x) = 0$ given that $y = 0$ when $x = 0$, $y = 120$ when $x = 6$ and $F(x)$ is given by the table:

x	1	2	3	4	5
$F(x)$	8	4	−7	2	−6.

2. Using an interval $h = 1$, solve the differential equation $y''+F(x) = 0$ given that $y = 0$ when $x = 0$, $y = 0$ when $x = 6$ and $F(x)$ is given by the table:

x	1	2	3	4	5
$F(x)$	20	25	25	30	10.

[Hint: use block relaxation initially.]

3. Solve the two-point boundary value problem of Exercise 2 above when the values of $F(x)$ are given by:

x	1	2	3	4	5
$F(x)$	22	12	−2	−22	−18.

4. A function y satisfies the differential equation $y''+(\pi^2 y/9) = 0$ and the boundary conditions $y = 1$ when $x = 0$, $y = \frac{1}{2}$ when $x = 1$. By taking successively intervals of $\frac{1}{2}$ and $\frac{1}{4}$, obtain by the relaxation method successive approximations to the value of y when $x = \frac{1}{2}$. Compare your approximations with the exact value of y for this value of x.

5. The sine integral $Si(x)$ is defined by $Si(x) = \int_0^x t^{-1} \sin t\, dt$ and it is known that $Si(0) = 0$, $Si(2) = 1{\cdot}605$. If $y = 10^3 Si(x)$, show that y satisfies the differential equation
$$\frac{d^2y}{dx^2}+10^3\left(\frac{\sin x - x\cos x}{x^2}\right) = 0,$$
with the boundary conditions $y = 0$ when $x = 0$, $y = 1605$ when $x = 2$. By using the relaxation method to solve this equation, find approximate values for $Si(x)$ when $x = 0{\cdot}25, 0{\cdot}50, 0{\cdot}75, 1{\cdot}00, 1{\cdot}25, 1{\cdot}50$ and $1{\cdot}75$.

6. The deflexion y at distance x from an end of a uniform beam of unit length and uniformly loaded with a weight w per unit length is given by $2EIy''+wx(1-x) = 0$ where E is Young's modulus of the material of the beam and I is the (constant) second moment of the cross-sectional area. If the beam is simply supported at its ends, use the relaxation method with $h = \frac{1}{8}$ to find the deflexion at the mid-point and compare your answer with the exact result.

7. Use an interval of $h = 0{\cdot}05$ to find the approximate value of y when $x = 0{\cdot}7$ satisfying the boundary-value problem $y''+F(x) = 0$, $y = 19$ when $x = 0{\cdot}3$, $y' = 98$ when $x = 0{\cdot}7$ and $F(x)$ given by:

x	0·35	0·40	0·45	0·50	0·55	0·60	0·65	0·70
$F(x)$	−630	−480	−270	0	330	720	1170	1680.

8. A beam of length 8 ft., carrying a uniform load of 10 lb. per ft. is simply supported at both ends. The deflexion y at distance x from an end is given by $2EIy''+10x(8-x) = 0$ where $E = 4\times 10^9$ lb./ft.^{2} and I is given in terms of x by

x	1, 7	2, 6	3, 4, 5	ft.
I	1	1·2	1·6	10^{-6} ft.^{4},

the cross-section of the beam being non-uniform but symmetrical about the middle point. Using an interval of 1 ft., find the deflexion at the centre.

9. The error function erf x is defined by $\operatorname{erf} x = (2/\sqrt{\pi})\int_0^x e^{-t^2}\,dt$. If

$$y = 10^3 \operatorname{erf} x,$$

show that y is given by the equation

$$\frac{d^2y}{dx^2}+\frac{4\times 10^3}{\sqrt{\pi}}xe^{-x^2} = 0,$$

with the boundary conditions $y = 0$ when $x = 0$, $y = 2\times 10^3\pi^{-1/2}e^{-1}$ when $x = 1$. Use the relaxation method with an interval of $h = \frac{1}{8}$ to solve this boundary-value problem and deduce approximate values for erf $\frac{1}{2}$ and erf 1.

10. Use an interval of 0·1 to solve the differential equation

$$\frac{d^2y}{dx^2}+\frac{1}{5}\frac{dy}{dx}+F(x) = 0,$$

if $y = 0$ both when $x = 0$ and when $x = \frac{1}{2}$, and $F(x)$ is given by

x	0·1	0·2	0·3	0·4
$F(x)$	6·63	7·33	8·10	8·95.

10.8 Extension to partial differential equations

The method outlined in the previous sections for the solution of ordinary differential equations can easily be extended to cover partial differential equations in two independent variables. The direct extension of the boundary-value problem $y''+F(x) = 0$ with y given at $x = a$ and $x = b$ is the solution of the two-dimensional form of Poisson's equation

$$\frac{\partial^2 V}{\partial x^2}+\frac{\partial^2 V}{\partial y^2}+F(x, y) = 0 \qquad (1)$$

with values of V prescribed on the boundary of a given plane area. This equation is one of the simplest from the relaxation point of view and, as it includes the important Laplace's equation as a special case, we shall use it to illustrate the procedure.

The plane area over which the integration is to be performed is subdivided by a uniform network and values of the wanted function V are calculated at the *nodal* or *mesh* points of this net. In practice a square network is usually employed although, occasionally, nets consisting of equilateral triangles or rectangles have proved useful. Here we confine ourselves to a square network of side h and we first write down the finite-difference approximation to equation (1) in terms of the values of V at the points labelled 0, 1, 2, 3 and 4 in Fig. 42. Using suffixes to denote values at corresponding points and neglecting terms of order h^4 and above, we have by equation (3) of §10.2,

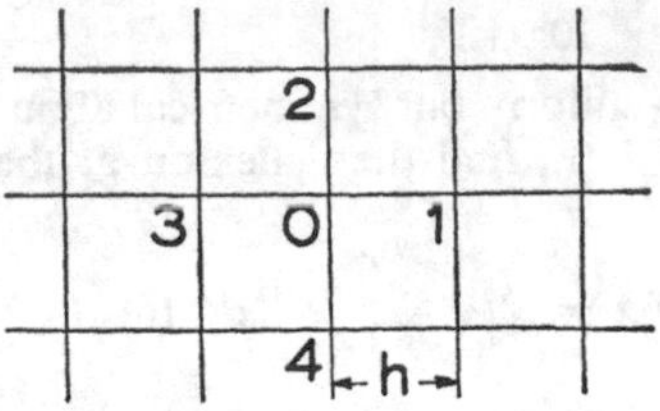

FIG. 42

$$V_1+V_3-2V_0 = h^2\left(\frac{\partial^2 V}{\partial x^2}\right)_0.$$

Similarly,

$$V_2+V_4-2V_0 = h^2\left(\frac{\partial^2 V}{\partial y^2}\right)_0,$$

and hence,

$$\left(\frac{\partial^2 V}{\partial x^2}\right)_0+\left(\frac{\partial^2 V}{\partial y^2}\right)_0 = \frac{1}{h^2}(V_1+V_2+V_3+V_4-4V_0). \tag{2}$$

Thus equation (1) can be replaced by the algebraic equation

$$V_1+V_2+V_3+V_4-4V_0+h^2F(x_0, y_0) = 0, \tag{3}$$

and a similar equation holds at each nodal point of the network.

The set of simultaneous equations of which equation (3) is typical is now solved by the relaxation process and it should be clear that the residual R_0 at the point 0 is given by

$$R_0 \equiv V_1+V_2+V_3+V_4-4V_0+h^2F(x_0, y_0), \tag{4}$$

while the relaxation pattern is that shown in Fig. 43. All the hints and devices previously given still apply, but the following additional comments may be useful:

(i) the square mesh, which should be drawn on the reverse side of the tracing cloth, should be large enough (at least one inch and preferably larger) to avoid too frequent erasure as the squares are filled up with repeated increments and adjustments of residuals:

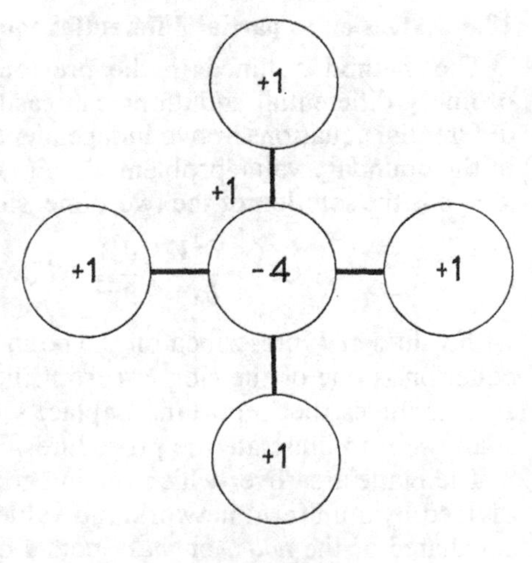

FIG. 43

(ii) when an increment is made at a given point, it is advisable to be systematic in adjusting the residuals, first adjusting the residual at the point 0 and then those at the points 1, 2, 3, 4 in that order:

(iii) the liquidation of residuals is now complete when the absolute magnitude of each individual residual is less than three, when the total of all the residuals is approximately zero and when individual positive and negative residuals are well distributed over the area of integration.

Example 5. *A square is enclosed by the lines $x = 0$, $x = 4$, $y = 0$, $y = 4$. The temperature θ degrees at the boundary is given to be x^3+y^2 and, throughout the region,*

$$\frac{\partial^2\theta}{\partial x^2}+\frac{\partial^2\theta}{\partial y^2} = 0.$$

Use the method of relaxation to determine the temperature at the nine internal points ($x = m$, $y = n$) where $m = 1, 2, 3$ and $n = 1, 2, 3$. (L.U.)

Here the network is shown in Fig. 44 and the boundary values have been calculated from the formula $\theta = x^3+y^2$. The equation for solution is that of Laplace so that $F(x, y) \equiv 0$ and the residuals are given by

$$R_0 \equiv \theta_1+\theta_2+\theta_3+\theta_4-4\theta_0. \tag{5}$$

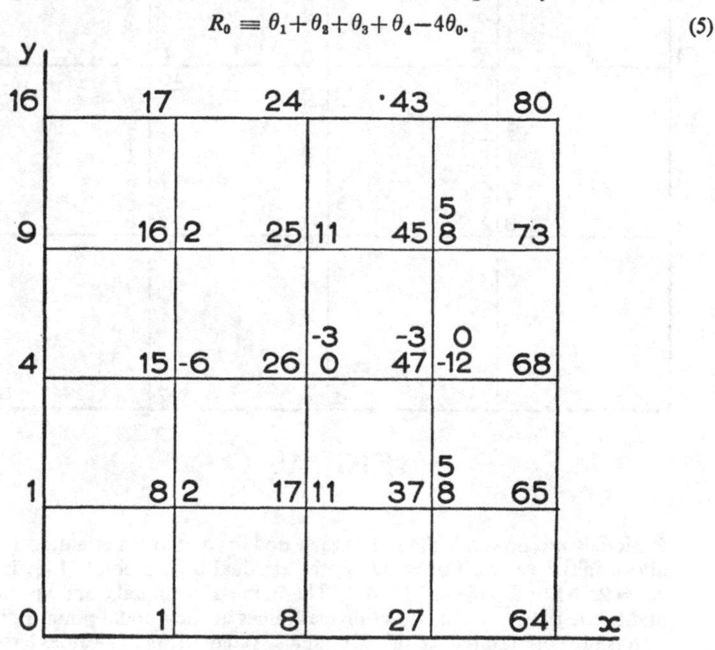

FIG. 44

A first estimate of the temperature at the central mesh point (2, 2) can be found by using a mesh length of 2 and equating to zero the residual given by equation (5) with $\theta_1 = 68$, $\theta_2 = 24$, $\theta_3 = 4$, $\theta_4 = 8$ so that $68+24+4+8-4\theta_0 = 0$, leading to $\theta_0 = 26$. Initial values at the points (1, 2), (3, 2) can be obtained by interpolating linearly between 4, 26 and 26, 68 respectively so that we have a complete set of starting values on the line $y = 2$. Initial values on the lines $y = 1$, $y = 3$ can now be obtained by interpolating between corresponding values on the lines $y = 0$, $y = 2$, and $y = 2$, $y = 4$ respectively. Fig. 44 shows such initial values above and to the left of the intersecting mesh lines at each nodal point.

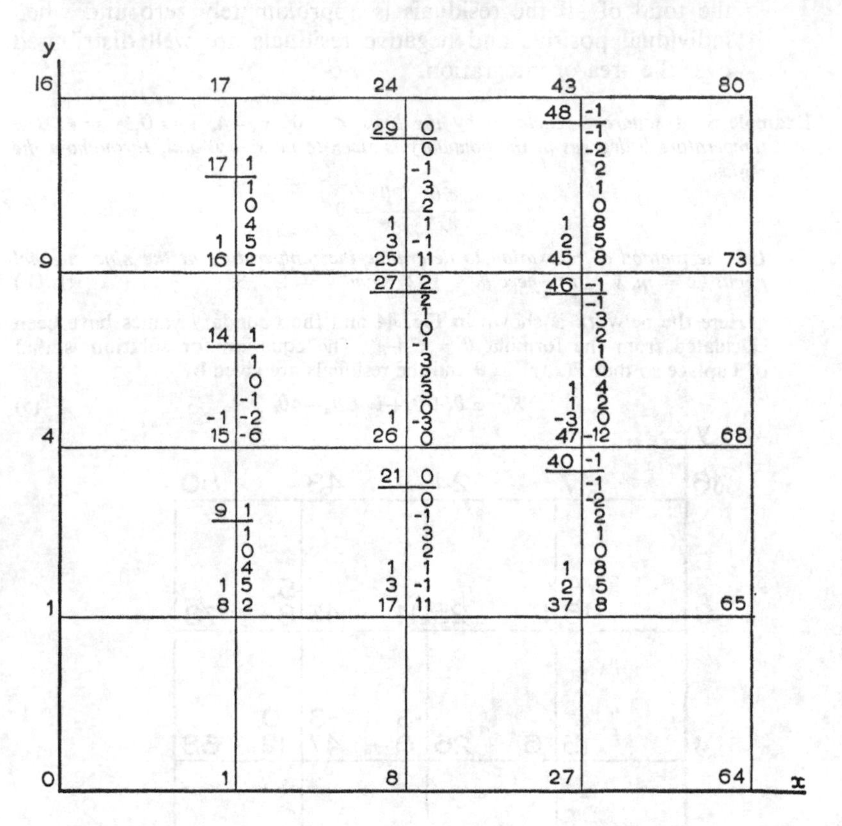

FIG. 45

Residuals are now calculated for each nodal point from equation (5) and the above initial values. For example, the residual at the point (1, 3) is given by $R_0 = 25+17+9+15-4\times 16 = 2$. These initial residuals are inserted above and to the right of the intersecting mesh lines at each nodal point. Point relaxation is now used to reduce these residuals systematically to values less than 3 in absolute magnitude. The largest residual being -12 at the point (3, 2), we first reduce this to zero by applying an increment of -3 $(-12/4)$ to the value of θ at this point. A glance at the relaxation pattern of Fig. 43 shows that such an increment decreases the residuals by 3 at each of the points (3, 3), (2, 2) and (1, 3), there being no alteration to make at the point (4, 2) as this is a boundary point at which the value of θ is prescribed. The effect of this increment is shown in Fig. 44 where the increment of -3 is shown above the initial value of 47 and the residuals at the point (3, 2) and those at the neighbouring points have been adjusted to their new values.

The next step is to deal similarly with the next largest residual (here we can either deal with that at the point (2, 1) or that at the point (2, 3) for the residual is 11 at each of these points). The liquidation process is systematically carried out in this way and Fig. 45 shows the relaxation diagram when the work has been completed. The figures underlined at the top of the left-hand column at each nodal point are the final values of θ obtained as the sum of the initial values and the increments made during the course of the work. Those underlined at the top of the right-hand columns are the residuals calculated from equation (5) and the final values of θ. These act as a check and will show up any errors made during the process. If, because of errors, these recalculated residuals are not sufficiently small, a fresh relaxation process must be carried out to reduce them to acceptable values.

10.9 Block relaxation in two-dimensional problems

Over- and under-relaxation play the same part in two-dimensional problems as they do in the solution of ordinary differential equations.

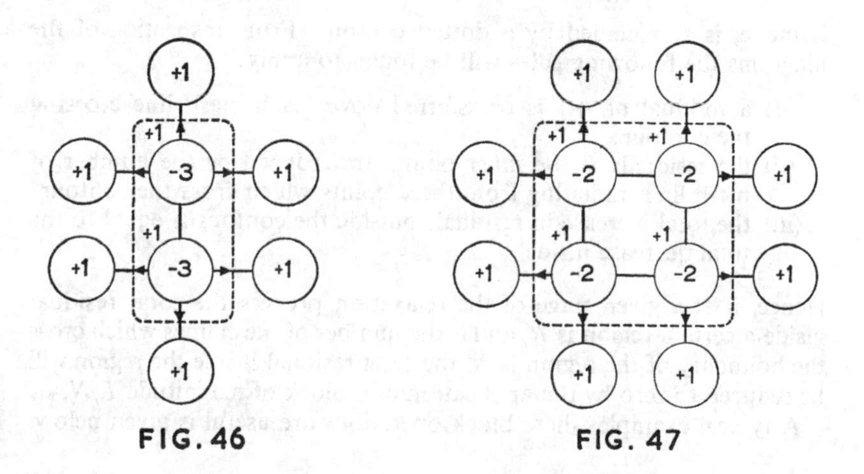

FIG. 46 FIG. 47

So too does block relaxation—an operation which is extremely useful in reducing the total residual to zero.

Relaxation patterns for blocks of points are easily found by combining two or more of the patterns for individual points given in Fig. 43. Typical cases are shown in Figs. 46 and 47 which show respectively the diagrams for two- and four-point blocks. The effect of a larger block is shown in Fig. 48.

In each diagram the block of points, at each of which a unit increment

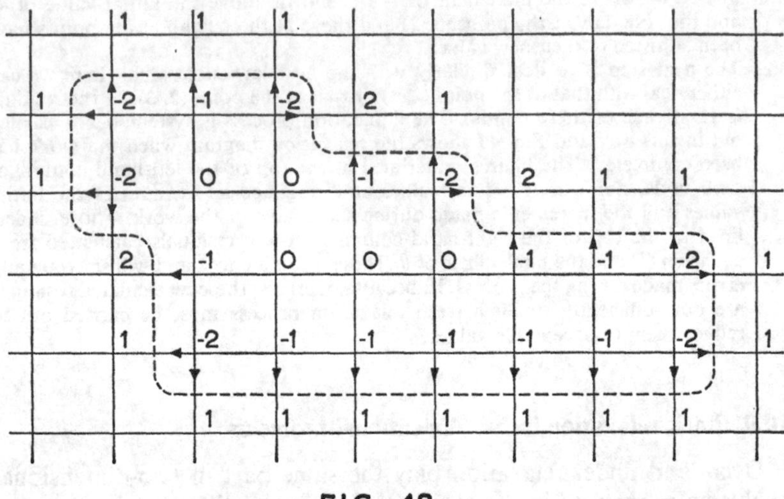

FIG. 48

is made, is surrounded by a dotted contour. From inspection of the diagrams the following rules will be found to apply:

(i) a residual of $+1$ is transferred down each mesh line crossing the contour;
(ii) the residuals at the inner points are reduced by the number of mesh lines radiating from these points which cross the contour;
(iii) the total increase in residuals outside the contour is equal to the total decrease inside.

Hence, if at a given stage of the relaxation process the total residual inside a certain region is R, and if the number of mesh lines which cross the boundary of the region is N, the total residual inside the region will be reduced to zero by the application of a block of magnitude R/N.

A typical example where block operations are useful is given below.

Example 6. *In a torsion problem the solution of Poisson's equation* $\nabla^2\chi+6400 = 0$ *is required inside a rectangle* 1 *ft.* 6 *in. by* 1 *ft., the values of* χ *on the boundaries of the rectangle being everywhere zero. Use mesh lengths of* 6 *in. and* 3 *in. successively to find approximate values of* χ.

For mesh length h, the residuals are given by

$$R_0 \equiv \chi_1+\chi_2+\chi_3+\chi_4-4\chi_0+6400h^2,$$

so that, taking $h = \frac{1}{2}$ ft. and zero initial values for χ, the residuals at the two mesh points of Fig. 49 are each 1600. Surrounding these two points by the dotted contour shown, the total residual inside the contour is 3200 and the

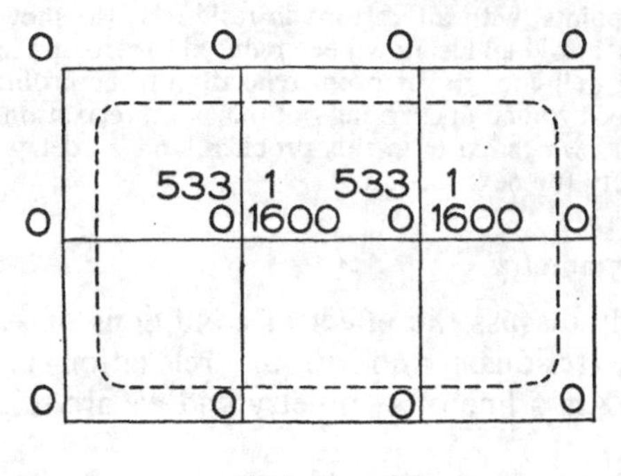

FIG. 49

number of mesh lines crossing it is 6. Hence, by applying a block of 533 (3200/6) and using the pattern of Fig. 46, the first approximate solution is that the values of χ at the mesh points shown in Fig. 49 are each 533 with residuals unity.

With $h = \frac{1}{4}$ and 15 mesh points, zero initial values of χ give initial residuals of 400 as shown in Fig. 50. The total initial residual is 15×400 and, if a block

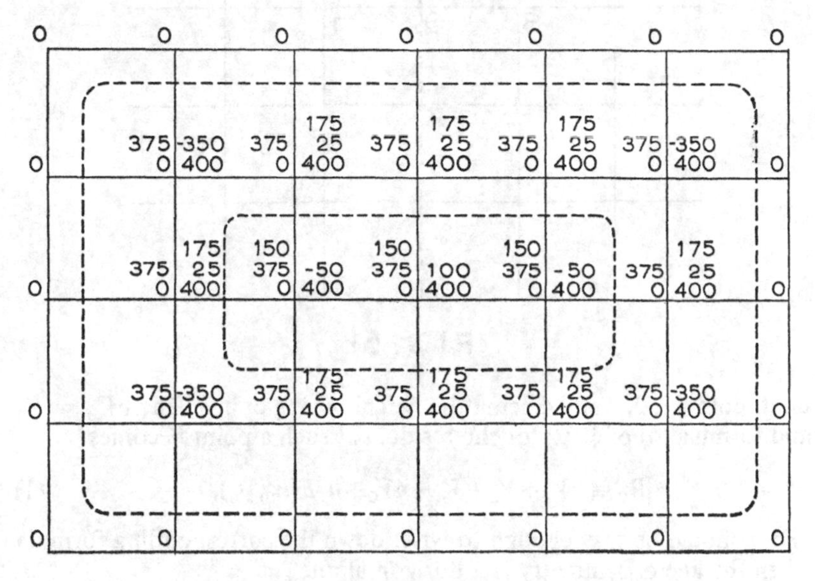

FIG. 50

operation is applied to all the nodal points inside the outer dotted contour, the number of mesh lines crossing the contour is 16. Hence the increment to be used in this block operation is $(15 \times 400)/16 = 375$ and Fig. 50 shows the values to which the residuals are altered after its application. Within the inner contour the total residual is 1200 and the number of mesh lines crossing this contour is 8. Hence we apply a block of magnitude $1200/8 = 150$ to the three central nodal points, with alterations in residuals also shown in Fig. 50. The large total initial residual has now been reduced to zero and individual residuals are distributed well enough for point relaxation to be profitable. A good deal of labour can be avoided in carrying out this point relaxation if full use is made of the symmetry which exists in this problem, and we delay the final stages of the solution until the next section.

10.10 Lines of symmetry

We now briefly discuss the effect of conditions of symmetry on the formula for the residuals and on the relaxation pattern. Suppose (Fig. 51) that YY is a line of symmetry and we are taking our point 0

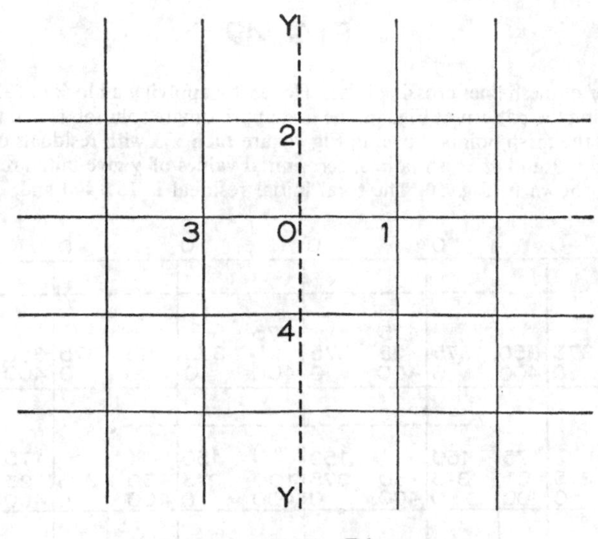

FIG. 51

on it, points 1, 2, 3 and 4 being the neighbouring points. Then $V_3 = V_1$ and formula (4) of §10.2 for the residual at such a point becomes

$$R_0 \equiv 2V_1 + V_2 + V_4 - 4V_0 + h^2F(x_0, y_0), \tag{1}$$

and it should be easy enough to write down the corresponding formula when the line of symmetry is a horizontal one.

The relaxation pattern for a point such as 1 now becomes that shown in Fig. 52 for it must be remembered that whenever an increment is applied to the point 1, the same increment is applied also to the point 3. With a little practice the reader will be able to use such patterns skilfully and will be able to take advantage of the great saving of labour which conditions of symmetry offer. As an example we take up the solution of Example 6 at the stage when point relaxation is profitable.

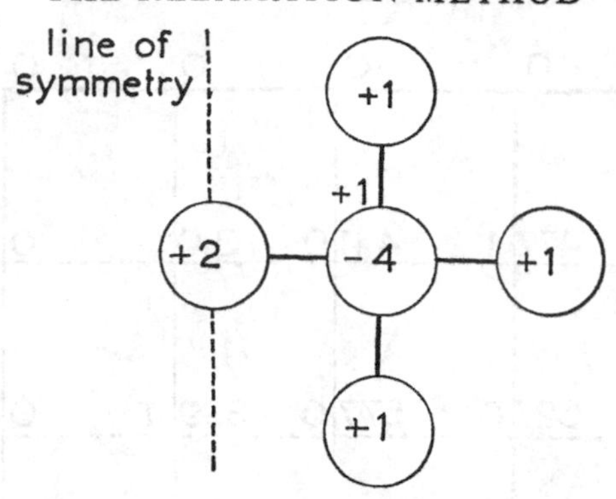

FIG. 52

Example 7. *Complete the solution of Example 6, using the values given on Fig. 50 as starting values.*

Here there are two lines of symmetry and we need only consider one-quarter of the rectangle shown in Fig. 50. Fig. 53 shows starting values and initial residuals taken from the previous diagram and the lines of symmetry are

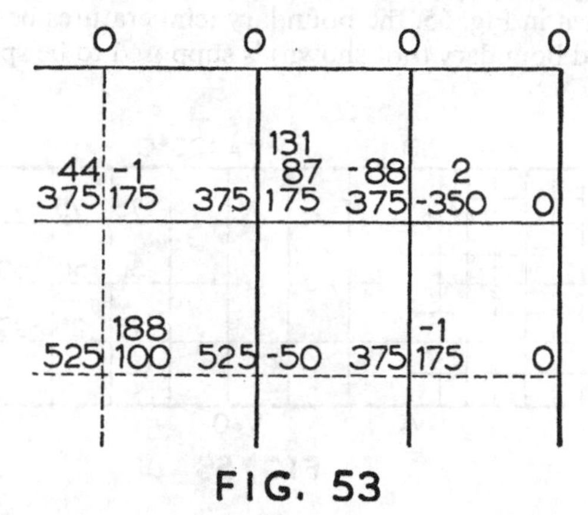

FIG. 53

shown dotted. The first step is to liquidate approximately the residual of -350 by applying an increment of -88. This reduces the residual at the point to the left by 88 and, because of the horizontal line of symmetry, the residual at the point below that at which we are working is reduced by 2×88. The next residual to liquidate is 175 and this is dealt with by applying an increment of 44. The effect of both these operations is shown in Fig. 53.

The solution is continued in this way and Fig. 54 shows the final values of χ and the final residuals. These were obtained by using point relaxation until individual residuals were less than 3 and then multiplying everything by 10 and carrying out a further liquidation process—the final results so obtained were then divided by 10 and rounded off to the nearest integer.

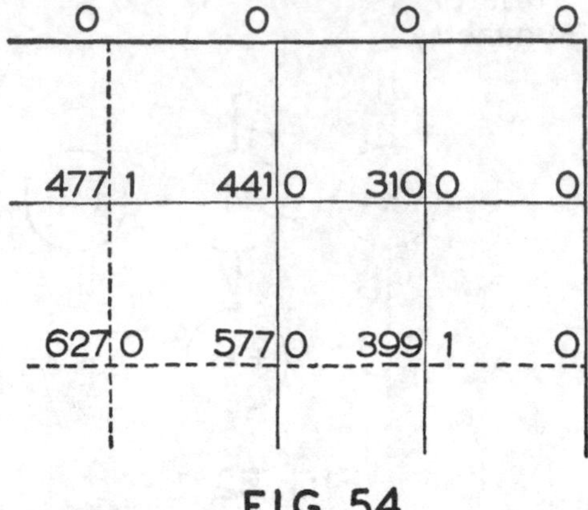

FIG. 54

10.11 Local effects

In some problems use can be made of the fact that disturbances in a physical field are often very local in their effect. Suppose, for example, we are finding the steady temperature V in a long bar whose cross-section is shown in Fig. 55, the boundary temperatures being as shown. The right-hand boundary (not shown) is supposed to be specified by the

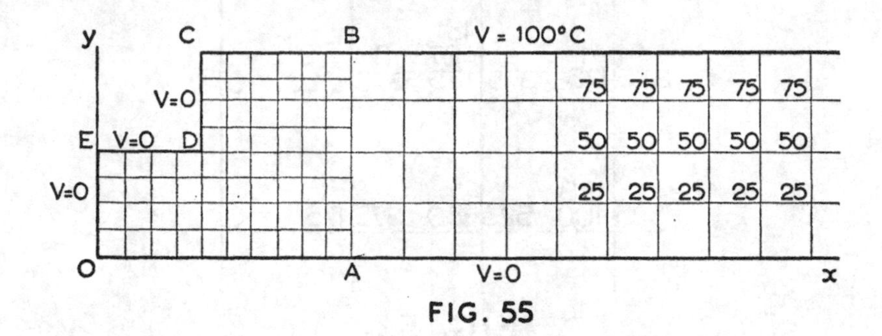

FIG. 55

line $x = a$ where a is large. When x is not too small or not too large, the influence of both left-hand and right-hand boundaries being local, the temperature will be approximately independent of x, and the equation satisfied by V (Laplace's equation) will reduce to $\partial^2 V/\partial y^2 = 0$. This has solution $V = Ay+B$ and, choosing A, B to satisfy the boundary conditions $V = 0$ when $y = 0$, $V = 100$ when $y = 4$, we find $V = 25y$, giving values of 25, 50 and 75°C. when $y = 1$, 2 and 3 respectively. Hence, in obtaining a relaxation solution for values of x which are not too large we can use these values as starting values, and we shall find that during the liquidation process the residuals have only to be adjusted in a certain region near the left-hand end. We may well find, in fact, that 25, 50 and 75 are perfectly good final values for the temperature in a region such as that shown on the right of the diagram and the number of nodal points in the working diagram can consequently be kept reasonably small.

Near the left-hand boundary, however, the temperature will no longer be independent of x nor linear in y, and it is often necessary to work on a much finer net over a limited region such as $OABCDE$. Special methods are available for "advancing to a finer net" in such a region without incurring the labour of using the finer net over the whole diagram. Lack of space precludes giving the details here but these can be found in the standard texts.*

10.12 Curved boundaries

In the examples so far considered the boundary of the area of integration has been rectilinear and, to make the method of universal application, it is clearly necessary to consider the modifications required to

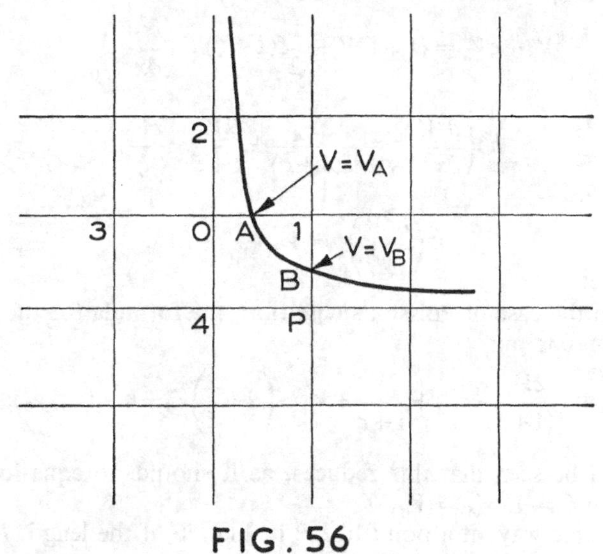

FIG. 56

deal with examples in which part or all of the boundary is a curved line. We continue to use a square network of mesh length h and consider first the case (Fig. 56), in which the point 0 is so situated that the point 1 lies outside the boundary.

The value of V at the point of intersection A of the boundary with

* See, for example, R. V. Southwell, *Relaxation Methods in Theoretical Physics*, Oxford University Press, 1946.

the mesh line 01 is supposed to be V_A and we take the length of $0A$ to be ξh where $0 < \xi < 1$. Expanding V in a Taylor's series at a point (x, y_0) near the point $0(x_0, y_0)$ we have

$$V = V_0+\left(\frac{\partial V}{\partial x}\right)_0 (x-x_0)+\frac{1}{2}\left(\frac{\partial^2 V}{\partial x^2}\right)_0 (x-x_0)^2+\frac{1}{6}\left(\frac{\partial^3 V}{\partial x^3}\right)_0 (x-x_0)^3+\ldots, \tag{1}$$

and, writing $V = V_A$, $x-x_0 = \xi h$, this gives

$$V_A = V_0+\left(\frac{\partial V}{\partial x}\right)_0 \xi h+\frac{1}{2}\left(\frac{\partial^2 V}{\partial x^2}\right)_0 \xi^2 h^2+\frac{1}{6}\left(\frac{\partial^3 V}{\partial x^3}\right)_0 \xi^3 h^3+\ldots. \tag{2}$$

Putting $V = V_3$, $x-x_0 = -h$ in (1), we have

$$V_3 = V_0-\left(\frac{\partial V}{\partial x}\right)_0 h+\frac{1}{2}\left(\frac{\partial^2 V}{\partial x^2}\right)_0 h^2-\frac{1}{6}\left(\frac{\partial^3 V}{\partial x^3}\right)_0 h^3+\ldots, \tag{3}$$

so that, multiplying (3) by ξ, adding to (2) and neglecting terms of order h^3 and above,

$$V_A+\xi V_3 = (1+\xi)V_0+\frac{1}{2}\xi(1+\xi)h^2\left(\frac{\partial^2 V}{\partial x^2}\right)_0,$$

giving

$$h^2\left(\frac{\partial^2 V}{\partial x^2}\right)_0 = \frac{2V_A}{\xi(1+\xi)}+\frac{2V_3}{1+\xi}-\frac{2V_0}{\xi}. \tag{4}$$

As in §10.8,

$$h^2\left(\frac{\partial^2 V}{\partial y^2}\right)_0 = V_2+V_4-2V_0,$$

so that, in the case of Poisson's equation, the formula for the residual at the point 0 is

$$R_0 \equiv \frac{2V_A}{\xi(1+\xi)}+V_2+\frac{2V_3}{1+\xi}+V_4-\left(2+\frac{2}{\xi}\right)V_0+h^2F(x_0, y_0), \tag{5}$$

and it will be seen that this reduces, as it should, to equation (4) of §10.8 when $\xi = 1$, $V_A = V_1$.

In the same way, at a point like P in Fig. 56, if the length PB is ηh, the residual would be given by

$$R_0 \equiv V_1+\frac{2V_B}{\eta(1+\eta)}+V_3+\frac{2V_4}{1+\eta}-\left(2+\frac{2}{\eta}\right)V_0+h^2F(x_0, y_0), \tag{6}$$

V_B being the given value of V at the point B. It may also happen that a mesh point 0 is so situated that two of the mesh lines radiating from it are intersected by the boundary as shown in Fig. 57. In this case, if

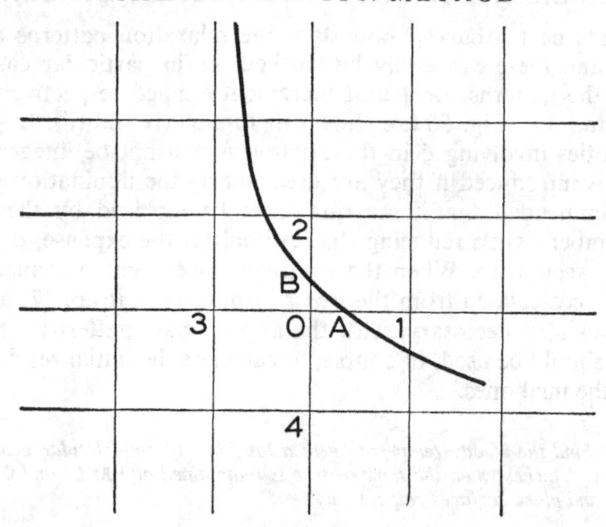

FIG. 57

$0A = \xi h$, $0B = \eta h$ and the given values of V on the boundary at A and B respectively are V_A and V_B, then the residual at 0 is given by

$$R_0 \equiv \frac{2V_A}{\xi(1+\xi)} + \frac{2V_B}{\eta(1+\eta)} + \frac{2V_3}{1+\xi} + \frac{2V_4}{1+\eta} - \left(\frac{2}{\xi} + \frac{2}{\eta}\right)V_0 + h^2 F(x_0, y_0). \quad (7)$$

The above should cover all the cases arising in practice for, if the network is such that three mesh lines are intersected by the boundary, this network is too coarse and a finer one should be substituted.

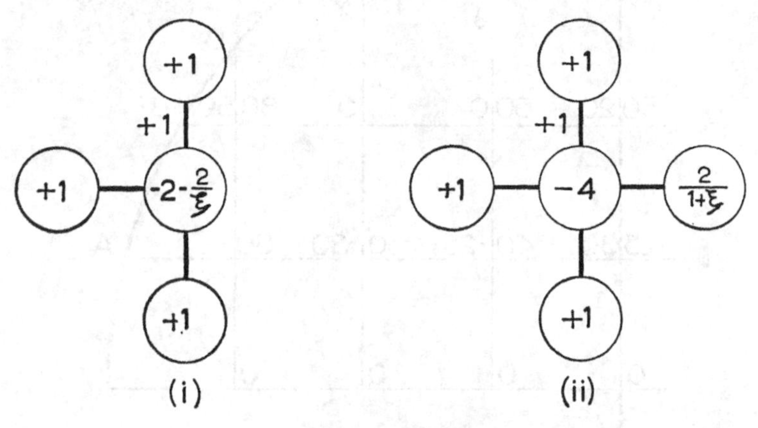

FIG. 58

At points near a curved boundary the relaxation patterns are also modified and these can easily be worked out in particular cases. For example, the patterns for a unit increment applied respectively at the points 0 and 3 of Fig. 56 are shown in Fig. 58 (i) and (ii). In general, the quantities involving ξ in these patterns will not be integers and a difficulty is introduced if they are used during the liquidation process. It is recommended that these quantities be replaced by the nearest whole numbers when reducing the residuals at the expense, of course, of some inaccuracies. When the residuals have been so reduced, they should be recalculated from the *exact* formulae (5), (6) or (7) and then reduced again, if necessary, with the approximate patterns. The exact formulae should be used, of course, to calculate the initial residuals and to check the final ones.

Example 8. *Find the steady temperature in a long bar of semicircular cross-section of radius* 4 *inches when the temperature is maintained at* 100°*C. and* 0°*C. on its curved and plane surfaces respectively.*

From reasons of symmetry we need only consider half the semicircular area and we propose to use a mesh length of 1 inch as shown in Fig. 59. Because of the curved boundary, the calculation of the residuals at the points α, β, γ, δ requires special treatment. At the point α, we find by elementary geometry (or

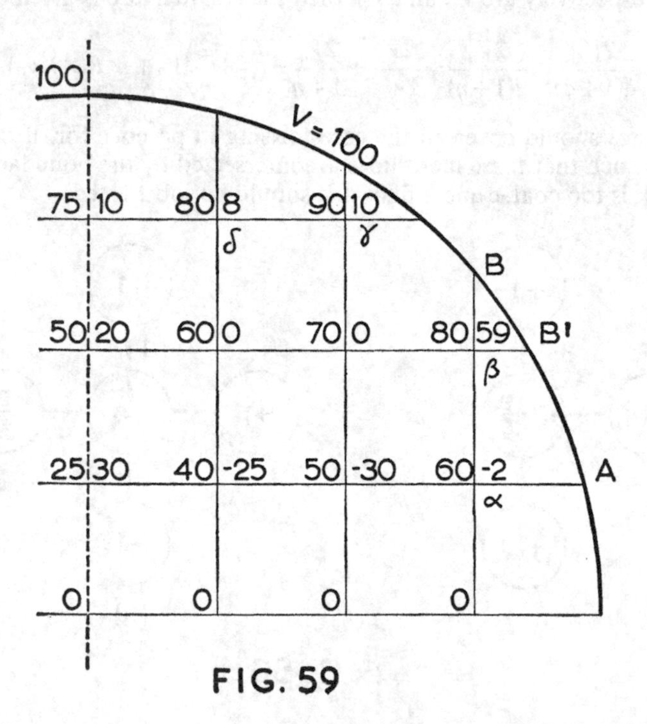

FIG. 59

by measurement) that $\alpha A = 0{\cdot}873 = \xi$, so that by equation (5) the residual at this point is given by

$$R_\alpha = \frac{2\times 100}{0{\cdot}873\times 1{\cdot}873}+V_2+\frac{2V_3}{1{\cdot}873}+V_4-\left(2+\frac{2}{0{\cdot}873}\right)V_0$$

$$= 122+V_2+1{\cdot}07V_3+V_4-4{\cdot}29V_0, \tag{8}$$

since $V_A = 100$ and $F(x_0, y_0) = 0$ as the temperature V satisfies Laplace's equation. At the point β, $\beta B' = 0{\cdot}464 = \xi$, $\beta B = 0{\cdot}646 = \eta$, and substitution in equation (7) gives, after a little reduction

$$R_\beta = 483+1{\cdot}37V_3+1{\cdot}22V_4-7{\cdot}41V_0. \tag{9}$$

In the same way we find

$$R_\gamma = 483+1{\cdot}22V_3+1{\cdot}37V_4-7{\cdot}41V_0, \tag{10}$$

and

$$R_\delta = 122+V_1+V_3+1{\cdot}07V_4-4{\cdot}29V_0. \tag{11}$$

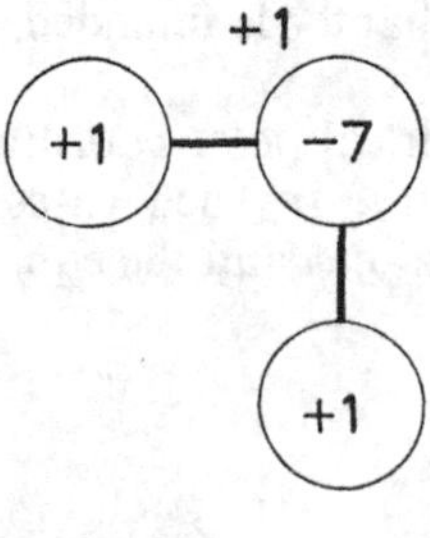

+1

+1 −7

+1

FIG. 60

At the three mesh points on the line of symmetry the residuals are given by

$$R_0 = 2V_1+V_2+V_4-4V_0, \tag{12}$$

while at the remaining four points,

$$R_0 = V_1+V_2+V_3+V_4-4V_0. \tag{13}$$

Guessed starting values for V as shown on Fig. 59 have been used with these formulae to calculate the initial residuals, and these are also shown on the diagram.

To the nearest integer, the approximate relaxation patterns are the usual ones for use with Laplace's equation with the exception of those at the points β and γ. Here the approxi-

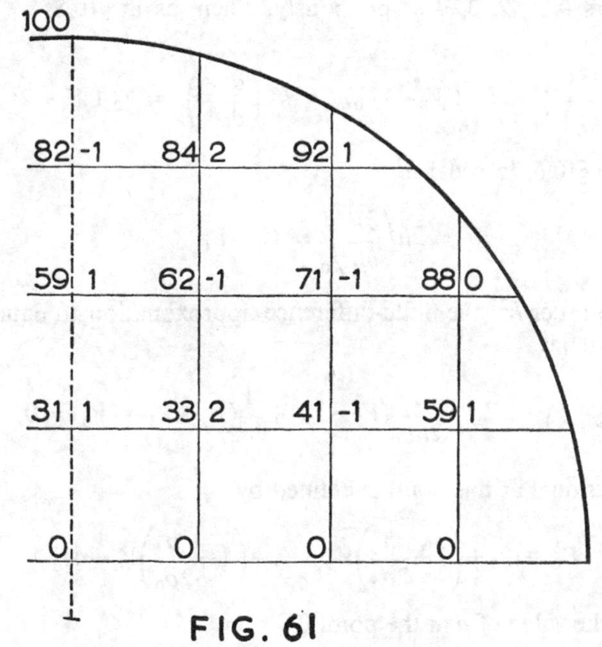

FIG. 61

mate pattern is that shown in Fig. 60. Liquidating the residuals with these approximate patterns, recalculating the residuals from the exact formulae (8)–(13) and repeating the process, we find the values for the temperature and the residuals shown in Fig. 61.

10.13 Extension to other partial differential equations

The two-dimensional form of Poisson's equation in *rectangular* coordinates is the simplest partial differential equation to solve by the relaxation technique but the method has been used to deal successfully with other equations. Here we shall mention only one or two other examples since, as has already been said, this chapter is intended only as a brief introduction to the method.

We first consider Laplace's equation in cylindrical polar coordinates (ρ, ϕ, z) in cases where there is axial symmetry. In such cases the wanted function V is independent of the coordinate ϕ and the equation (see equation (4) of §7.6) reduces to

$$\frac{\partial^2 V}{\partial \rho^2}+\frac{1}{\rho}\frac{\partial V}{\partial \rho}+\frac{\partial^2 V}{\partial z^2} = 0. \tag{1}$$

If we require a solution of this equation when values of V are given for prescribed values of z and ρ, we take a square network of mesh length h as shown in Fig. 62 and number a typical nodal point and its four neighbours 0, 1, 2, 3, 4 as previously. Then, as in §10.8, we have to order h^3,

$$h^2\left(\frac{\partial^2 V}{\partial z^2}\right)_0 = V_1+V_3-2V_0, \quad h^2\left(\frac{\partial^2 V}{\partial \rho^2}\right)_0 = V_2+V_4-2V_0,$$

and, as in §10.6, to order h^2,

$$2h\left(\frac{\partial V}{\partial \rho}\right)_0 = V_2-V_4.$$

Hence, to order h^2, the finite-difference approximation to equation (1) at the point 0 is

$$\frac{1}{h^2}(V_2+V_4-2V_0)+\frac{1}{2h\rho_0}(V_2-V_4)+\frac{1}{h^2}(V_1+V_3-2V_0) = 0,$$

and the residual at this point is defined by

$$R_0 \equiv V_1+\left(1+\frac{h}{2\rho_0}\right)V_2+V_3+\left(1-\frac{h}{2\rho_0}\right)V_4-4V_0, \tag{2}$$

ρ_0 being the value of ρ at the point 0.

The relaxation pattern is now that shown on the right of Fig. 62, ρ_2 and ρ_4 being the values of ρ at the points 2 and 4 respectively. Thus the pattern remains constant when residuals are being liquidated on a line of mesh points possessing the same value of ρ but changes when

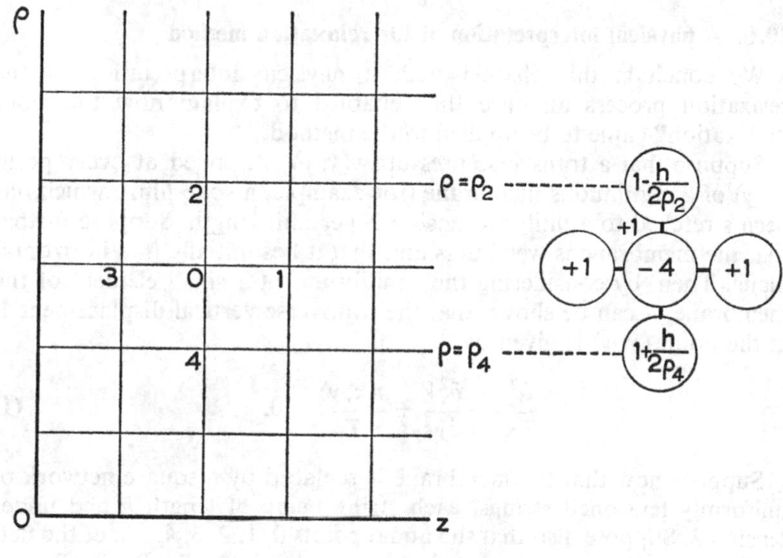

FIG. 62

points specified by a different value of ρ are dealt with. The relaxation process is therefore a good deal more complicated than in the cases considered in the previous sections but, with a little practice, the solution of this type of problem can be effectively carried through.

Equations of higher order than the second have also yielded to the relaxation method. For example, several problems involving the *biharmonic equation* in two variables, that is,

$$\nabla^4 V \equiv \frac{\partial^4 V}{\partial x^4} + 2\frac{\partial^4 V}{\partial x^2\,\partial y^2} + \frac{\partial^4 V}{\partial y^4} = 0,$$

have been successfully solved. Here the relaxation pattern will be found to involve thirteen nodal points instead of the five points in the case of second order equations but this pattern, although complicated, has not prevented solutions being obtained by experienced workers. Boundary conditions specifying the normal gradient of the wanted function instead of the function itself have also been brought within the scope of the method as have also problems involving unknown boundaries. Quite recently, a technique has been devised to deal with partial differential equations in three variables, and there is no doubt that the relaxation method has indeed proved a most powerful tool in the solution of practical problems.

10.14 A physical interpretation of the relaxation method

We conclude this chapter with a physical interpretation of the relaxation process and are thus enabled to explain how the word "relaxation" came to be applied to the method.

Suppose that a transverse pressure $p(x, y)$ is applied at every point (x, y) of a continuous membrane (for example, a soap film), which has been stretched to a uniform tension T per unit length. Suppose further that the membrane is weightless and that it lies initially in a horizontal plane. Then, by considering the equilibrium of a small element of the membrane, it can be shown that the transverse vertical displacement V at the point (x, y) is given by

$$\frac{\partial^2 V}{\partial x^2}+\frac{\partial^2 V}{\partial y^2}+\frac{p(x, y)}{T}=0. \tag{1}$$

Suppose now that the membrane is replaced by a square network of uniformly tensioned strings, each string being of length h and under tension λ. Suppose also that the nodal points 0, 1, 2, 3, 4, . . . of the network are subject to concentrated transverse loads $P_0, P_1, P_2, P_3, P_4, \ldots$. By writing down the condition for equilibrium of the nodal point 0, we have

$$\lambda\left(\frac{V_1-V_0}{h}\right)+\lambda\left(\frac{V_2-V_0}{h}\right)+\lambda\left(\frac{V_3-V_0}{h}\right)+\lambda\left(\frac{V_4-V_0}{h}\right)+P_0=0,$$

where V_0, V_1, V_2, V_3, V_4 are the (small) transverse displacements of the nodal points 0, 1, 2, 3, 4. This equation can be written

$$V_1+V_2+V_3+V_4-4V_0+\frac{hP_0}{\lambda}=0 \tag{2}$$

and this is identical with the finite-difference approximation used in the relaxation solution of equation (1) if $hP_0/\lambda = h^2p(x_0, y_0)/T$.

The relaxation process can now be interpreted in terms of the string network as follows:

(i) Give each nodal point of the string net an arbitrary displacement. With these arbitrary displacements we cannot expect the net to be in equilibrium and there will, in fact, be unbalanced forces at the nodal points. These arbitrary displacements correspond to the initial values of the wanted function assumed in the relaxation solution.

(ii) To maintain the string net in equilibrium, constraints equal in magnitude to these unbalanced forces will have to be applied to the nodal points. These constraints correspond to the residuals in the relaxation solution.

(iii) Point relaxation is then equivalent to altering the displacement at the point at which the largest constraint is applied. Such an alteration affects, of course, the magnitude of the constraints at the four neighbouring points and gives an interpretation of the relaxation pattern.

This gives a physical picture of the liquidation process which was originally devised for the computation of the stresses in braced frameworks and which was based on the notion of *the systematic relaxation of constraints.* It suggests also a physical reason for the convergence of the process for, in using it, we are moving steadily to the equilibrium configuration—a configuration of minimum potential energy.

Exercises 10(*b*)

1. Solve Laplace's equation for the region of Fig. 63, using the square network and boundary values shown.

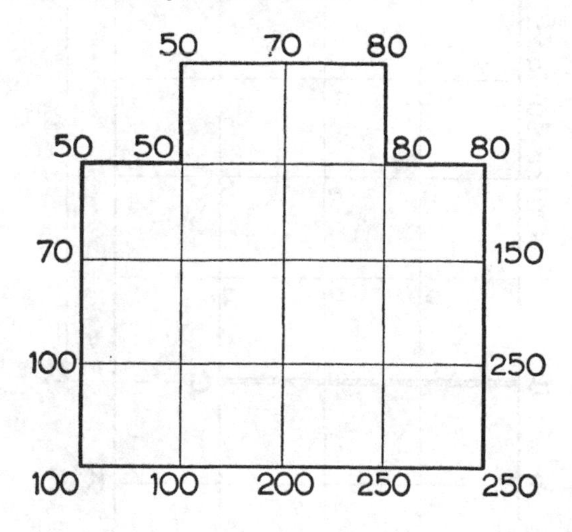

FIG. 63

2. The rectangular cross-section of a long metal bar is bounded by the lines $x = 0$, $x = 6$, $y = 0$, $y = 4$ and the temperature V at its surfaces is given by $V = 2x^2 + y^2$. Find the steady temperature at the two internal nodal points when a square network of mesh length 2 is used. Find also

the temperature at these two nodal points when the above solution is refined by halving the mesh length.

3. A function χ satisfies the equation $\nabla^2\chi+100 = 0$ inside a square region of side 4 in., and vanishes on the boundaries. With a square network of mesh length 1 in. and assuming zero starting values for χ, use block and point relaxation to find an approximate value of χ at the centre of the square.

4. The walls of a refrigerator consist of a homogeneous insulating material

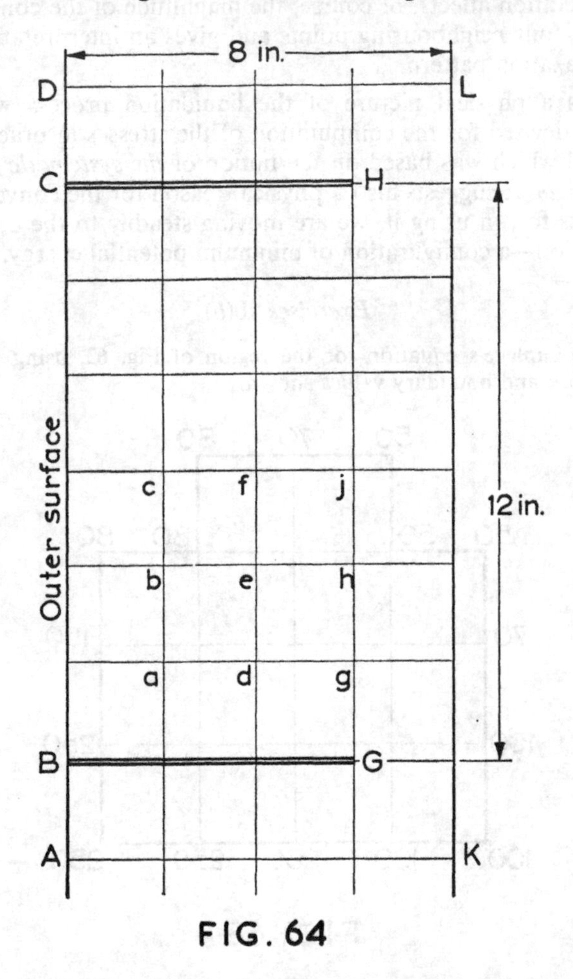

FIG. 64

of thickness 8 in., supported by internal ribs of length 6 in. In Fig. 64, *ABCD* is the outer surface at constant temperature 60°F. *BG* and *CH* are two of the ribs, 12 in. apart and assumed to be at a constant temperature of 60°F.; *KL* is the inner surface at a constant temperature of 20°F.

The temperature θ at all points within the insulating material may be assumed to satisfy the equation

$$\frac{\partial^2\theta}{\partial x^2}+\frac{\partial^2\theta}{\partial y^2}=0.$$

Using a square net of mesh length 2 in. find, by the method of relaxation, the values of θ at each of the nine internal nodes $a, b, c, d, e, f, g, h, j$. (L.U.)

5. If $f(x, y)$ satisfies $\frac{\partial^2 f}{\partial x^2}+\frac{\partial^2 f}{\partial y^2}=0$ when f has the values 100 on AB, CD and EF, and is zero on PQ and RS (Fig. 65), find the approximate value of f at O.

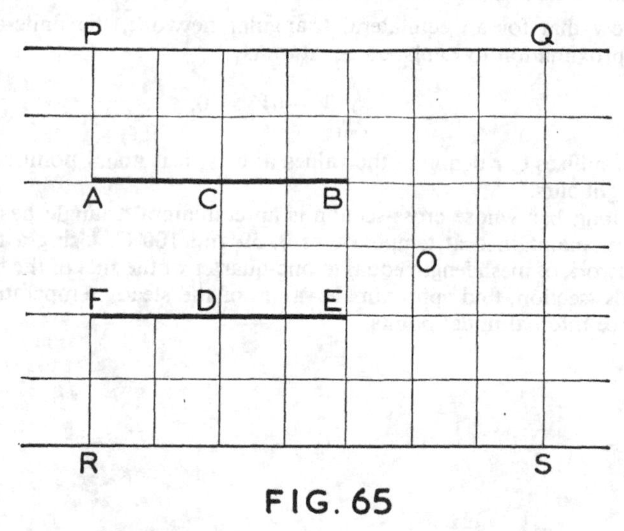

FIG. 65

[The lines PQ and RS extend to infinity in both directions; CA and DF extend to infinity on the left.] (L.U.)

6. A function z satisfies the equation

$$\frac{\partial^2 z}{\partial x^2}+\frac{\partial^2 z}{\partial y^2}=-100$$

inside the trapezium defined by the lines $x=0$, $y=0$, $y=4$, $2x+2y=13$ and takes zero values on its boundaries. Use the square network given by $x=r$ $(r=1, 2, 3, 4, 5, 6)$ and $y=s$ $(s=1, 2, 3)$ to find approximate values of z at the points $(t, 2)$ where $t=1, 2, 3$ and 4.

7. A function satisfies Laplace's equation inside a quadrant of the ellipse $(x/6)^2+(y/4)^2=1$, vanishes on the ellipse and takes a constant value C on both the semi-major and minor axes. Use a square network of mesh length one-twelfth of the major axis to find approximate values of the function at the three internal nodal points which lie on a line making an angle of 45° with the axes of the ellipse.

8. A function ϕ satisfying the differential equation $\nabla^2\phi = \phi$ inside a square of side π, takes the value 2000 on one side of the square and vanishes on the other three. Find an approximate value for ϕ at the centre of the square (i) with a square network of mesh length $\pi/4$, (ii) when the mesh length is $\pi/8$.

9. A hollow right circular cylinder is bounded by the surfaces $\rho = 4$, $\rho = 8$, $z = 0$ and $z = 4$, ρ and z being cylindrical polar coordinates. The temperature θ of the surface $z = 0$ is given by $\theta = 25\rho$ $(4 < \rho < 8)$ and the other surfaces of the cylinder are maintained at temperature zero. Use the square mesh given by $\rho = 4+t$, $z = t$ where t takes the values 1, 2 and 3 to find approximate values of the steady temperature at the nine internal nodal points.

10. Show that for an equilateral triangular network, the finite-difference approximation to Laplace's equation is

$$\sum_{r=1}^{6} V_r - 6V_0 = 0,$$

the suffixes 0, r denoting the values at a typical nodal point and its six neighbours.

A long bar whose cross-section is an equilateral triangle has its three faces maintained at temperatures 0, 50 and 100°C. Using a triangular network of mesh length equal to one-quarter of the side of the triangular cross-section, find approximate values of the steady temperature at the three internal nodal points.

CHAPTER 11

NON-LINEAR EQUATIONS

11.1 Introductory

In this chapter we give a brief account of some of the methods used in the solution of non-linear differential equations. This is a subject in which there has been considerable development in recent years. Formerly the engineer who had to deal with a vibrational problem in a design linearized the problem by making approximations so that the differential equation which had to be solved was in a simple form such as

$$\ddot{x}+2k\dot{x}+\omega^2 x = E\cos pt.$$

The solution obtained was a first approximation to the actual vibration and adequate for design purposes. More recently it has been found that a small non-linear term can sometimes cause very large changes in the vibration and that the solution of the linearized problem may be inadequate.

By using numerical methods, such as those considered in Chapter 9, a solution of most equations can be found to satisfy given initial conditions. This is often not enough, for two reasons. Firstly, the equation to be solved in the design of structures or in electronics often contains constants whose numerical values cannot be decided upon until the nature of the response, that is, the solution of the differential equation, is known. Secondly, the arbitrary constant appears in the solution of non-linear equations in a more complicated way than in the solution of linear equations. Thus if y_1 be a solution of the linear equation $y'+P(x)y = 0$, the complete solution is $y = Ay_1$, where A is a constant. This is not true for non-linear equations. For example, the non-linear equation

$$y = x\frac{dy}{dx}-\left(\frac{dy}{dx}\right)^3$$

has solution $y = Ax-A^3$, where A is arbitrary. Thus one has not found the complete solution of a non-linear equation when it has been solved for one set of initial conditions.

In non-linear equations, some of the constants in the differential equation can be reduced by a change of variable. Thus in the equation $\ddot{x}+ax+bx^3 = 0$, if we write $x = \lambda z$ and $t = \mu\tau$, we have

$$\frac{\lambda}{\mu^2}\frac{d^2z}{d\tau^2}+a\lambda z+b\lambda^3 z^3 = 0,$$

and taking $\mu^2 = 1/a$ and $\lambda^2 = a/b$, the equation becomes

$$\ddot{z}+z+z^3 = 0.$$

This sometimes makes the equation look deceptively simple; often, however, the algebra involved in the solution is less complicated.

11.2 Non-linear equations which are integrable

In the earlier chapters of this book we considered several types of non-linear differential equations which can be solved in terms of known functions. These include:

(i) Equations with one variable explicitly absent, §1.6 (*b*), (*c*).

Example $$\left(\frac{dy}{dx}\right)^2 - y^2 = 1.$$

(ii) Exact equations, §1.7.

Example $$(2x/y)\,dx+(1-x^2/y^2)\,dy = 0.$$

(iii) Equations with variables separable, §2.2.

Example $$\frac{dy}{dx} = (x+y)^2.$$

(iv) Bernoulli's equation, §2.5.

Example $$x\frac{dy}{dx}+y = x^4y^3.$$

(v) Homogeneous equations, §2.7.

Examples $$x(x-2y)\frac{dy}{dx}-y(y-2x) = 0,$$

$$(x-y+3)\frac{dy}{dx}-(3x-y-1) = 0.$$

(vi) Equations soluble for *p*, §2.9.

Example $$\left(\frac{dy}{dx}-y\right)^2-(x-y)^2 = 0.$$

(vii) Equations in which *y* is a function of *x* and *p*, §2.9.

Example $$y = px+p^2x^2.$$

(viii) Clairaut's equation, §2.10.

Example $$y = px+p^4.$$

We also considered several cases in which the order of an equation can be reduced by a change of variable (§4.6). In non-linear theory, equations of the form $\ddot{x}+f(x, \dot{x}) = 0$ are of frequent occurrence, and a change of variable from x to v, where $v = \dot{x}$, reduces them to the first order equation $v\,(dv/dx)+f(x, v) = 0$.

Any soluble differential equation can be transformed into a complicated non-linear equation by a change of variables, but the reverse process is seldom possible. We may note, however, the transformation of the Riccati equation (§2.13) by which the solution of the non-linear equation

$$\frac{dy}{dx}+y^2+Py+Q = 0, \tag{1}$$

is made to depend on that of the linear equation

$$\frac{d^2z}{dx^2}+P\frac{dz}{dx}+Qz = 0, \tag{2}$$

by the change of variable $zy = dz/dx$. Using the same transformation of the linear equation

$$\frac{d^3z}{dx^3}+P\frac{d^2z}{dx^2}+Q\frac{dz}{dx}+Rz = 0, \tag{3}$$

we have $z'/z = y$, $z''/z = y'+y^2$, $z'''/z = y''+3yy'+y^3$, and hence the non-linear equation

$$y''+y'(3y+P)+(y^3+Py^2+Qy+R) = 0, \tag{4}$$

can be solved whenever a solution of equation (3) can be obtained.

Most of the work done in recent years on the solution of non-linear equations concerns those which represent vibrations of one kind or another. We may also mention, however, the *Lane-Emden* equation which occurs in astronomy, namely,

$$\frac{d^2y}{dx^2}+\frac{2}{x}\frac{dy}{dx}+y^n = 0. \tag{5}$$

Solutions of this equation such that $y = 1$ and $dy/dx = 0$ when $x = 0$ are known as Lane-Emden functions of order n. The equation (5) can be solved explicitly when $n = 0$, 1 and 5 and numerical solutions for $n = 1{\cdot}5$, 2, $2{\cdot}5$, . . ., $4{\cdot}5$ have been tabulated.* The complicated procedure necessary in the case $n = 5$ to obtain the simple solution $y = (1+x^2/3)^{-\frac{1}{2}}$ is indicated in Exercise 11(*a*) 5.

Exercises 11(*a*) contain some non-linear differential equations which can be solved by the methods indicated in this section.

* British Association Mathematical Tables, **2**, 1932.

Exercises 11(*a*)

1. Show that the transformation $X = p = dy/dx$, $Y = xp - y$, transforms the equation $f(x, y, p) = 0$ into the equation $f(P, PX - Y, X) = 0$, where $P = dY/dX$. Hence find the solution of the equation $(y - px)x = y$.
2. Verify that a solution of the differential equation $(1+x^3)(y')^2 = (1+y^3)$ is
$$x^2y^2+2axy(x+y)+a^2(x-y)^2-4(x+y)+4a = 0,$$
where a is an arbitrary constant.
3. By means of the substitution $xy' - y = v$, or otherwise, show that a solution of the equation $f(x)y'' - (xy' - y)^2 = 0$ is
$$y = x\int (v/x^2)\,dx + Ax, \quad \text{where } 1/v = B - \int \{x/f(x)\}\,dx,$$
and A and B are arbitrary constants,
4. By means of the substitution $y^2 = x^2 + u$ reduce the equation
$$(x^2+y^2-2pxy)^2 = 4a^2y^2(1-p^2)$$
to the equation
$$(u - xu' - 2a^2)^2 = a^2\{4a^2 - (u')^2\},$$
and find (i) the solution, (ii) the singular solution.
5. The Lane-Emden equation of order 5 is $y'' + (2/x)y' + y^5 = 0$. By means of the substitution $y = (1/\sqrt{2})e^{t/2}z$, $x = e^{-t}$, reduce this equation to the form $4\ddot{z} - z + z^5 = 0$. Hence show that the Lane-Emden function of order 5, with $y = 1$ and $y' = 0$ when $x = 0$, is $y = (1+x^2/3)^{-1/2}$.
6. Solve the equation $y'' + (y')^2 + y'/x = 0$.
7. Show that the complete solution of the equation $y'' + 3yy' + y^3 + \omega^2 y = 0$ is $y = \omega(\cos\omega x - A\sin\omega x)/(\sin\omega x + A\cos\omega x + B)$, where A and B are constants.
8. Find a solution of the equation $y'' + \omega^2 y = 2(\omega^2/a^2)y^3$, such that $y' = 0$ when $y = a$.

11.3 Jacobian elliptic functions

The solutions of certain non-linear differential equations can be expressed simply in terms of *Jacobian elliptic functions.* These are generalizations of the ordinary sine and cosine functions.

The Jacobian sine function of t is written as

$$x = \operatorname{sn}(t, k), \tag{1}$$

where k is a number between 0 and 1. The number k is called the *modulus* and there is a separate function for each value of k. When k is known, the values of x corresponding to different values of t can be read off from tables* in much the same way as ordinary sines and

* For example, *Tables of Elliptic Functions*, Smithsonian Institute, 1939; E. Jahnke and F. Emde, *Tables of Functions*, Leipzig, 1933; L. M. Milne-Thompson, *Jacobian Elliptic Function Tables,* Dover Publications Inc., 1950.

cosines. When a particular modulus is specified it is convenient to write the function as $x = \text{sn}\, t$, omitting the modulus. The modulus is often expressed as the sine of an angle by writing $k = \sin\alpha$, and α is called the *modular angle.*

The function is defined by the inverse relationship

$$t = \text{sn}^{-1}x = \int_0^x \frac{du}{\{(1-u^2)(1-k^2u^2)\}^{\frac{1}{2}}}, \tag{2}$$

and, for a particular modulus, corresponding values of x and t can be computed from this integral. The integral is clearly a generalization of the inverse sine integral

$$\sin^{-1}x = \int_0^x \frac{du}{(1-u^2)^{\frac{1}{2}}},$$

and the Jacobian function reduces to the sine function when $k = 0$.

From the integral (2) it follows that, if $x = \text{sn}(t, k)$, and t is real,

(i) $-1 \leqq x \leqq 1$, and $x = 0$ when $t = 0$,

(ii) $-t = \text{sn}^{-1}(-x)$, that is $\text{sn}\,(-t) = -\text{sn}\,(t)$,

(iii) $dt/dx = 1/\{(1-x^2)(1-k^2x^2)\}^{\frac{1}{2}}$ and hence $x = \text{sn}\,t$ is a solution of the differential equation

$$\left(\frac{dx}{dt}\right)^2 = (1-x^2)(1-k^2x^2). \tag{3}$$

The complete solution of the equation (3) is therefore

$$x = \text{sn}\,(\pm t + A),$$

where A is constant.

(iv) If K be the value of t corresponding to $x = 1$,

$$\begin{aligned} K &= \int_0^1 \frac{du}{\{(1-u^2)(1-k^2u^2)\}^{\frac{1}{2}}} \\ &= \int_0^{\frac{1}{2}\pi} \frac{d\phi}{(1-k^2\sin^2\phi)^{\frac{1}{2}}} \\ &= \tfrac{1}{2}\pi\, {}_2F_1(\tfrac{1}{2}; \tfrac{1}{2}; 1; k^2), \end{aligned} \tag{4}$$

from §5.9(*a*).

(v) The function $\text{sn}\,t$ is thus defined for $-K \leqq t \leqq K$. The definition is extended to all real values of t by the relation

$$\text{sn}\,(t \pm 2K) = -\text{sn}\,t, \tag{5}$$

and hence

$$\text{sn}\,(t \pm 4K) = \text{sn}\,t. \tag{6}$$

The function $x = \text{sn}\,t$ thus defined is a continuous differentiable and periodic function of t with period $4K$ and $x = \text{sn}\,(t+A)$

provides a periodic solution of the differential equation (3) for all real values of t. The graph of $x = \text{sn}\, t$ for $k^2 = \frac{1}{2}$ and $K = 1{\cdot}180\pi/2$ is shown in Fig. 66, and for comparison the values of $x = \sin(t/1{\cdot}18)$, which has the same period, are shown on the figure by a broken line.

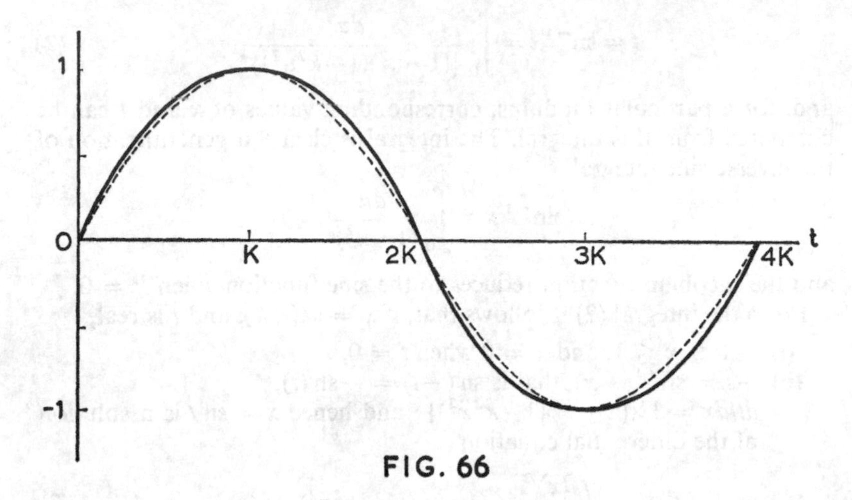

FIG. 66

The Jacobian elliptic functions cn (t, k) and dn (t, k) are defined by the equations

$$\text{cn}^2 t + \text{sn}^2 t = 1, \tag{7}$$

$$\text{dn}^2 t + k^2 \text{sn}^2 t = 1, \tag{8}$$

with dn 0 = cn 0 = 1, and with the further condition that the functions and their derivatives shall be continuous. These conditions determine the signs of cn t and dn t. As can be seen from the graphs of cn t and dn t with $k^2 = \frac{1}{2}$ (Fig. 67), these are periodic functions of t, the period of cn t being $4K$ and that of dn t, which is always positive and not less than $1-k^2$, $2K$. From equation (3) using (7) and (8), it follows that

$$\frac{d}{dt}(\text{sn}\, t) = \text{cn}\, t\, \text{dn}\, t, \tag{9}$$

and from equations (7) and (8),

$$\frac{d}{dt}(\text{cn}\, t) = -\text{sn}\, t\, \text{dn}\, t, \tag{10}$$

$$\frac{d}{dt}(\text{dn}\, t) = -k^2\, \text{sn}\, t\, \text{cn}\, t. \tag{11}$$

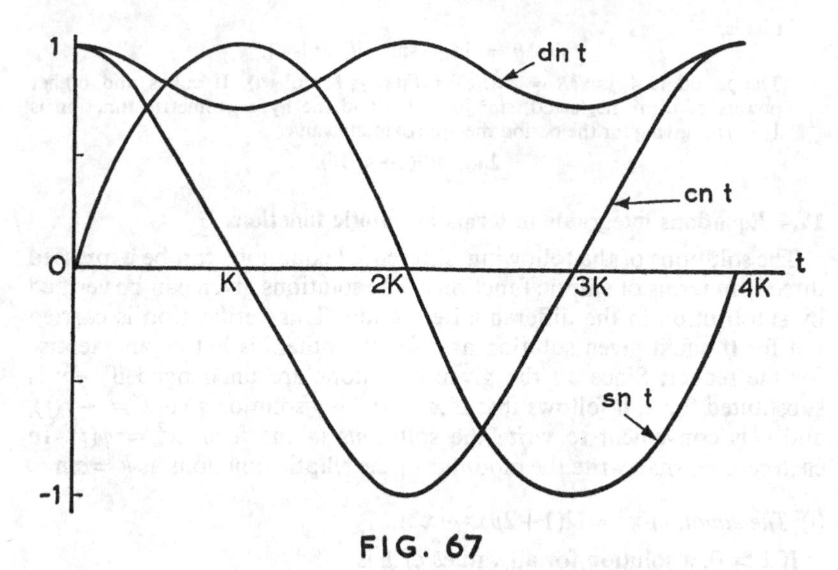

FIG. 67

Example 1. *Solve the equation of the simple pendulum* $l\ddot{\theta}+g\sin\theta = 0$, *given that* $\theta = \alpha$ and $\dot{\theta} = 0$ *when* $t = 0$.

Integrating the given equation with respect to θ we have

$$\tfrac{1}{2}l\dot{\theta}^2 - g\cos\theta = \text{constant} = -g\cos\alpha,$$

and hence

$$\dot{\theta}^2 = \frac{4g}{l}(\sin^2\tfrac{1}{2}\alpha - \sin^2\tfrac{1}{2}\theta).$$

Thus

$$\dot{\theta} = -2(g/l)^{1/2}(\sin^2\tfrac{1}{2}\alpha - \sin^2\tfrac{1}{2}\theta)^{1/2},$$

the negative sign being taken since $\dot{\theta}$ is initially negative. Hence

$$2(g/l)^{1/2}t = -\int_{\alpha}^{\theta}\frac{du}{(\sin^2\tfrac{1}{2}\alpha - \sin^2\tfrac{1}{2}u)^{1/2}}.$$

Writing $k = \sin\tfrac{1}{2}\alpha$ and $\sin\tfrac{1}{2}u = kv$, we have $du = 2k\,dv/(1-v^2)^{1/2}$ and hence

$$(g/l)^{1/2}t = -\int_{1}^{x}\frac{dv}{\{(1-v^2)(1-k^2v^2)\}^{1/2}},$$

where $kx = \sin\tfrac{1}{2}\theta$. Thus

$$\begin{aligned}(g/l)^{1/2}t &= \int_0^1\frac{dv}{\{(1-v^2)(1-k^2v^2)\}^{1/2}} - \int_0^x\frac{dv}{\{(1-v^2)(1-k^2v^2)\}^{1/2}}\\ &= K - \operatorname{sn}^{-1}x,\end{aligned}$$

the modulus of the Jacobian function being k and its period $4K$. Then

$$\begin{aligned}x &= \operatorname{sn}\{-(g/l)^{1/2}t + K\}\\ &= -\operatorname{sn}\{(g/l)^{1/2}t - K\}\\ &= \operatorname{sn}\{(g/l)^{1/2}t + K\},\end{aligned}$$

that is,

$$\sin \tfrac{1}{2}\theta = \sin \tfrac{1}{2}\alpha \,\text{sn}\,\{(g/l)^{1/2}t+K\}.$$

The period is $4(l/g)^{1/2}K = 2\pi(l/g)^{1/2}\,{}_2F_1(\tfrac{1}{2}, \tfrac{1}{2}; 1; \sin^2 \tfrac{1}{2}\alpha)$. If cubes and higher powers of α are neglected, $\sin^2 \tfrac{1}{2}\alpha = \tfrac{1}{4}\alpha^2$ and the hypergeometric function is $1+\alpha^2/16$, giving for the period the approximate value

$$2\pi(l/g)^{1/2}(1+\alpha^2/16).$$

11.4 Equations integrable in terms of elliptic functions

The solutions of the following differential equations can be expressed directly in terms of elliptic functions. The solutions given can be verified by substitution in the differential equation. This verification is carried out for the first given solution and for the others is left as an exercise for the reader. Since all the given equations are unchanged if $-x$ is substituted for x, it follows that if $x = f(t)$ is a solution so is $x = -f(t)$, and it is convenient to write the solutions in the form $x^2 = \phi(t)$. In each case we shall write the modulus of the elliptic functions as $k = \sin \alpha$.

(i) *The equation* $\dot{x}^2 = \lambda(1+2\mu x^2-x^4)$

If $\lambda > 0$, a solution for all values of μ is

$$x^2 = \tan \alpha \,\text{cn}^2\,(\omega t+\beta), \tag{1}$$

where $\cot 2\alpha = -\mu$, $\omega^2 = 2\lambda/\sin 2\alpha$ and β is an arbitrary constant. To verify this result we differentiate equation (1), giving

$$2x\dot{x} = -2\omega \tan \alpha \,\text{cn}\,(\omega t+\beta)\,\text{sn}\,(\omega t+\beta)\,\text{dn}\,(\omega t+\beta),$$

and hence

$$\begin{aligned} x^2\dot{x}^2 &= \omega^2 \tan^2 \alpha \,\text{cn}^2(\omega t+\beta)\{1-\text{cn}^2(\omega t+\beta)\}\{\cos^2\alpha+\sin^2\alpha\,\text{cn}^2(\omega t+\beta)\} \\ &= \omega^2 x^2(\tan \alpha - x^2)(\cos^2 \alpha + x^2 \sin \alpha \cos \alpha). \end{aligned}$$

Thus

$$\begin{aligned} \dot{x}^2 &= \tfrac{1}{2}\omega^2 \sin 2\alpha(\tan \alpha - x^2)(\cot \alpha + x^2) \\ &= \tfrac{1}{2}\omega^2 \sin 2\alpha(1-2x^2 \cot 2\alpha - x^4). \end{aligned}$$

If $\lambda < 0$, a solution for all values of μ is

$$x^2 = \cot \alpha/\text{cn}^2\,(\omega t+\beta), \tag{2}$$

where $\cot 2\alpha = \mu$, $\omega^2 = -2\lambda/\sin 2\alpha$ and β is arbitrary.

ii) *The equation* $\dot{x}^2 = \lambda(1+2\mu x^2+x^4)$

The substitution $x = 1/z$ in this equation leaves the equation unchanged and hence if $x^2 = \phi(t)$ is a solution so also is $x^2 = 1/\phi(t)$.

If $\lambda > 0$ and $\mu < -1$, we have solutions

$$x^2 = \sin \alpha \,\text{sn}^2\,(\omega t+\beta), \tag{3}$$

and

$$x^2 = \operatorname{cosec}\alpha/\operatorname{sn}^2(\omega t+\beta), \tag{4}$$

where $-\mu = (1+\sin^2\alpha)/(2\sin\alpha)$, $\omega^2 = \lambda/\sin\alpha$ and β is arbitrary.

If $\lambda > 0$ and $-1 < \mu < 1$, we have solutions

$$x^2 = \frac{1-\operatorname{cn}(\omega t+\beta)}{1+\operatorname{cn}(\omega t+\beta)}, \tag{5}$$

and

$$x^2 = \frac{1+\operatorname{cn}(\omega t+\beta)}{1-\operatorname{cn}(\omega t+\beta)}, \tag{6}$$

where $\mu = 1-2\sin^2\alpha$, $\omega^2 = 4\lambda$ and β is arbitrary.

If $\lambda > 0$ and $\mu > 1$, we have solutions

$$x^2 = \cos\alpha\operatorname{sn}^2(\omega t+\beta)/\operatorname{cn}^2(\omega t+\beta), \tag{7}$$

and

$$x^2 = \sec\alpha\operatorname{cn}^2(\omega t+\beta)/\operatorname{sn}^2(\omega t+\beta), \tag{8}$$

where $\mu = (1+\cos^2\alpha)/(2\cos\alpha)$, $\omega^2 = \lambda/\cos\alpha$ and β is arbitrary.

If $\lambda < 0$ and $\mu < -1$, we have solutions

$$x^2 = \sec\alpha\operatorname{dn}^2(\omega t+\beta), \tag{9}$$

and

$$x^2 = \cos\alpha/\operatorname{dn}^2(\omega t+\beta), \tag{10}$$

where $-\mu = \frac{1}{2}(\sec\alpha+\cos\alpha)$, $\omega^2 = -\lambda\sec\alpha$ and β is arbitrary.

11.5 The equation $\ddot{x}+ax+bx^3 = 0$

This equation describes motion in which the force on a particle varies as the distance and as the cube of the distance from the origin. In the case of a mass m controlled by a spring of uniform stiffness the equation of motion is $\ddot{x}+ax = 0$ and the stiffness is given by $s = ma$. The stiffness may, however, vary with the extension so that an equation of the above form is appropriate. The stiffness is then defined as the derivative of the restoring force, that is $s = ma+3mbx^2$. If b is positive the spring is said to be *hard*, that is the stiffness increases with the extension, whereas if b is negative the spring is said to be *soft*. We shall consider subsequently the case where the stiffness is a linear function of x, so that $s = ma+2mbx$.

We proceed to find solutions in terms of Jacobian elliptic functions of the equations $\ddot{x}\pm ax\pm bx^3 = 0$, where a and b are positive, and in each case we shall assume that $x = x_0 > 0$ and $\dot{x} = 0$ when $t = 0$.

(i) *The cn oscillation* $\ddot{x}+ax+bx^3 = 0$

A first integral of this equation obtained by writing $\ddot{x} = (d/dx)(\dot{x}^2/2)$ and integrating with respect to x is

$$\tfrac{1}{2}\dot{x}^2 = \tfrac{1}{2}a(x_0{}^2-x^2)+\tfrac{1}{4}b(x_0{}^4-x^4),$$

giving

$$\dot{x}^2 = \tfrac{1}{2}b(x_0{}^2 - x^2)(x^2 + x_0{}^2 + 2a/b). \tag{1}$$

This equation is of the type considered in §11.4 (i) and has solution

$$x = x_0 \operatorname{cn} \omega t. \tag{2}$$

To find the modulus $k = \sin\alpha$ and the value of ω the simplest method is to form the differential equation for x. Differentiating

$$\dot{x} = -x_0\omega \operatorname{sn}\omega t \operatorname{dn}\omega t.$$

Then

$$\dot{x}^2 = x_0{}^2\omega^2(1 - \operatorname{cn}^2\omega t)(\cos^2\alpha + \sin^2\alpha \operatorname{cn}^2\omega t)$$

$$= \frac{\omega^2\sin^2\alpha}{x_0{}^2}(x_0{}^2 - x^2)(x^2 + x_0{}^2\cot^2\alpha).$$

Comparing with (1) we find

$$x_0{}^2\cot^2\alpha = x_0{}^2 + 2a/b, \quad \text{and} \quad \omega^2 = \tfrac{1}{2}bx_0{}^2\operatorname{cosec}^2\alpha.$$

The motion given by (2) is oscillatory and the period is

$$T = 4K/\omega = (2\pi/\omega)\,{}_2F_1(\tfrac{1}{2}, \tfrac{1}{2}; 1; \sin^2\alpha).$$

(ii) *The sn oscillation* $\ddot{x} + ax - bx^3 = 0$

The initial motion will be towards the origin if $a > bx_0{}^2$ and this is the condition that the motion should be oscillatory. The first integral of the equation is now

$$\dot{x}^2 = \tfrac{1}{2}b(x^2 - x_0{}^2)(x^2 + x_0{}^2 - 2a/b). \tag{3}$$

This is an equation of the type considered in §11.4 (ii) equation (3) and the solution is

$$x = x_0 \operatorname{sn}(\omega t + K). \tag{4}$$

Differentiation gives

$$\dot{x} = x_0\omega \operatorname{cn}(\omega t + K)\operatorname{dn}(\omega t + K),$$

and

$$\dot{x}^2 = x_0{}^2\omega^2(1 - x^2/x_0{}^2)(1 - x^2\sin^2\alpha/x_0{}^2)$$

$$= \frac{\omega^2\sin^2\alpha}{x_0{}^2}(x^2 - x_0{}^2)(x^2 - x_0{}^2\operatorname{cosec}^2\alpha).$$

Hence, by comparison with (3),

$$x_0{}^2\operatorname{cosec}^2\alpha = 2a/b - x_0{}^2, \qquad \omega^2 = \tfrac{1}{2}bx_0{}^2\operatorname{cosec}^2\alpha,$$

and the period is $4K/\omega$.

(iii) *The dn oscillation* $\ddot{x}-ax+bx^3 = 0$

In this case the first integral is

$$\dot{x}^2 = \tfrac{1}{2}b(x_0^2-x^2)(x^2+x_0^2-2a/b), \tag{5}$$

and if $bx_0^2 > 2a$, this is a cn oscillation similar to that considered in (i), so that

$$x = x_0 \operatorname{cn} \omega t, \tag{6}$$

with $x_0^2 \cot^2 \alpha = x_0^2-2a/b$ and $\omega^2 = \tfrac{1}{2}bx_0^2 \operatorname{cosec}^2 \alpha$.

If $bx_0^2 < 2a$ the solution is of the type considered in §11.4 (ii), equation (9), but we must distinguish between the cases in which the initial motion is towards and away from the origin.

If $a < bx_0^2 < 2a$ the initial motion is towards the origin and the solution is

$$x = x_0 \operatorname{dn} \omega t. \tag{7}$$

Proceeding as before we find

$$\dot{x}^2 = x_0^2\omega^2k^4 \operatorname{sn}^2 \omega t \operatorname{cn}^2 \omega t$$

$$= \frac{\omega^2}{x_0^2}(x_0^2-x^2)(x^2-x_0^2\cos^2\alpha),$$

and comparison with (5) gives $\cos^2\alpha = (2a/b-x_0^2)/x_0^2$ and $\omega^2 = \tfrac{1}{2}bx_0^2$. This oscillation is between x_0 and $x_0 \cos\alpha$ and the period is $2K/\omega$.

If $bx_0^2 < a$ the initial motion is away from the origin and the appropriate solution is

$$x = x_0 \sec\alpha \operatorname{dn}(\omega t+K). \tag{8}$$

Since $\operatorname{dn} K = \cos\alpha$ this gives the correct initial value of x. To find the modulus we have

$$\dot{x}^2 = x_0^2\omega^2\sec^2\alpha\, k^4 \operatorname{sn}^2(\omega t+K)\operatorname{cn}^2(\omega t+K)$$

$$= (\omega^2\cos^2\alpha/x_0^2)(x_0^2-x^2)(x^2-x_0^2\sec^2\alpha).$$

Hence, by comparison with (5),

$$\sec^2\alpha = (2a/b-x_0^2)/x_0^2 \quad \text{and} \quad \omega^2 = \tfrac{1}{2}bx_0^2\sec^2\alpha.$$

This oscillation is between x_0 and $x_0 \sec\alpha$ and the period is $2K/\omega$.

(iv) *Non-oscillatory motion*

The equation $\ddot{x}-ax-bx^3 = 0$ has a first integral

$$\dot{x}^2 = \tfrac{1}{2}b(x^2-x_0^2)(x^2+x_0^2+2a/b), \tag{9}$$

and this is the type considered in §11.4 (i) with solution

$$x = x_0/\operatorname{cn}\omega t \tag{10}$$

Proceeding as before we find

$$\dot{x}^2 = \frac{x_0{}^2\omega^2}{\text{cn}^4\,\omega t}(1-\text{cn}^2\,\omega t)(\cos^2\alpha+\sin^2\alpha\,\text{cn}^2\,\omega t)$$

$$= \frac{\omega^2\cos^2\alpha}{x_0{}^2}(x^2-x_0{}^2)(x^2+x_0{}^2\tan^2\alpha).$$

Hence $\tan^2\alpha = (x_0{}^2+2a/b)/x_0{}^2$ and $\omega^2 = \frac{1}{2}bx_0{}^2\sec^2\alpha$.

Similarly the equation $\ddot{x}+ax-bx^3 = 0$ whose first integral is given in (3) has the solution

$$x = x_0/\text{cn}\,\omega t, \tag{11}$$

where $\tan^2\alpha = (x_0{}^2-2a/b)/x_0{}^2$ and $\omega^2 = \frac{1}{2}bx_0{}^2\sec^2\alpha$, in the non-oscillatory case when $bx_0{}^2 > 2a$. If, however, $a < bx_0{}^2 < 2a$ the solution is still non-oscillatory but the solution is that of §11.4, equation (4) and

$$x = x_0/\text{sn}\,(\omega t+K). \tag{12}$$

This gives

$$\dot{x}^2 = \frac{\omega^2}{x_0{}^2}(x^2-x_0{}^2)(x^2-x_0{}^2\sin^2\alpha)$$

and hence $\sin^2\alpha = (2a/b-x_0{}^2)/x_0{}^2$ and $\omega^2 = \frac{1}{2}bx_0{}^2$. In each of these cases x becomes infinite after time K/ω.

Example 2. *A non-linear spring exerts a restoring force $mg(x/l+x^3/l^3)$ when it is extended a distance x. Find the period of oscillation of a mass m controlled by the spring when the amplitude is $\frac{1}{2}l$. Find also the maximum velocity of the mass.*

The equation of motion $\ddot{x}+(g/l)x+(g/l^3)x^3 = 0$ gives a cn oscillation as described in §11.5 (i) and the solution is

$$x = \tfrac{1}{2}l\,\text{cn}\,(\omega t, k),$$

where

$$k^2 = \frac{\frac{1}{2}(g/l^3)(l^2/4)}{g/l+(g/l^3)(l^2/4)} = \tfrac{1}{10},$$

and

$$\omega^2 = \frac{g}{l}+\frac{g}{4l} = \frac{5g}{4l}.$$

Since

$$_2F_1(\tfrac{1}{2}, \tfrac{1}{2}; 1; k^2) = 1+(\tfrac{1}{2})^2k^2+(\tfrac{3}{8})^2k^4+(\tfrac{5}{16})^2k^6+\ldots$$
$$= 1{\cdot}0265,$$

the period is

$$T = 1{\cdot}0265(2\pi/\omega)$$
$$= 2\pi(l/g)^{1/2}(2{\cdot}053/5^{1/2}) = 0{\cdot}918\{2\pi(l/g)^{1/2}\}.$$

The velocity at any time is

$$\dot{x} = -\tfrac{1}{2}l\omega\,\text{sn}\,(\omega t)\,\text{dn}\,(\omega t),$$

and the maximum velocity, when $\omega t = K$, is

$$-\tfrac{1}{2}l(5g/4l)^{1/2}(1-k^2)^{1/2} = -\tfrac{3}{8}(2gl)^{1/2}.$$

11.6 The equation $\ddot{x}+ax+bx^2 = 0$

This equation which may represent an oscillation or a movement from the initial point towards infinity can be solved in terms of Jacobian elliptic functions. If initially $x = x_0 > 0$ and $\dot{x} = 0$, a first integral obtained by writing $\ddot{x} = (d/dx)(\dot{x}^2/2)$ and integrating is

$$\dot{x}^2 = a(x_0{}^2-x^2)+(2b/3)(x_0{}^3-x^3)$$
$$=\tfrac{2}{3}b(x_0-x)\{(x_0-x)^2-3(x_0+a/2b)(x_0-x)+3x_0(x_0+a/b)\}. \quad (1)$$

To reduce this equation to a standard form we make the substitution $x_0-x = z^2$ if the initial acceleration is towards the origin and $x_0-x = -z^2$ if it is away from the origin. We shall find the solution for the case where a and b are positive, other cases being dealt with in much the same way.

Since a and b are positive, the initial acceleration is towards the origin and the substitution $x_0-x = z^2$ gives

$$4z^2\dot{z}^2 = \tfrac{2}{3}bz^2\{z^4-3(x_0+a/2b)z^2+3x_0(x_0+a/b)\},$$
$$\dot{z}^2 = \tfrac{1}{6}b\{z^4-3(x_0+a/2b)z^2+3x_0(x_0+a/b)\}. \quad (2)$$

This gives zero velocity at the points where

$$z^2 = (x_0-x) = [3(a+2bx_0)\pm\{3(a-2bx_0)(3a+2bx_0)\}^{\frac{1}{2}}]/4b.$$

Hence if $a > 2bx_0$ there are stopping points x_1 and x_2, the corresponding values of z^2 being positive and equal to $z_1{}^2$ and $z_2{}^2$. If $|z_1{}^2| < |z_2{}^2|$ the equation (2) can be written as

$$\dot{z}^2 = \tfrac{1}{6}b(z^2-z_1{}^2)(z^2-z_2{}^2),$$

and, putting $z = z_1 y$,

$$\dot{y}^2 = \tfrac{1}{6}bz_2{}^2(1-y^2)(1-k^2y^2),$$

where $k^2 = z_1{}^2/z_2{}^2$. Thus if $\omega^2 = \tfrac{1}{6}bz_2{}^2$

$$y = \operatorname{sn}\omega t,$$

and

$$z^2 = z_1{}^2\operatorname{sn}^2\omega t,$$

that is,

$$x_0-x = (x_0-x_1)\operatorname{sn}^2\omega t,$$

or,

$$x = x_0\operatorname{cn}^2\omega t+x_1\operatorname{sn}^2\omega t.$$

Thus the motion is oscillatory with period $2K/\omega$.

If $a < 2bx_0$ there are no stopping points, and putting $z^2 = \{3x_0(x_0+a/b)\}^{\frac{1}{2}}y^2$ in (2) we find

$$\dot{y}^2 = \tfrac{1}{6}b\{3x_0(x_0+a/b)\}^{\frac{1}{2}}(y^4+2\mu y^2+1)$$

where

$$-\mu = \frac{3(x_0+a/2b)}{2\{3x_0(x_0+a/b)\}^{\frac{1}{2}}}.$$

It is easily seen that μ^2-1 is negative and hence that $-1 < \mu < 1$, so that the solution is, §11.4(ii) (5),

$$y^2 = \frac{1-\text{cn}\,\omega t}{1+\text{cn}\,\omega t},$$

with $\cos 2\alpha = \mu$ and $\omega^2 = \frac{2}{3}b\{3x_0(x_0+a/b)\}^{\frac{1}{2}}$. Hence

$$z^2 = x_0 - x = \{3x_0(x_0+a/b)\}^{\frac{1}{2}}\left(\frac{1-\text{cn}\,\omega t}{1+\text{cn}\,\omega t}\right).$$

11.7 The equation $\ddot{x}+ax+bx|x| = 0$

This equation can represent the motion of a particle of unit mass controlled by a spring of stiffness $s = a+2b|x|$, the modulus ensuring that the restoring force changes sign at the origin. Thus if a and b are positive the motion is bound to be oscillatory, the motion for the first quarter period being identical with that given by $\ddot{x}+ax+bx^2 = 0$, §11.6.

Hence if $a > 2bx_0$ the motion between x_0 and the origin is given by

$$x_0 - x = (x_0-x_1)\,\text{sn}^2\,\omega t,$$

and at the origin $\text{sn}\,\omega t = \{x_0/(x_0-x_1)\}^{\frac{1}{2}}$, so that the period

$$(4/\omega)\,\text{sn}^{-1}\{x_0/(x_0-x_1)\}^{\frac{1}{2}}$$

can be found from the tables.

If $a < 2bx_0$ the motion from x_0 to the origin is given by

$$x_0 - x = \{3x_0(x_0+a/b)\}^{\frac{1}{2}}\left(\frac{1-\text{cn}\,\omega t}{1+\text{cn}\,\omega t}\right),$$

whence the value of $\text{cn}\,\omega t$ at the origin and the period of the oscillation can be found.

Example 3. *Find the periodic time of oscillation of a particle of unit mass controlled by a spring of stiffness $(g/l)(1+|x|)$ where x is the extension and the initial extension is $\frac{1}{2}l$.*

The equation of motion is $\ddot{x}+gx/l+\frac{1}{2}gx|x|/l^2 = 0$. Disregarding the modulus sign we have the first integral

$$\dot{x}^2 = (g/l)(\tfrac{1}{4}l^2-x^2)+(g/3l^2)(\tfrac{1}{8}l^3-x^3)$$

$$= \frac{g}{3l^2}(\tfrac{1}{2}l-x)\{(\tfrac{1}{2}l-x)^2-(9l/2)(\tfrac{1}{2}l-x)+15l^2/4\}.$$

The velocity is zero when $(\frac{1}{2}l-x) = 0$ or $\{9\pm\sqrt{(21)}\}l/4$. Putting $z^2 = \frac{1}{2}l-x$, we can write

$$\dot{z}^2 = \frac{g}{12l^2}(z^2-z_1^2)(z^2-z_2^2),$$

where
$$z_1^2/z_2^2 = \{9-\sqrt{(21)}\}/\{9+\sqrt{(21)}\} = (0{\cdot}5702)^2 = k^2.$$
Then
$$z^2 = (\tfrac{1}{2}l-x) = z_1^2\,\mathrm{sn}^2\,\omega t,$$
where
$$\omega^2 = (g/12l^2)z_2^2 = 0{\cdot}283(g/l), \qquad \omega = 0{\cdot}532(g/l)^{1/2}.$$
The particle reaches the origin in time t where
$$\mathrm{sn}^2\omega t = 2l/\{9-\sqrt{(21)}\}l = 0{\cdot}4528,$$
$$\mathrm{sn}\,\omega t = 0{\cdot}6729,$$
and for this with $k = 0{\cdot}5702 = \sin 34°46'$, the tables give $\omega t = 0{\cdot}670$. Thus $4t = 5{\cdot}04(l/g)^{1/2}$ and this is the period, which may be written in order to show the effect of non-linearity, as
$$T = 0{\cdot}80\{2\pi(l/g)^{1/2}\}.$$

Exercises 11(*b*)

1. Prove that if
$$t = \int_0^\theta \frac{d\phi}{(1-k^2\sin^2\phi)^{1/2}},$$
then $\sin\theta = \mathrm{sn}\,(t, k)$.

2. Prove that if the Jacobian elliptic functions are expanded in series of powers of their arguments,
$$\mathrm{sn}\,t = t-\tfrac{1}{6}(1+k^2)t^3+O(t^5),$$
$$\mathrm{cn}\,t = 1-\tfrac{1}{2}t^2+O(t^4),$$
$$\mathrm{dn}\,t = 1-\tfrac{1}{2}k^2t^2+O(t^4).$$

3. Prove that, if
$$t = \int_x^1 (1-u^2)^{-1/2}(\cos^2\alpha+u^2\sin^2\alpha)^{-1/2}\,du,$$
then $x = \mathrm{cn}\,(t, k)$, where $k = \sin\alpha$.

4. Prove that, if
$$t = \int_x^1 (1-u^2)^{-1/2}(u^2-\cos^2\alpha)^{-1/2}\,du,$$
then $x = \mathrm{dn}\,(t, k)$, where $k = \sin\alpha$.

5. Prove that the solution of the equation $\ddot{x}+gx/a+gx^3/a^3 = 0$ with $x = a$ and $\dot{x} = 0$ when $t = 0$ represents an oscillation of amplitude a and period $1{\cdot}52\pi(a/g)^{1/2}$.

6. Solve the equation $2\ddot{x} = 1+12x^2$, with $x = 10$ and $\dot{x} = 0$ when $t = 0$. Given that for $k = \sin 15°$, $K = 1{\cdot}598$, and that $\mathrm{sn}\,\omega t = 0{\cdot}736$ when $\omega t = 2{\cdot}527$, show that $x = 100$ when $t = 0{\cdot}30$ approximately.

7. Show that the equation $\ddot{x}-\tfrac{2}{3}bcx+bx^2 = 0$, $b>0$, $c>0$, with $x = x_0$ and $\dot{x} = 0$ when $t = 0$, has a solution of the form
$$x = x_1\,\mathrm{cn}^2\,(\omega t+K)+x_0\,\mathrm{sn}^2\,(\omega t+K)$$
if $0 < x_0 < 2c/3$, where $2x_1 = c-x_0+\{(c-x_0)(c+3x_0)\}^{1/2}$. Show also that the modulus of the elliptic functions is given by
$$k^2 = (x_1-x_0)/(2x_1+x_0-c), \quad \text{and} \quad \omega^2 = \tfrac{1}{6}b(2x_1+x_0-c).$$

8. Show that the equation $\ddot{x}+\frac{2}{3}bcx-bx^2 = 0$, $b > 0$, $c > 0$ with $x = x_0$ and $\dot{x} = 0$ when $t = 0$, has a solution of the form

$$x = x_1 \operatorname{cn}^2(\omega t+K)+x_0 \operatorname{sn}^2(\omega t+K)$$

if $0 < x_0 < \frac{2}{3}c$, where $2x_1 = c-x_0-\{(c-x_0)(c+3x_0)\}^{1/2}$. Show also that the modulus of the elliptic function is given by

$$k^2 = (x_0-x_1)/(c-2x_1-x_0), \quad \text{and} \quad \omega^2 = \tfrac{1}{6}b(c-2x_1-x_0).$$

9. Show that the equation $\ddot{x}+\frac{2}{3}bcx-bx^2 = 0$, $b > 0$, $c > 0$, with $x = x_0$ and $\dot{x} = 0$ when $t = 0$ has solutions

(i) if $x_0 < c < \frac{3}{2}x_0$,

$$x-x_0 = \lambda \operatorname{sn}^2 \omega t/\operatorname{cn}^2 \omega t,$$

(ii) if $0 < c < x_0$,

$$x-x_0 = \lambda'(1-\operatorname{cn} \omega t)/(1+\operatorname{cn} \omega t),$$

where λ, λ' and ω are constants.

10. Show that the equation $\ddot{x}-\frac{2}{3}bcx-bx^2 = 0$, $b > 0$, $c > 0$, with $x = x_0\,(>0)$ and $\dot{x} = 0$ when $t = 0$ has solutions

(i) if $c > 3x_0$,

$$x-x_0 = \lambda \operatorname{sn}^2\omega t/\operatorname{cn}^2 \omega t,$$

(ii) if $c < 3x_0$,

$$x-x_0 = \lambda'(1-\operatorname{cn} \omega t)/(1+\operatorname{cn} \omega t),$$

where λ, λ' and ω are constants.

11. The differential equation of the *elastica*, that is a thin strut with a large deflexion, is $EI\,d\psi/ds = -Py$, where y is the deflexion at a point at a distance s along the strut and ψ is the inclination of the tangent at the point to the line of thrust. P is the axial thrust, EI the modulus of rigidity and $P = EIn^2$. Show that

$$\frac{d^2\psi}{ds^2}+n^2 \sin \psi = 0,$$

and hence, if α be the slope at the ends of the strut, that

$$\sin \tfrac{1}{2}\psi = \sin \tfrac{1}{2}\alpha \operatorname{sn}(ns+K),$$

where $K = \frac{1}{2}\pi\, {}_2F_1(\frac{1}{2}, \frac{1}{2}; 1; \sin^2 \frac{1}{2}\alpha)$. Deduce that the length of the strut is $2K/n$ and the maximum deflexion $(2/n) \sin \frac{1}{2}\alpha$.

12. The equations of motion of a symmetrical top are

$$\dot{\phi} \sin^2 \theta = (Cn/B)(\cos \alpha-\cos \theta),$$
$$\dot{\phi}^2 \sin^2 \theta+\dot{\theta}^2 = (2\mu/B)(\cos \alpha-\cos \theta),$$

where θ is the angle of nutation, ϕ the azimuth angle and $\theta = \alpha$, $\dot{\theta} = \phi = \dot{\phi} = 0$ when $t = 0$, C, n, B, μ being constants. Show that

$$\dot{\theta}^2 \sin^2 \theta = (2\mu/B)(\cos \alpha-\cos \theta)(\cos \theta-\cos \beta)(c-\cos \theta),$$

where $\cos \beta = s-(s^2-2s \cos \alpha+1)^{1/2}$, $c = s+(s^2-2s \cos \alpha+1)^{1/2}$ and $s = C^2n^2/4B\mu$. Hence show that

$$\cos \theta = \cos \alpha \operatorname{sn}^2(\omega t+K)+\cos \beta \operatorname{cn}^2(\omega t+K)$$

where $(c-\cos \beta)k^2 = \cos \alpha-\cos \beta$ and $B\omega^2 = \mu(s^2-2s \cos \alpha+1)^{1/2}$.

11.8 Damped oscillations

Much of recent work in the theory of non-linear equations is concerned with oscillations in which damping occurs. An oscillation will, of course, eventually die away when there is a damping force opposing motion; this is called *positive damping*. In certain cases, however, the damping may be positive for one part of the motion and negative for another part, due to energy being received from an outside source, and in such cases the motion can be a sustained oscillation.

A general form of second order equation representing a damped oscillation is

$$\ddot{x}+\psi(\dot{x}, x) = 0, \tag{1}$$

but this can often be written in the form

$$\ddot{x}+\phi(\dot{x})+f(x) = 0, \tag{2}$$

where $f(x)$ represents the restoring force and $\phi(\dot{x})$ the damping force. In the particular case of *Coulomb damping* $\phi(\dot{x})$ is constant but changes sign with $\dot{x}$ so that equation (2) takes the form

$$\ddot{x}+c \operatorname{sgn} \dot{x}+f(x) = 0, \tag{3}$$

$\operatorname{sgn} \dot{x}$ being $+1$ when $\dot{x}$ is positive and -1 when $\dot{x}$ is negative.

In equations of the type (1), (2) and (3), the time t is explicitly absent and the equation can be reduced to one of the first order by writing $\dot{x} = v$. Thus (2) becomes

$$v\frac{dv}{dx}+\phi(v)+f(x) = 0. \tag{4}$$

If this equation can be solved to give v in terms of x, a second integration will give a relation between x and t. If an explicit solution of (4) cannot be obtained, graphical and numerical methods can be used to obtain a solution and there are other methods by which information can be obtained about the nature of the motion and its period and amplitude.

We now consider some cases in which equation (4) can be solved.

(*a*) *Coulomb damping*

(i) *The linear equation* $\ddot{x}+c \operatorname{sgn} \dot{x}+\omega^2 x = 0$

We have, in effect, two equations, namely

when $\dot{x} < 0$,

$$\ddot{x}-c+\omega^2 x = 0,$$

and when $\dot{x} > 0$,

$$\ddot{x}+c+\omega^2 x = 0.$$

These equations represent harmonic oscillations centred at c/ω^2 and $-c/\omega^2$ respectively. If initially $\dot{x} = 0$ and $x = x_0 > 0$, for the first phase of the motion $x-c/\omega^2 = (x_0-c/\omega^2)\cos\omega t$, and the particle comes to rest after time π/ω at $x_1 = -x_0+2c/\omega^2$. For the second phase we have $x+c/\omega^2 = (x_1+c/\omega^2)\cos\omega t_1$, where $t_1 = t-\pi/\omega$, and the particle comes to rest after a further time π/ω at

$$x_2 = -x_1-2c/\omega^2 = x_0-4c/\omega^2.$$

Thus the motion continues with the distance travelled diminishing by $2c/\omega^2$ each time, until the particle comes to rest at x_n where $|x_n| < c/\omega^2$, when the friction will prevent it starting again. The total time of motion will then be $n\pi/\omega$.

(ii) *The equation* $\ddot{x}+c\operatorname{sgn}\dot{x}+ax+bx^2 = 0,\ a,\ b,\ c > 0$

If initially $\dot{x} = 0$ and $x = x_0 > 0$, we have for the first phase

$$\ddot{x}+ax+bx^2-c = 0,$$

and writing $x-\lambda_1 = z$, where $b\lambda_1^2+a\lambda_1-c = 0$, this becomes

$$\ddot{z}+(a+2b\lambda_1)z+bz^2 = 0.$$

This equation can be solved in terms of elliptic functions by the method of §11.6. For the second phase, we have the equation

$$\ddot{x}+ax+bx^2+c = 0,$$

and the substitution $x-\lambda_2 = z$, where $b\lambda_2^2+a\lambda_2+c = 0$, gives

$$\ddot{z}+(a+2b\lambda_2)z+bz^2 = 0,$$

which can be solved similarly.

(iii) *The equation* $\ddot{x}+c\operatorname{sgn}\dot{x}+ax+bx^3 = 0,\ a,\ b,\ c > 0$

If initially $\dot{x} = 0$ and $x = x_0 > 0$, we have for the first phase of the motion

$$\ddot{x}-c+ax+bx^3 = 0,$$

and a first integral of this equation is

$$\dot{x}^2 = \tfrac{1}{2}b(x_0-x)\{x^3+x_0x^2+(x_0^2+2a/b)x+x_0^3+2ax_0/b-4c/b\}.$$

This equation gives a stop at a *negative* value x_1 if $x_0^3+2ax_0/b-4c/b>0$, and hence

$$\dot{x}^2 = \tfrac{1}{2}b(x_0-x)(x-x_1)\{x^2+(x_0+x_1)x+d^2\},$$

where $d^2 > 0$.

The substitution $x = (\lambda z+\mu)/(z+1)$ reduces this equation to one of the form

$$\dot{z}^2 = Az^4+Bz^2+C,$$

if λ and μ are chosen to make the coefficients of z^3 and z zero, and hence the solution can be expressed in terms of elliptic functions. The equation $\ddot{x}+c+ax+bx^3 = 0$ for the second phase can be solved in a similar way.

(*b*) *Damping proportional to the square of velocity*

(i) *Linear restoring force:* $\ddot{x}+c\dot{x}|\dot{x}|+\omega^2 x = 0,\ c > 0$

If initially $x = x_0 > 0$ and $\dot{x} = 0$, the equation for the first phase of the motion is

$$\ddot{x}-c\dot{x}^2+\omega^2 x = 0.$$

Using the integrating factor e^{-2cx} and integrating with respect to x we find

$$\dot{x}^2 e^{-2cx} = (\omega^2/2c^2)\{f(x)-f(x_0)\},$$

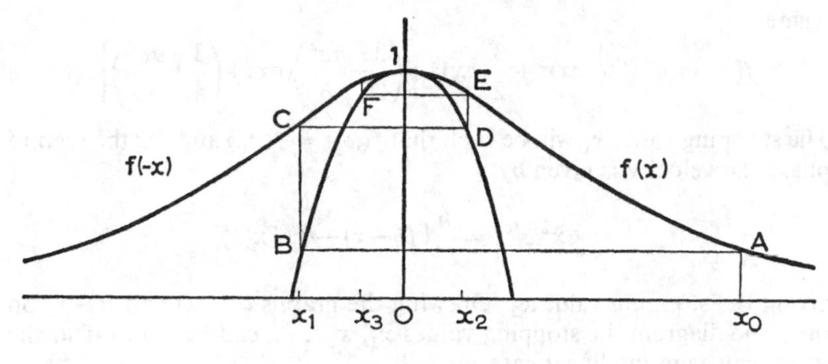

FIG. 68

where $f(x) = (1+2cx)e^{-2cx}$. This gives a stopping point at $x = x_1$, where $f(x_1) = f(x_0)$, and for the second phase we have a similar expression for $\dot{x}^2$ obtained by replacing c by $-c$ and x_0 by x_1, namely

$$\dot{x}^2 e^{2cx} = (\omega^2/2c^2)\{f(-x)-f(-x_1)\}.$$

This gives a stopping point at $x = x_2$, where $f(-x_2) = f(-x_1)$, and for the third phase

$$\dot{x}^2 e^{-2cx} = (\omega^2/2c^2)\{f(x)-f(x_2)\},$$

and so on.

The motion can be represented graphically by drawing the graphs of $f(x) = (1+2cx)e^{-2cx}$ and $f(-x) = (1-2cx)e^{2cx}$ on the same diagram (Fig. 68). Thus A and B, where AB is parallel to the x-axis, are points at which $f(x)$ has the same value and the abscissa of B is x_1. Similarly, C and D are points at which $f(-x)$ has the same value and the abscissa of D is x_2. Thus the set of values $x_1, x_2, x_3, \ldots$, is determined and the diminishing amplitude of the oscillation is clearly shown. It is evident that whatever may be the value of x_0 all the values $x_1, x_2, \ldots,$ lie within the range bounded by the zeros of $f(x)$ and $f(-x)$, that is $-1/2c$ to $1/2c$.

(ii) *Non-linear restoring force:* $\ddot{x}+c\dot{x}|\dot{x}|+ax+bx^3 = 0$

Assuming that a, b and c are positive and that initially $\dot{x} = 0$ and $x = x_0 > 0$, the equation for the first phase is

$$\ddot{x}-c\dot{x}^2+ax+bx^3 = 0.$$

Using the integrating factor e^{-2cx} and integrating with respect to x we find

$$\dot{x}^2 e^{-2cx} = \frac{b}{c^4}\{f(x)-f(x_0)\},$$

where

$$f(x) = e^{-2cx}\left\{(cx)^3+\frac{3}{2}(cx)^2+\left(\frac{3}{2}+\frac{ac^2}{b}\right)(cx)+\left(\frac{3}{4}+\frac{ac^2}{2b}\right)\right\}.$$

The stopping value x_1 will be such that $f(x_1) = f(x_0)$ and for the second phase the velocity is given by

$$\dot{x}^2 e^{2cx} = \frac{b}{c^4}\{f(-x)-f(-x_1)\},$$

giving the stopping value x_2. Drawing the graphs of $f(x)$ and $f(-x)$ on the same diagram the stopping values $x_1, x_2, \ldots,$ can be read off in the same way as in the linear case.

11.9 Use of the phase-plane diagram

We have seen that equations of the type $\ddot{x}+f(\dot{x}, x) = 0$ can be expressed as first order equations in the form

$$v\frac{dv}{dx}+f(v, x) = 0 \qquad (1)$$

When this equation cannot be integrated directly, numerical or graphical methods may be used to study the variation of v with x. By the method of isoclinals integral curves of the equation (1) can be drawn with differing initial conditions to give a representation of the motion on the phase-plane diagram, see §9.3, from which the amplitude of an oscillation and the maximum velocity can be seen at once. When an oscillation is ultimately damped out the phase-plane diagram will, in general, be a spiral ending in a point of stable equilibrium, whereas if the motion is unstable the velocity will increase indefinitely. In the case of a sustained oscillation where sufficient energy is received from an external force to counter the damping, the phase-plane diagram will be a closed curve. The period of such an oscillation may be found by evaluating the integral $\int dt = \int \frac{dx}{v}$, the integral being taken along the closed curve in the phase-plane diagram.

As an example, consider the equation

$$\ddot{x}+c\dot{x}+ax+bx^3 = 0, \tag{2}$$

with a, b and c positive, that is

$$\frac{dv}{dx}+c+\frac{ax+bx^3}{v} = 0. \tag{3}$$

With $c = 0$ the solution is (§11.5(1).)

$$v^2 = \tfrac{1}{2}b({x_0}^2-x^2)(x^2+{x_0}^2+2a/b),$$

where initially $\dot{x} = 0$ and $x = x_0$. The phase-plane diagram in this case is a closed curve resembling an ellipse as shown in Fig. 69. When

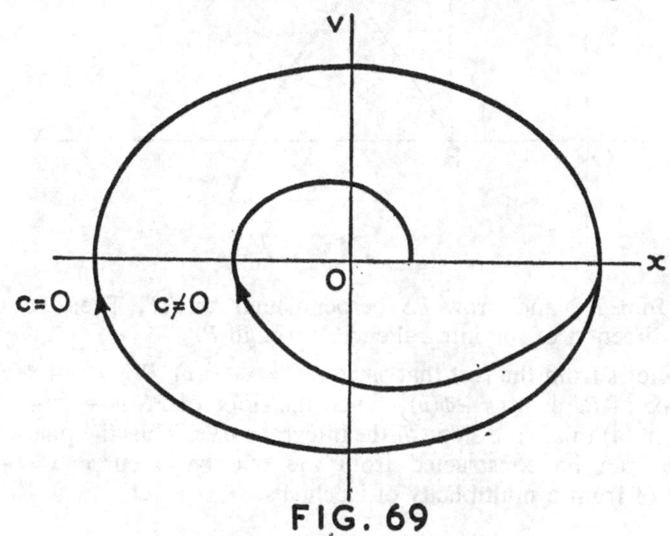

FIG. 69

$c \neq 0$ the equation (3) cannot be integrated but the phase-plane diagram will be a spiral towards the origin as shown in the figure.

When the restoring force is linear and the damping force is a function of velocity only, equation (1) can be written in the form

$$v\frac{dv}{dx}+\phi(v)+x = 0, \tag{4}$$

by making, if necessary, a change of variable to absorb the constant in the restoring force. In this case a graphical method known as *Liénard's construction* can be used to draw the phase-plane diagram. The procedure is as follows:

(i) Draw the curve $x = -\phi(v)$ on the phase plane (Fig. 70).

(ii) At any point $P(x, v)$ draw a parallel to the x-axis to meet the curve in the point Q.

(iii) From Q drop a perpendicular to meet the x-axis at R.

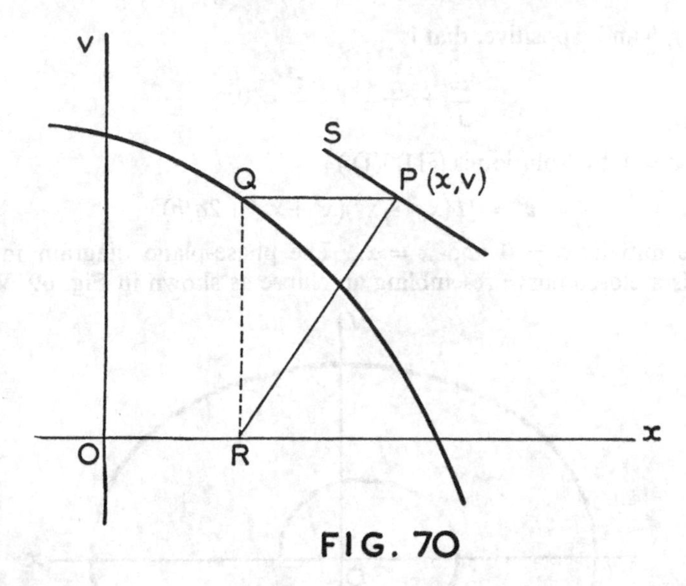

FIG. 70

(iv) Join RP and draw PS perpendicular to RP. Then PS is the direction of the integral curve through P.

This follows from the fact that since $OR = -\phi(v)$, $PQ = x+\phi(v)$ and the slope of RP is $v/\{x+\phi(v)\}$. Then the slope of PS is $-\{x+\phi(v)\}/v$ and from (4) this is the slope of the integral curve. Thus the phase-plane diagram can be constructed from the one basic curve $x = -\phi(v)$ instead of from a multiplicity of isoclinals.

Example 4. *Draw the phase-plane diagram of the solution of the equation*

$$\ddot{x}+\dot{x}|\dot{x}|+x = 0 \text{ with } x = 2 \text{ and } \dot{x} = 0 \text{ when } t = 0.$$

The first order equation is

$$v\frac{dv}{dx}-v^2+x = 0, \quad \text{when } v \text{ is negative,}$$

$$v\frac{dv}{dx}+v^2+x = 0, \quad \text{when } v \text{ is positive.}$$

In the first stage of the motion v is negative so the parabola $v^2 = x$ is drawn in the phase plane (Fig. 71). At $A(x = 2)$ the construction gives a line AB per-

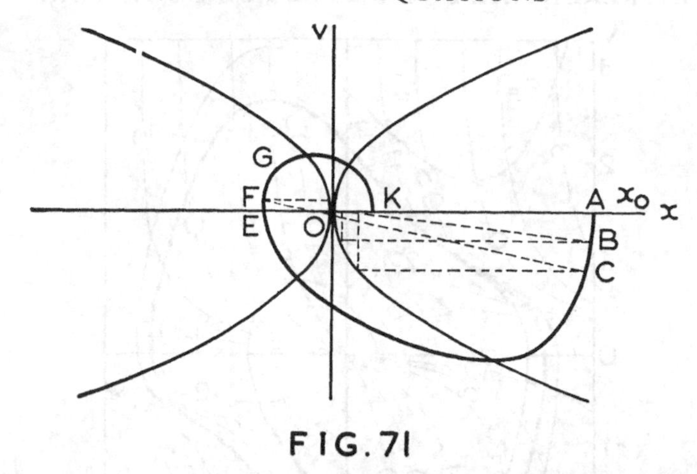

FIG.71

pendicular to OX and at B, a little way along this line, we find a direction BC, and so on until the point E on the x-axis is reached.

At this stage v changes sign so that the second equation is now appropriate and the parabola $v^2 = -x$ is drawn and points $F, G, \ldots$, found. When the x-axis is again reached at K the parabola $v^2 = x$ comes into use again. Thus a polygon is constructed which is smoothed out to give the integral curve.

11.10 Van der Pol's equation

Van der Pol's equation is

$$\ddot{x} - \varepsilon(1-x^2)\dot{x} + x = 0, \tag{1}$$

with $\varepsilon > 0$. This equation represents a self-oscillatory thermionic valve circuit, x being a function of the current. It is clear that in an oscillation in which $|x|$ is sometimes greater and sometimes less than unity the damping coefficient $\varepsilon(1-x^2)$ will alternate in sign so that compensation for the energy dissipated in the circuit can be received from an external source. Thus the equation represents a sustained periodic oscillation. This was shown by van der Pol* by using the method of isoclinals to draw the phase-plane diagram of the motion for different values of ε. The diagram for $\varepsilon = 1$ is shown in Fig. 72. It is seen that if x is initially large the positive damping reduces the amplitude and the oscillation quickly settles down to the steady oscillation shown by the heavy black line. If x is initially small the negative damping increases the amplitude until the same steady state is reached. The solution for $\varepsilon = 1$, giving x in terms of t following a small initial disturbance, is shown in Fig. 73. For all positive values of ε the phase-plane diagram shows that there is eventually a sustained steady oscillation. When ε is large the motion has the characteristics of what is called a *relaxation oscillation*, with sudden changes between the maximum and

* *Phil. Mag.* **2**, 1926, 978.

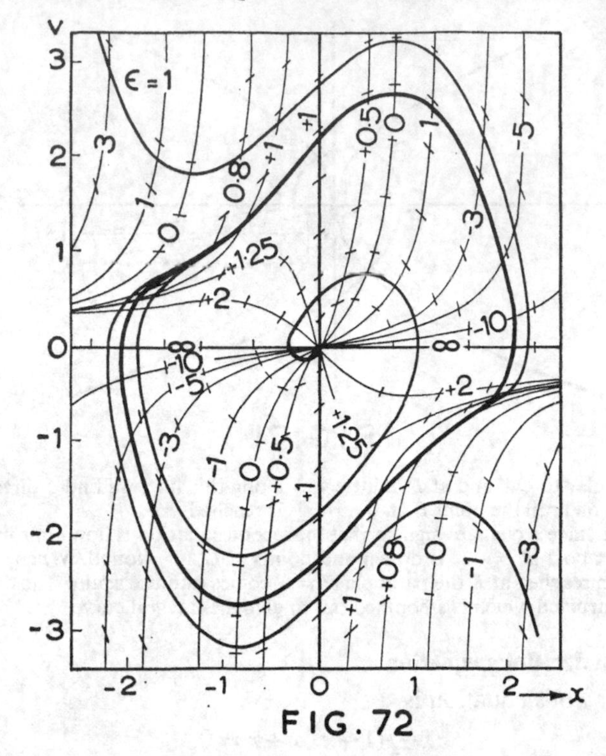

FIG. 72

minimum positions. This is exemplified by the solution for $\varepsilon = 10$ shown in Fig. 74. When ε is small the phase-plane diagram is approximately circular and the oscillation differs little from simple harmonic motion.

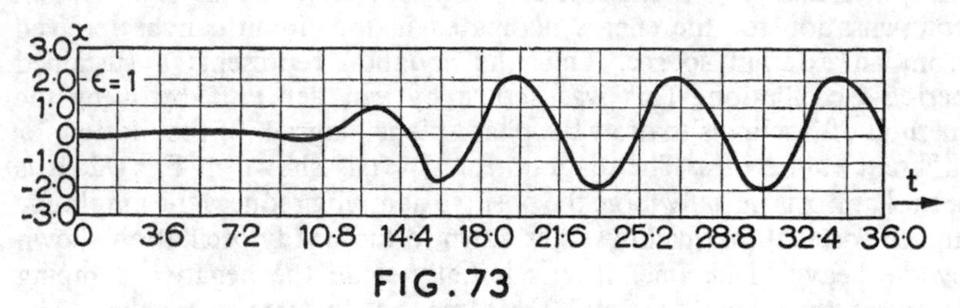

FIG. 73

The equation

$$\ddot{u} - \varepsilon(\dot{u} - \dot{u}^3/3) + u = \upsilon \tag{2}$$

is an alternative form of van der Pol's equation, since by differentiating this equation with respect to t we find

$$\dddot{u} - \varepsilon(1 - \dot{u}^2)\ddot{u} + \dot{u} = 0,$$

and setting $\dot{u} = x$ this reduces to equation (1).

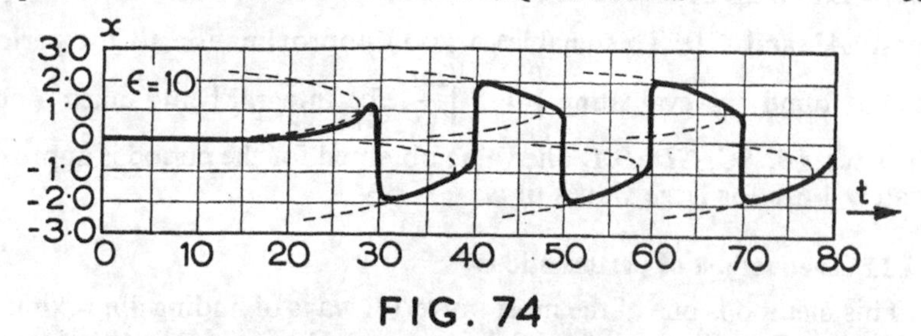

FIG. 74

Liénard's construction is particularly suitable for drawing the phase-plane diagram of equation (2) and determining the limit cycle of the steady oscillation. Writing $\dot{u} = w$ the first order equation is

$$w\frac{dw}{du} - \varepsilon(w - w^3/3) + u = 0,$$

so that the basic curve for the construction is $u = \varepsilon(w - w^3/3)$. This cubic curve is drawn in the u, w plane (Fig. 75) showing turning values

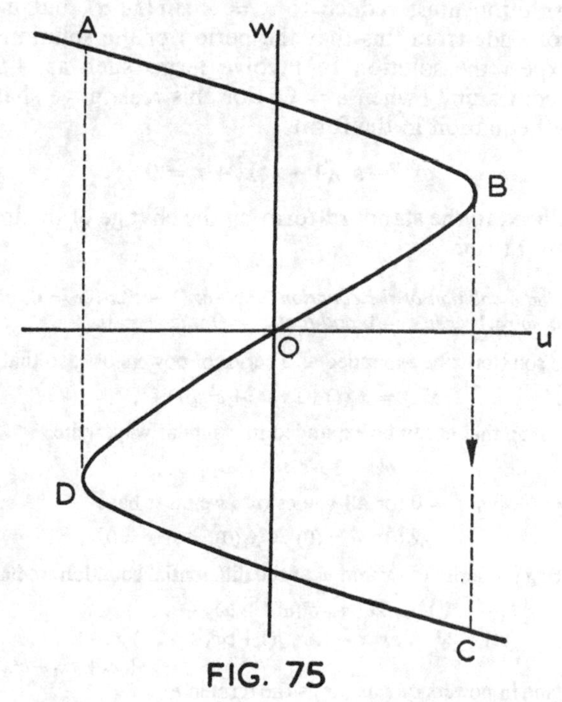

FIG. 75

for u at $B(\frac{2}{3}\varepsilon, 1)$ and $D(-\frac{2}{3}\varepsilon, -1)$. It is found that when ε is large (>10, say) the limit cycle approximates very closely to the figure *ABCD* bounded by the vertical lines *BC* and *DA* and the arcs of the curve AB and CD. This enables a good approximation to the period to be found by evaluating $\int dt = \int \frac{du}{w}$, the integral being taken along the lines AB, BC, CD, DA. The value obtained for the period is approximately $1{\cdot}61\varepsilon$ for large values of ε.

11.11 The method of perturbations

This method is one of the most powerful ways of finding approximate solutions of non-linear differential equations. Due originally to Poincaré, who applied the method to the solution of problems of celestial mechanics, it involves expanding the solution and some parameters of the equation in a series of powers of some small quantity and building up a solution step by step. Each step usually involves the solution of a linear differential equation.

In the following example we apply this method to find a periodic solution of van der Pol's equation $\ddot{x}-\varepsilon(1-x^2)\dot{x}+x = 0$ when ε is small and obtain the solution as a power series in ε. Obviously when $\varepsilon = 0$ the solution must reduce to $x = A \sin (t+\alpha)$, but it would be wrong to conclude from this that the period of the solution is 2π. We therefore expect the solution to involve terms such as $A \sin (\omega t+\alpha)$ where ω becomes unity when $\varepsilon = 0$. For this reason we shall consider van der Pol's equation in the form

$$\omega^2\ddot{x}-\varepsilon\omega(1-x^2)\dot{x}+x = 0, \tag{1}$$

which is reduced to the standard form by the change of the independent variable from t to ωt.

Example 5. *Find a solution of the equation $\omega^2\ddot{x}-\varepsilon\omega(1-x^2)\dot{x}+x = 0$, given that ε is small, that $\omega = 1$ when $\varepsilon = 0$ and that $x = 0$ when $t = 0$.*

Let the solution x be expanded as a series of powers of ε, so that

$$x(t) = x_0(t)+\varepsilon x_1(t)+\varepsilon^2x_2(t)+ \ldots,$$

and, assuming that ω can be expanded in a similar way, write

$$\omega = 1+\varepsilon\omega_1+\varepsilon^2\omega_2+ \ldots.$$

Since $x = 0$ when $t = 0$ for all values of ε we must have

$$x_0(0) = x_1(0) = x_2(0) = \ldots = 0.$$

Substituting the series for x and ω in the differential equation we have

$$\begin{aligned}(\ddot{x}_0+\varepsilon\ddot{x}_1+\varepsilon^2\ddot{x}_2+ \ldots)\{1+2\omega_1\varepsilon+\varepsilon^2(\omega_1^2+2\omega_2)+ \ldots\}&\\ -\varepsilon(1-x_0^2-2\varepsilon x_0x_1- \ldots)(1+\varepsilon\omega_1+ \ldots)(\dot{x}_0+\varepsilon\dot{x}_1+ \ldots)&\\ +x_0+\varepsilon x_1+\varepsilon^2x_2+ \ldots &= 0.\end{aligned}$$

Rearranging in powers of ε as far as the term in ε^2,

$$\begin{aligned}(\ddot{x}_0+x_0)+\varepsilon\{\ddot{x}_1+x_1+2\omega_1\ddot{x}_0-(1-x_0^2)\dot{x}_0\}&\\ +\varepsilon^2\{\ddot{x}_2+x_2+2\omega_1\ddot{x}_1+\ddot{x}_0(\omega_1^2+2\omega_2)-(1-x_0^2)(\dot{x}_1+\omega_1\dot{x}_0)+2x_0x_1\dot{x}_0\}&\\ + \ldots &= 0.\end{aligned}$$

This expression must be zero for all values of ε, hence the quantity multiplying each power of ε must be zero. Therefore, firstly,

$$\ddot{x}_0+x_0 = 0,$$

and hence, since x_0 is initially zero,

$$x_0 = A\sin t.$$

Secondly,

$$\begin{aligned}\ddot{x}_1+x_1 &= -2\omega_1\ddot{x}_0+(1-x_0^2)\dot{x}_0\\ &= 2\omega_1 A\sin t+A(1-A^2\sin^2 t)\cos t\\ &= 2\omega_1 A\sin t+A(1-\tfrac{1}{4}A^2)\cos t+\tfrac{1}{4}A^3\cos 3t.\end{aligned}$$

The solution of this equation for x_1 is

$$x_1 = B\cos t+C\sin t-\omega_1 At\cos t+\tfrac{1}{2}A(1-\tfrac{1}{4}A^2)t\sin t-(A^3/32)\cos 3t.$$

If the solution is to be periodic the terms involving $t\sin t$ and $t\cos t$ must vanish, and hence we must have

$$A = 2,\quad \omega_1 = 0.$$

Then

$$x_0 = 2\sin t,$$

$$x_1 = B\cos t+C\sin t-\tfrac{1}{4}\cos 3t,$$

and, since $x_1(0) = 0$,

$$x_1 = \tfrac{1}{4}\cos t+C\sin t-\tfrac{1}{4}\cos 3t.$$

Thirdly, since $\omega_1 = 0$,

$$\begin{aligned}\ddot{x}_2+x_2 &= -2\omega_2\ddot{x}_0+(1-x_0^2)\dot{x}_1-2x_0x_1\dot{x}_0\\ &= 4\omega_2\sin t+(1-4\sin^2 t)(-\tfrac{1}{4}\sin t+C\cos t+\tfrac{3}{4}\sin 3t)\\ &\qquad -8\sin t\cos t(\tfrac{1}{4}\cos t+C\sin t-\tfrac{1}{4}\cos 3t)\\ &= (4\omega_2+\tfrac{1}{4})\sin t-2C\cos t+3C\cos 3t-\tfrac{3}{2}\sin 3t+\tfrac{5}{4}\sin 5t.\end{aligned}$$

The solution of this equation will not be periodic unless the coefficients of $\cos t$ and $\sin t$ on the right-hand side are zero. Therefore, we must have

$$C = 0,\quad \omega_2 = -\tfrac{1}{16}.$$

Then

$$\ddot{x}_2+x_2 = -\tfrac{3}{2}\sin 3t+\tfrac{5}{4}\sin 5t,$$

$$x_2 = D\cos t+E\sin t+\tfrac{3}{16}\sin 3t-\tfrac{5}{96}\sin 5t,$$

and, since $x_2(0) = 0$,

$$x_2 = E\sin t+\tfrac{3}{16}\sin 3t-\tfrac{5}{96}\sin 5t.$$

The next step, equating to zero the coefficient of ε^3, shows that $E = -\frac{29}{96}$. Using this value we have the solution as far as the term in ε^2

$$x = 2\sin t+\tfrac{1}{4}\varepsilon(\cos t-\cos 3t)-\tfrac{1}{96}\varepsilon^2(29\sin t-18\sin 3t+5\sin 5t).$$

This is the solution of the equation (1). Reverting to the original form of van der Pol's equation, $\ddot{x}-\varepsilon(1-x^2)\dot{x}+x = 0$, we have the solution

$$x = 2\sin\omega t+\tfrac{1}{4}\varepsilon(\cos\omega t-\cos 3\omega t)-\tfrac{1}{96}\varepsilon^2(29\sin\omega t-18\sin 3\omega t+5\sin 5\omega t),$$

where $\omega^2 = 1-\varepsilon^2/16$, the values of x and ω being accurate as far as the term in ε^2.

11.12 Forced oscillations

Under this heading we consider some equations of the type

$$\ddot{x}+f(x,\dot{x}) = E\cos qt, \tag{1}$$

in which the oscillation is forced by an external agency. The linear forced oscillation given by $\ddot{x}+2k\dot{x}+\omega^2x = E\cos qt$ was considered in §3.10 and the solution was seen to consist of a transient part $C\,e^{-kt}\cos(pt+\varepsilon)$, $p^2 = \omega^2-k^2$, and a persisting oscillation

$$E\{(\omega^2-q^2)^2+4k^2q^2\}^{-\frac{1}{2}}\cos(qt-\alpha), \quad \tan\alpha = 2kq/(\omega^2-q^2).$$

In the non-linear field we are led to consider equations such as

$$\ddot{x}+k\dot{x}+ax+bx^3 = E\cos qt, \tag{2}$$

which is known as *Duffing's equation* and represents a forced and damped oscillation such as arises with a non-linear spring. We may also have equations such as van der Pol's with a forcing term, that is,

$$\ddot{x}-\varepsilon(1-x^2)\dot{x}+x = E\cos qt. \tag{3}$$

The solutions of such equations will, as in the linear case, usually contain a transient term and the more important persisting oscillation maintained by the applied force.

In the linear case the persisting oscillation has the same period as the forcing term. In non-linear equations, the situation is complicated by the occurrence of *subharmonics*, which are oscillations whose frequency is a fraction of that of the forcing term. We may also have *ultraharmonics* whose frequency is a multiple of that of the term giving the forced oscillations.

Solutions of equations of this type can often be obtained by the method of perturbations. As an example, consider Duffing's equation without the damping term, that is,

$$\ddot{x}+ax+bx^3 = E\cos qt. \tag{4}$$

Replacing x by $(E/a)x$ and t by (t/q) this equation becomes

$$\omega^2\ddot{x}+x+\varepsilon x^3 = \cos t, \tag{5}$$

where

$$\omega^2 = q^2/a \quad \text{and} \quad \varepsilon = bE^2/a^3.$$

Assuming a solution

$$x = x_0+\varepsilon x_1+\varepsilon^2x_2+\ldots,$$

with

$$\omega^2 = \omega_0+\varepsilon\omega_1+\varepsilon^2\omega_2+\ldots,$$

we take

$$x_0(0) = c, \qquad x_1(0) = x_2(0) = \ldots = 0,$$

and

$$\dot{x}_0(0) = \dot{x}_1(0) = \ldots = 0.$$

Substituting in the equation (5) we find successively

$$x_0 = c\cos t,$$

$$\omega_0 = 1-1/c,$$

$$x_1 = \tfrac{1}{4}c^4(\cos t-\cos 3t)/(9-8c),$$

$$\omega_1 = \tfrac{3}{4}c^2+\tfrac{1}{4}c^2/(9-8c),$$

and

$$\omega^2 = \omega_0+\varepsilon\omega_1+\ldots.$$

The detail of this work is left as an exercise for the reader.

Alternatively, if the constants in the equation (4) are such that no subharmonics exist we may assume a Fourier series solution and determine successive coefficients by substitution in the differential equation. The nature of the disturbing force $ax+bx^3$ leads us to try the solution

$$x = A_1\cos qt+A_3\cos 3qt+A_5\cos 5qt+\ldots.$$

Substitution in equation (4), which involves some rather heavy algebra, gives

$$\{-q^2A_1+aA_1+\tfrac{3}{4}bA_1(A_1{}^2+A_1A_3+2A_3{}^2)-E\}\cos qt$$
$$+\{-9q^2A_3+aA_3+\tfrac{1}{4}b(A_1{}^3+6A_1{}^2A_3+3A_3{}^3)\}\cos 3qt+\ldots = 0. \quad (6)$$

If we now equate to zero the coefficients of $\cos qt$ and $\cos 3qt$ we have two equations to determine A_1 and A_3.

In particular, if $b = 0$ these equations give $A_3 = 0$, $A_1 = E/(a-q^2)$, for the linear equation.

We may use the preceding analysis to show the possibility of the existence of a subharmonic solution. If we write Duffing's equation as

$$\ddot{x}+ax+bx^3 = E\cos 3qt \quad (7)$$

and assume a solution

$$x = A_1\cos qt+A_3\cos 3qt+\ldots,$$

then the term $A_1\cos qt$ is a subharmonic of order $\tfrac{1}{3}$, that is, its frequency is one-third of that of the forcing frequency. Substitution in the equation (7) gives as before equations determining A_1 and A_3 which now are

$$-q^2A_1+aA_1+\tfrac{3}{4}bA_1(A_1{}^2+A_1A_3+2A_3{}^2) = 0,$$

$$-9q^2A_3+aA_3+\tfrac{1}{4}b(A_1{}^3+6A_1{}^2A_3+3A_3{}^3)-E = 0,$$

the only change being that the quantity E moves to the second equation. This does not prove the existence of subharmonics but they are known to occur and mechanisms have been constructed which exhibit oscillations of this type.

Exercises 11(*c*)

1. The equation of motion of a particle is $\ddot{x}+c\dot{x}|\dot{x}|+ax+bx|x| = 0$, where a, b, c are positive, and initially $x = x_0 > 0$ and $\dot{x} = 0$. Show that if the particle reaches the origin it comes to rest when $x = x_1$ (<0) where
$$f(x_1)-f(x_0)+(b/c^3)\{1-e^{-2cx_0}(2c^2x_0^2+2cx_0+1)\} = 0,$$
and $f(x) = (e^{-2cx}/2c^3)\{-2bc^2x^2+2(ac-b)cx+(ac-b)\}$.

2. Assuming a solution of van der Pol's equation $\ddot{x}-\varepsilon(1-x^2)\dot{x}+x = 0$ of the form $x = A \sin \omega t+$ harmonics of higher frequency, show by substitution that $A = 2$ and $\omega = 1$.

3. Taking van der Pol's equation with $\varepsilon = 1$ in the form
$$\ddot{u}-(\dot{u}-\dot{u}^3/3)+u = 0,$$
use Liénard's construction to draw the integral curve in the phase plane which passes through the point $u = 1$, $\dot{u} = 0$, and determine the limit cycle.

4. Use the solution of van der Pol's equation obtained by the method of perturbations (§11.11) to show that if $\varepsilon = 0{\cdot}1$ the period of the sustained oscillation is approximately 2π and the amplitude 2. Show also that the phase-plane diagram is approximately circular.

5. Assuming a solution of Duffing's equation $\ddot{x}+k\dot{x}+x+\varepsilon x^3 = \cos \omega t$ of the form $x = A \cos \omega t+B \sin \omega t+$harmonics of higher frequency, show by substitution that
$$A(1-\omega^2)+Bk\omega+\tfrac{3}{4}\varepsilon A(A^2+B^2) = 1,$$
$$B(1-\omega^2)-Ak\omega+\tfrac{3}{4}\varepsilon B(A^2+B^2) = 0.$$

6. The equation $\ddot{x}+ax+bx^2 = E \cos qt$ is reduced by the substitution $ax = Ey$, $qt = \theta$, to the form $\omega^2y''+y+\varepsilon y^2 = \cos \theta$, where $a\omega^2 = q^2$, and $\varepsilon = Eb/a^2$. Use the method of perturbations to obtain the approximate solution with $y = c$ and $y' = 0$ when $\theta = 0$,
$$y = c \cos \theta+\varepsilon\left\{-\frac{c^2}{2}+\frac{c^2(c-2)}{3c-4}\cos \theta+\frac{c^3}{6c-8}\cos 2\theta\right\}$$
with $\omega^2 = 1-1/c+\varepsilon(c-2)/(3c-4)$.

7. Use the method of perturbations to obtain the solution of the equation $\omega^2(d^2x/d\theta^2)+x+\varepsilon x^3 = \cos \theta$, with $x = c$ and $\dot{x} = 0$ initially, in the form
$$x = c \cos \theta+\left(\frac{\varepsilon c^4}{36-32c}\right)(\cos \theta-\cos 3\theta),$$
with $\omega^2 = 1-1/c+\varepsilon\{3c^2/4+c^2/(36-32c)\}$.

8. An iteration method applied to Duffing's equation
$$\ddot{x}+ax+bx^3 = E \cos qt, \quad b \text{ being small},$$
consists in taking the first approximation as $x_0 = A \cos qt$. A second approximation x_1 is obtained as the periodic solution of the equation
$$\ddot{x}_1+ax_0+bx_0^3 = E \cos qt.$$
Show that $q^2 = a+\frac{3}{4}bA^2-E/A$, and $x_1 = A \cos qt+(bA^3/36q^2) \cos 3qt$.

ANSWERS TO THE EXERCISES

Exercises 1(a)

1. (*a*) 1, 2. (*b*) 2, 1. (*c*) 4, 1. (*d*) 2, 1.
2. (*a*) $y''-n^2y = 0$. (*b*) $\ddot{y}+2k\dot{y}+(k^2+n^2)y = 0$.
   (*c*) $y''-2my'+my = 0$. (*d*) $yy''+(y')^2 = 0$.
   (*e*) $4y(y')^2+2xy'-y = 0$
4. $(x^2+y^2)y''-2(xy'-y)\{1+(y')^2\} = 0$.
7. $y = A\,e^{nx}$.
10. $x+(2/3)x^3+(8/15)x^5$.

Exercises 1(b)

1. $3y = 4-2(x+2)(1-x)^{1/2}$.
2. $y = \tan(1+y-x)$.
6. $y^2 = \sin^2 x$.
7. $4(y-1)^2 = \{x(1-x^2)^{1/2}+\sin^{-1}x\}^2$.
8. $(x-A)^2(y^2+1) = 1$.
9. $y-1 = \pm[\log_e x-\log_e\{1+(1+x^2)^{1/2}\}+\log_e(1+\sqrt{2})]$.
10. $x^2 = \cos^2(x-y)$.
11. $8x^3 = 27(1-y)^2$.
12. $(2y-x^2)^2 = [x(1+x^2)^{1/2}+\log_e\{x+(1+x^2)^{1/2}\}]^2$.
15. 80 ft./sec., 1440 ft., 85·3 ft./sec.

Exercises 1(c)

1. $xy = A$.
2. $x\cos y = A$.
3. $x(x^2+y^2) = A$.
4. $\sinh x\cosh y = A$.
5. $x\log_e y = A$.
6. $x^4+x^2y^2+y^4 = A$.
7. $x^2+y^2 = A$.
8. $x^3+y^2-2xy^2 = A$.
9. $ax^2+dy^2+2bxy+2cx+2ey = A$.
10. $2a_3x-2a_2y+b_3x^2-c_2y^2+(c_3-b_2)xy = A$.
11. $\log_e(x^2+y^2)+\tan^{-1}(y/x) = A$.
12. $\sin x\sin y = A$.
13. $x^2+y^2 = A(x+y)$.
14. $x^{-2}e^x$; $y = Ax\,e^{-x}$.
15. $y = Ax$.
16. $x^2+y^2 = A(x-y)$.
17. x; $x^4-4x^2y = A$.
18. -2; $y^2 = Ax^2$.

Exercises 2(a)

1. $(1+y)^{1/2}/(1-y)^{1/2} = Ax$.
2. $(y+b)/(x+a) = A$.
3. $y = Ax\,e^x$.
4. $(y^2-1)x^2e^{x^2} = A$.
5. $(x-1)(y+1) = A(x+1)(y-1)\,e^{2y-2x}$.
6. $y^2 = Ax^2/(x^2+1)-1$.
7. $\sin^{-1}x+\sin^{-1}y = A$, or $y(1-x^2)^{1/2}+x(1-y^2)^{1/2} = \sin A$.
8. $\sin y\sin x = A$.
9. $x+y+\cot y-\frac{1}{2}\tan 2x = A$.
10. $(\log_e y)^2+(x^2-1)\log_e(1+x)+x-x^2/2 = A$.
11. $y = \log_e\tan(\cosh x-x\sinh x+A)$.
12. $4y = 1-2x+C\,e^{-2x}$.
13. $8y = \coth(2x+A)-2x$.
14. $y^2 = A\,e^{2x}-x^2$.
15. $(1+x^2)^{1/2}+e^{-y} = 2$.

Exercises 2(b)

6. $r = (T/\pi f)^{1/2}\,e^{-wx/2f}$.

Exercises 2(c)

1. $6y(x+1)^2 = 2x^3+3x^2+A$.
2. $2y = A(x+1)^2-2x-1$.
3. $x^2y = e^x(x-1)+A$.
4. $30(x-1)^3y = 6x^5-15x^4+10x^3+A$.

5. $3y\,e^{2x} = A(x+1)^3-1$.
6. $4xy = \sin 2x-2x\cos 2x+A$.
7. $2x^3y = x^2+A$.
8. $(x^2+1)^{1/2}(y-1) = A$.
9. $y \sin x = x+A$.
10. $2(x-1)y = 2x\log_e x-4x^2+x^3+Ax$.
11. $y = 1+x^2+A(1+x^2)^{1/2}$.
12. $y(1+x^2)^{1/2} = \tan^{-1} x+A$.
13. $y = 2(e^{-\sin x}+\sin x-1)$.
14. $2(x-1)\sin y = 2x\log_e x-4x^2+x^3+Ax$.
15. $x^{1/2}(x-1)^{1/2}y = \log_e \{x^{1/2}+(x-1)^{1/2}\}+A$.
16. $3x\,e^{2y} = A(y+1)^3-1$.
17. $4xy = 1+Ay^4$.
18. $y^2(4x+A) = x$.
19. $y^3\{e^x(x-1)+A\} = x^2$.
20. $2x^3 = y^{3/2}(x^2+A)$.
21. $3(1+x^2)^{1/2} = y^5\{(1+x^2)^{3/2}+A\}$.

Exercises 2(d)

7. $c = c_0q_0t(2v_0+q_0t)/(v_0+q_0t)^2$.
8. $a(q/p)^{p/(p-q)}$, $a(q/p)^{q/(p-q)}$, $a-a(1+p/q)(q/p)^{p/(p-q)}$.

Exercises 2(e)

1. $2\tan^{-1}(y/x) = \log_e A(x^2+y^2)$.
2. $x^2-xy-y^2 = A$.
3. $(\sqrt{2})\tan^{-1}(\sqrt{2}y/2x)+\log_e A(2x^2+y^2) = 0$.
4. $\log_e y-x^2/2y^2 = A$.
5. $y-2x = Ax^2y$.
6. $x^4+4xy^3-y^4 = A$.
7. $(x+y)^3 = A(x-y)$.
8. $xy^2 = A(x-y)^2$.
9. $4\log_e(y/x)+x^4/y^4 = A$.
10. $\tan(y/2x) = Ax$.
11. $x-2y-4\log_e(2x-5y+10) = A$.
12. $(y-x+3)^4 = A(y+2x-3)$.
13. $\log_e(y-2)-(x-1)^2/2(y-2)^2 = A$.
14. $24x = 2(x-y)(x-y+5)+9\log_e(2x-2y-1)+A$.

Exercises 2(f)

1. $(x^2-y+A)(2x^2+y+A) = 0$.
2. $(y+e^{2x}+A)(y-e^{-2x}+A) = 0$.
3. $x = 3p^2/2-3p+3\log_e(1+p)+A$, $y = p^3-x$.
4. $x = 4\log_e p+6p^2+A$, $y = 4p+4p^3$.
5. $\sqrt{(p^2-1)}x = p\log_e\{p+\sqrt{(p^2-1)}\}+Ap$,
   $\sqrt{(p^2-1)}(y-p) = \log_e\{p+\sqrt{(p^2-1)}\}+A$.
6. $2x = 9p^2+1/p^2+A$, $y = 3p^3+1/p$.
7. $2x = 2p+\log_e(2p-1)+A$, $4y = 2p^2+2p-2+\log_e(2p-1)+A$.
8. $y = Ax+1/A$, $y^2 = 4x$.
9. $y = Ax+(b^2+a^2A^2)^{1/2}$, $x^2/a^2+y^2/b^2 = 1$.
10. $y = Ax+A^n$, $\{y/(n-1)\}^{n-1}+\{x/n\}^n = 0$.
11. $y = Ax-A/(1+A^2)^{1/2}$, $x^{2/3}-y^{2/3} = 1$.
12. $y = Ax-\log_e A$, $y = 1+\log_e x$.

Exercises 2(g)

1. $x^2-y^2 = A$.
2. $y = Ax^4$.
3. $\theta^2+(\log_e r)^2 = A$.
4. $(x+yp)(xp-y) = (a^2-b^2)p$.
5. $x^2+y^2 = A\,e^{-2cx}$.
6. $x/(x^2+y^2) = cx+A$.
7. $xy = (x^2-A)/(x^2+A)$.
8. $z = Ax+Bx^{-1}$.
9. $z = Ax^3+Bx(x-1)$.
10. $y = -3/x+1/(Ax+x\log_e x)$, $z = x^{-3}(A+B\log_e x)$.

Exercises 3(a)

1. $y = A\,e^{-t}+B\,e^{-4t}$.
2. $y = A\,e^{3t}+B\,e^{-7t}$.
3. $y = (A+Bx)\,e^{-5x/2}$.
4. $y = C\,e^{-2x}\cos(3x-\varepsilon)$.
5. $y = C\,e^{-t/2}\cos(t-\varepsilon)$.
6. $y = C\,e^{-3t}\cos(t/2-\varepsilon)$.

7. $y = 4\cos 2t$.
8. $\theta = a\cos\{(g/l)^{1/2}t\}$.
9. $y = e^{2x}$.
10. $y = (1+6x)e^{-5x}$.
11. $y = 4e^{-3t} - 3e^{-5t}$.
12. $y = e^{-5x}$.
13. $y = e^{-t}(2\cos 2t + \sin 2t)$.
14. $y = e^{-at}\sin 3at$.
15. $x = e^{-kt}[a\cos pt + \{(v+ka)/p\}\sin pt]$.
16. $q = Q\,e^{-Rt/2L}(1+Rt/2L)$.

Exercises 3(b)

1. $y = e^{-3x}(A + B\cos 2x + C\sin 2x)$.
2. $y = A\,e^{-2x} + B\cos 2x + C\sin 2x$.
3. $y = A + B\,e^{7x} + C\,e^{-13x}$.
4. $y = A\cosh 2x + B\sinh 2x + E\cos 2x + F\sin 2x$.
5. $y = e^{x}(A\cos x + B\sin x) + e^{-x}(E\cos x + F\sin x)$.
6. $y = A\,e^{-2x} + e^{x}(B\cos\sqrt{3}x + C\sin\sqrt{3}x)$.
7. $y = e^{-t}(A\cos t + B\sin t + E\cos 2t + F\sin 2t)$.
8. $y = A\cosh t + B\sinh t + t(E\cosh t + F\sinh t)$.
9. $y = e^{t}(A\cos 2t + B\sin 2t) + e^{-t}(E\cos 2t + F\sin 2t)$.
10. $y = e^{-ax}(A + Bx + Ex^2 + Fx^3)$.

Exercises 3(c)

1. $y = A\,e^{x} + B\,e^{2x} + e^{3x}$.
2. $y = A\,e^{x} + (B+x)\,e^{2x}$.
3. $y = A\,e^{x} + B\,e^{2x} + x + 1$.
4. $y = A\,e^{x} + B\,e^{2x} + (2x^2+6x+7)/4$.
5. $y = (A + Bx + x^2)\,e^{-2x}$.
6. $y = (A + Bx)\,e^{-x} + e^{-2x}$.
7. $y = (A + Bx)\,e^{-3x} + 2x - 4/3$.
8. $y = e^{-x}(A\cos 2x + B\sin 2x) + 2\cos x + \sin x$.
9. $y = (\omega^2 - p^2)^{-1}\cos pt + C\cos(\omega t + \varepsilon)$.
10. $y = (A + Bx + Cx^2 + x^3/6)\,e^{x}$.
11. $y = A\,e^{x} + B\,e^{-x} + C\,e^{2x} - x\,e^{x}$.
12. $y = A\,e^{x} + B\,e^{-x} + C\,e^{2x} + 2x^2 + 2x + 5$.

Exercises 3(d)

1. $y = A\,e^{-2x} + B\,e^{-4x} + 2\,e^{3x}$.
2. $y = A\,e^{-2x} + B\,e^{-4x} + 4x\,e^{-2x}$.
3. $y = e^{2x}(A + Bx + 2x^2)$.
4. $y = (A + 3x)\sinh 2x + B\cosh 2x$.
5. $y = e^{-2x}(A\cos 3x + B\sin 3x + 3)$.
6. $y = A + e^{-3x}(B + Cx - 2x^2)$.
7. $y = A + B\cosh 3x + (C + 6x)\sinh 3x$.
8. $y = (A + Bx + 2x^2)\cosh 2x + (E + Fx)\sinh 2x$.
9. $y = A\cos 3x + B\sin 3x + 3\cos 2x$.
10. $y = e^{-x}(A\cos 2x + B\sin 2x) + 5\cos 2x + 20\sin 2x$.
11. $y = A\cos 2x + (B + 2x)\sin 2x$.
12. $y = e^{-2x}(A\cos 3x + B\sin 3x) + 2\cos 3x + 6\sin 3x$.
13. $y = e^{-kt}(A\cos\omega t + B\sin\omega t) + (1/2kp)\sin pt$, $\omega^2 = p^2 - k^2$.
14. $y = e^{-\lambda t}(A\cos\omega t + B\sin\omega t) + \{2E/(R\lambda^2 + 4R\omega^2)\}(\lambda\cos\omega t + 2\omega\sin\omega t)$, $\lambda = R/2L$.
15. $y = e^{-x}(A\cos x + B\sin x) + e^{2x}(2\cos 3x + 36\sin 3x)$.
16. $y = A + Be^{-4x} + e^{2x}(\sin 3x - 8\cos 3x)$.
17. $y = -2\,e^{-t/2}\cos 3t + 3\cos t$.
18. $y = (1 + e^{-2x})(\cos 3x + 3\sin 3x)/8$.

Exercises 3(e)

1. $y = A\cos 2x + B\sin 2x + 2x^2 - 1$.
2. $y = A\,e^{x} + B\,e^{3x} + 9x^2 + 24x + 26$.
3. $y = (A + Bx)\,e^{2x} + 4x^3 + 12x^2 + 18x + 12$.
4. $y = e^{-3x}(A\cos 2x + B\sin 2x) + x^2 - x + 2$.
5. $y = A + B\cos 2x + C\sin 2x + 4x^3 - 6x$.
6. $y = A + Bx + E\cosh 2x + F\sinh 2x - 2x^4 - 6x^2$.
7. $y = A\,e^{x} + B\,e^{3x} - e^{2x}(x^2 + 2)$.

8. $y = e^{-2x}(A\cos 3x + B\sin 3x) + e^{-2x}(3x^3 - 2x)$.
9. $y = A\cos x + B\sin x - x^2\cos x + x\sin x$.
10. $y = e^{-2x}(A\cos 2x + B\sin 2x) + (50x^2 - 20x - 1)\cos 2x$
$+(100x^2 - 140x + 68)\sin 2x$.
11. $y = A\cos 2x + B\sin 2x + (8x^3 - 3x)\sin 2x + 6x^2\cos 2x$.
12. $y = e^{-3x}\{A\cos 3x + B\sin 3x + (6x^3 - x)\sin 3x + 3x^2\cos 3x\}$.

Exercises 3(f)

1. $y = Ae^x + Be^{2x} + 2x^2 + 6x + 7$.
2. $y = e^x(A + x^2) + Be^{-2x}$.
3. $y = A\cos 2x + B\sin 2x + x^2\sin 2x$.
4. $y = Ae^{-x} + e^x(B\cos x + C\sin x) + x^2 - 4x + 5$.
5. $y = (w/n^2P)\{\frac{1}{2}n^2x(x-l) - 1 + \cos nx + \sin nx\tan nl/2\}$, $P = EIn^2$.
6. $y = Ae^{-x}\cos(2x+\varepsilon) + \sqrt{(17)}\cos(2x - \tan^{-1}4)$.
7. $y = Ae^{-x}\cos(2x+\varepsilon) + \{5/\sqrt{(17)}\}\cos(2x + \tan^{-1}3/4 - \tan^{-1}4)$.
8. $y = Ae^x + Ce^{x/2}\cos(ax+\varepsilon) + \cos(2x - \tan^{-1}4/3)$, $4a^2 = 19$.

Exercises 3(g)

1. $y = x(A + B\log_e x)$.
2. $y = A/x + B/x^4$.
3. $y = x^{-5/2}(A + B\log_e x)$.
4. $y = Ax^3 + B/x^3$.
5. $y = Ax + Bx^2 + 2x^4$.
6. $y = A + B\log_e x + C(\log_e x)^2 + x^3$.
7. $y = x(A\cos\log_e x + B\sin\log_e x) + x\log_e x$.
8. $y = x^2\{A + B\log_e x + (\log_e x)^2\}$.
9. $y = (A + B\log_e x)/x^2 + x/9 + 1/x$.
10. $y = x^3(A + B\log_e x) + 3\log_e x + 2$.
11. $y = A(x-1) + B(x-1)^4 - (x-1)^2/2 - (2/3)(x-1)\log_e(x-1) + 1/4$.
12. $y = Ax + B/x + Ex^2 + F/x^2 + x^3$.

Exercises 3(h)

1. $3nx = (4an - u)e^{-nt} + (u - an)e^{-4nt}$.
3. $\ddot{x} + k\dot{x} + \omega^2 x = ku$.
4. $\ddot{x} + 2\dot{x} + 32x = 32\sin 4t$; $(8\sin 4t - 4\cos 4t)/5$.
5. $a(k^2p^2 + \lambda^2)^{1/2}\{m^2p^4 + (k^2 - 2\lambda m)p^2 + \lambda^2\}^{-1/2}$.
8. $\theta = \{12al\pi^2n^2/(3gl - 8\pi^2n^2l^2)\}\sin 2\pi nt$.

Exercises 3(i)

6. $W(a+b-y)$; $b = a(\sec nl - 1)$; $Wa\sec nl$; $\pi^2EI/9l^2$, $W = EIn^2$.
10. $(gEI/m)^{1/2}(1{\cdot}506\pi)^2/l^2$.

Exercises 3(j)

1. $CR^2 = 4L$, $EC(1 + 2t/CR)e^{-2t/CR}$, $(Et/L)e^{-2t/CR}$.
4. $v = 800e^{-1000t}(\cos 1000t + \sin 1000t)$.
6. $q = EC\{(\lambda\pi + \mu\beta)e^{\alpha t_1} - (\lambda\pi + \mu\alpha)e^{\beta t_1}\}/(\alpha-\beta)(\lambda^2+\mu^2)$, $\lambda = 1 - LC\pi^2$,
$\mu = RC\pi$, $t_1 = t - 2T$, $\alpha\beta = 1/LC$, $\alpha + \beta = -R/L$.
7. 0·0001, 3·1.
8. $(E/R)\cos\phi\sin(\omega t + \phi)$, $\tan\phi = (1 - LC\omega^2)/RC\omega$.
9. $CR^2 = 4L$.
10. $E\{1 - e^{-\omega t}(\cos 2\omega t + \frac{1}{2}\sin 2\omega t)\}$, $(5CE\omega/2)\sin 2\omega t$.

Exercises 4(a)

1. $x = 5A\cos 4t + 5B\sin 4t$, $y = (3A - 4B)\cos 4t + (4A + 3B)\sin 4t$.
2. $x = Ae^{4t} + 5Be^{-3t}$, $y = -Ae^{4t} + 2Be^{-3t}$.
3. $x = e^{2t}(\cos 3t - \sin 3t)$, $y = e^{2t}(\cos 3t + 2\sin 3t)$.
4. $2x = e^{-t} + e^{-3t}$, $2y = -e^{-t} + e^{-3t}$.

5. $40x = -5\,e^{t}+8\,e^{-2t}-3\,e^{-7t}$, $120y = 25\,e^{t}-16\,e^{-2t}-9\,e^{-7t}$.
6. $x = -(1+t)e^{-t}+6\,e^{-t/6}+t-5$, $y = (t-2)\,e^{-t}+9\,e^{-t/6}+t-7$.
7. $4x = 3\sinh t+t\,e^{t}$, $4y = 4\cosh t+5\sinh t+e^{t}(3t-4)$.
8. $27x = -(6t+1)\,e^{-3t}+3t+1$, $27y = -(6t+4)\,e^{-3t}-6t+4$.
9. $4x = (2-\pi+2t)\sin t$, $4y = (2-\pi+2t)\cos t+(\pi-2t)\sin t$.
10. $60x = -45\,e^{-t}-e^{-5t}+36+10\,e^{t}$, $60y = -45\,e^{-t}+e^{-5t}+24+20\,e^{t}$.
11. $x = e^{2t}$, $y = 2\,e^{2t}$, $z = 4\,e^{2t}$.
12. $y = A/z^2+Bz^3-(3\cos\log_e z-21\sin\log_e z)/50$,
    $x = A/z^2-2Bz^3/3-(8\cos\log_e z-6\sin\log_e z)/50$.

Exercises 4(b)

1. $x = t^2$, $y = 2t$.
2. $x = c(1-\cos\omega t)$, $y = c\omega t-c\sin\omega t$.
3. $5x = 52\cos t+64\sin t-12\cosh 3t-(64/3)\sinh 3t$,
   $y = -20\cos t+80\sin t+20\cosh 3t-(80/3)\sinh 3t$.
4. $x = A\cos t+B\sin t+E\cos at+F\sin at-(75\cos 3t)/424$,
   $y = -A\cos t-B\sin t+\frac{2}{3}E\cos at+\frac{2}{3}F\sin at+(3\cos 3t)/424$, $6a^2 = 1$.
5. $x = 2a\cos at+4ab\sin at$, $y = -a\cos at-2ab\sin at$, $2a^2 = 1$.
6. $x = A\cos 3t+B\sin 3t+3C\cos 4t+3E\sin 4t-2$,
   $y = -2A\cos 3t-2B\sin 3t+C\cos 4t+E\sin 4t$.
7. $x = -2\,e^{t}+2\,e^{2t}$, $y = -4\,e^{t}+5\,e^{2t}$.
8. $36x = 6t+11$, $36y = -48t-88$.
9. $7ax = 8\sin at+20at$, $7ay = -20\sin at+20at$, $a^2 = 336/5$.
10. $x = a\omega^2t^2-2a(1-\cos\omega t)$, $y = a\omega^2t^2+2a(1-\cos\omega t)$.

Exercises 4(c)

3. $m\ddot{x} = s(y-x)$, $m\ddot{y} = s(a+x-y)$.
5. $i_1 = (E/R)(1-e^{-Rt/L})$, $i_2 = (E/R)\,e^{-t/CR}$.
7. $E(R^2+L^2\omega^2)^{1/2}(9R^2+4L^2\omega^2)^{-1/2}\cos(\omega t+\beta)$; $\cot\beta = (3R^2+2L^2\omega^2)/RL\omega$.
8. $i = A\,e^{-Rt/L}+E(R^2+L^2\omega^2)^{-1/2}\cos(\omega t-\tan^{-1}L\omega/R)$.
10. $i_1 = 2-e^{-25t}-e^{-100t}$, $i_2 = -e^{-25t}+e^{-100t}$.

Exercises 4(d)

1. $2x = 2\cos t+\sin t+\frac{1}{2}\sin 2t$, $2y = 2\cos t+\sin t-\frac{1}{2}\sin 2t$.
2. $5x = 6\cos t+4\cos 2t$, $5y = 12\cos t-2\cos 2t$.
3. $13x = 15\cos 2t-2\cos 3t$, $13y = 10\cos 2t+3\cos 3t$.
4. $\lambda^2 = 1+2k$, $C = b/2$.
6. $4\theta = 4a\cos(3g/10a)^{1/2}t+3a\cos(6g/a)^{1/2}t$,
   $4\phi = 5a\cos(3g/10a)^{1/2}t-a\cos(6g/a)^{1/2}t$.
7. $2\theta+\phi$, $4\theta-3\phi$, $2\pi(17a/6g)^{1/2}$, $2\pi(a/3g)^{1/2}$.
8. $2\pi(6a/g)^{1/2}$, $2\pi(a/g)^{1/2}$.
9. $x+y+z$, $x+2y+z$, $x-y-z$; 2π, π, $\sqrt{2}\pi$.
12. $x = y = \{as_1/(s_1-m\omega^2)\}[\cos\omega t-\cos\{(s_1/m)^{1/2}t\}]$.

Exercises 4(e)

1. $y = x^2/6+2Ax^{1/2}+B$.
2. $y = x^2/2+Ax(\log_e x-1)+Bx+C$.
3. $y = Ax-(1+A^2)\log_e(x+A)+B$.
4. $9y^2 = 4(x-1)^3$.
5. $y = A\log_e\tan x/2+B$.
6. $y = (\sin^{-1}x)^2/2+A\sin^{-1}x+B$.
7. $y^2 = \sqrt{2}\sinh 2x+\cosh 2x$.
8. $y = 1-x^2/4$.
9. $y = 1+x^2$.
10. $y = \cosh(x/\sqrt{2})$.
11. $y = B\sec^2(x+C)$.
12. $y = \cosh x$.
13. $y = A\log_e(y+A)+x+B$.
14. $\{y(y+A)\}^{1/2} = A\log_e\{(y)^{1/2}+(y+A)^{1/2}\}\pm x+B$.

15. $\sin^2 y = 4x^2$.
16. $\{y(y-1)\}^{1/2} + \log_e \{(y)^{1/2} + (y-1)^{1/2}\} = \pm x/\sqrt{2}$.
17. $A \log_e x = y/x + (1/A) \log_e (Ay/x - 1) + B$.
18. $y = \sec x$.
19. $y^2 = Cx^{n+1} + A$.
20. $y = Bx(1+Ay)^2$.
21. $a \cos^{-1} (x^{1/2}/a^{1/2}) + x^{1/2}(a-x)^{1/2} = (2gR^2/a)^{1/2}t$.
22. $x^2 = a^2 \cos (2\mu t)$.

Exercises 5(a)

1. $$y = a_0\left(1+\frac{2}{1!1}x+\frac{2^2x^2}{2!1.4}+\frac{2^3x^3}{3!1.4.7}+\dots\right)$$
$$+b_0x^{2/3}\left(1+\frac{2}{1!5}x+\frac{2^2x^2}{2!5.8}+\frac{2^3x^3}{3!5.8.11}+\dots\right).$$
2. $$y = a_0x^3\left(1-\frac{4}{7}\frac{x}{1!}+\frac{4.5}{7.8}\frac{x^2}{2!}-\frac{4.5.6}{7.8.9}\frac{x^3}{3!}+\dots\right)+b_0x^{-3}\left(1-\frac{2}{5}x+\frac{x^2}{20}\right).$$
3. $$y = a_0\left(1-\frac{x}{3!}+\frac{x^2}{5!}-\frac{x^3}{7!}+\dots\right)+b_0x^{-1/2}\left(1-\frac{x}{2!}+\frac{x^2}{4!}-\frac{x^3}{6!}+\dots\right).$$
4. $$y = a_0\left(1-\frac{2}{1!}x+\frac{3}{2!}x^2-\frac{4}{3!}x^3+\dots\right).$$
5. $y = a_0x^2(1+3x+6x^2+10x^3+\dots)$.
6. $$y = a_0x^{-1}\left(1-\frac{1}{1!}x^{-1}+\frac{1}{2!}x^{-2}-\frac{1}{3!}x^{-3}+\dots\right).$$
7. $$y = a_0\left\{1-\frac{n(n+1)}{(1!)^2}\left(\frac{1-x}{2}\right)+\frac{(n-1)n(n+1)(n+2)}{(2!)^2}\left(\frac{1-x}{2}\right)^2-\dots\right\}.$$
10. $y_1 = a_0(1+3x^2+\frac{3}{5}x^4-\frac{1}{16}x^6+\dots)$, $y_2 = b_0x^{3/2}(1+\frac{3}{8}x^2-\frac{3}{128}x^4+\frac{5}{1024}x^6-\dots)$.
11. $$C\left\{(x-1)\log_e\left(\frac{x}{x-1}\right)-1\right\}.$$
12. $y_1 = x(1-x)^{-2}$; $y_2 = c(1-x)^{-2}(1+x\log_e x)$.
13. $$y = a_0\left[1+\frac{c}{2!}\left(\frac{x}{a}\right)^2+\frac{c\{c+2(b+1)\}}{4!}\left(\frac{x}{a}\right)^4+\dots\right]$$
$$+b_0\frac{x}{a}\left[1+\frac{c+b}{3!}\left(\frac{x}{a}\right)^2+\frac{(c+b)\{c+3(b+2)\}}{5!}\left(\frac{x}{a}\right)^4+\dots\right];\ |x| < |a|.$$
14. $$y_1 = a_0x^{-3}\left(1+\frac{3.4}{2.7}x^{-2}+\frac{3.4.5.6}{2.4.7.9}x^{-4}+\dots\right);\ y_2 = b_0\left(x^2-\frac{1}{3}\right).$$

Exercises (7a)

1. $\sum A \sin mx \sin mct$.
2. $\sum A e^{-mx} \sin my$.
3. $\sum e^{-a^2m^2t}(A \cos mx + B \sin mx)$.
4. $\sum A \cos m(ax-y)$.
5. $R''+r^{-1}R'+n^2R = 0$.
6. $F''+\cot\theta . F'+n(n+1)F = 0$.
7. $(k/n) \cos (nx/a) \sin nt$.
8. $(1/2a)\, e^{2ax} \sin 2ay$.
9. $c^2F''+\omega^2F = 0$; $F = A\cos(\omega r/c)+B\sin(\omega r/c)$.
10. $f(r) = A \sin kr + B \cos kr$; $u = (b/r)^{1/2} \sin k(r-a) \operatorname{cosec} k(b-a) \cos \frac{1}{2}\theta$.
11. $\phi = (a^3/2r^2)\cos\theta$.
12. $x^2X''+xX'-m^2X = 0$, $Y''+m^2Y = 0$; $X = Ax^m+Bx^{-m}$,
$Y = C \cos my + D \sin my$; $u = (a^3/2x^2)\cos 2y$.

13. $A = 0, B = 1, m = 1\ \omega = 1.$
14. $f''+r^{-1}f'+(\alpha^2-m^2\pi^2r^{-2})f = 0.$
15. $X = A\cosh(n^2-c^2)^{1/2}x + B\sinh(n^2-c^2)^{1/2}x,\ (c < n);$
$X = A\cos(c^2-n^2)^{1/2}x+B\sin(c^2-n^2)^{1/2}x,\ (c > n);$
$X = Ax+B,\ (c = n);$

$$v = \left(\cosh 2x+\frac{1-e}{1+e}\sinh 2x\right)\cos 1{\cdot}5t.$$

16. $V = \frac{1}{3}(1-e^{-6x})\sin 3t.$
17. $z = xe^{-kx}\sin kt.$

Exercises 7(b)

5. $al = 1{\cdot}506\pi.$
7. 49·0, 23·3 per sec.
9. $(4V_0/\pi)\{e^{-kt/4}\cos(x/2)-\frac{1}{3}e^{-9kt/4}\cos(3x/2)+\frac{1}{5}e^{-25kt/4}\cos(5x/2)-\ldots\}.$
11. 42° (approx.).

Exercises 7(c)

6. $\dfrac{\partial^2F}{\partial\rho^2}+\dfrac{1}{\rho}\dfrac{\partial F}{\partial\rho}+\dfrac{1}{\rho^2}\dfrac{\partial^2F}{\partial\phi^2};\ F = A\tan^{-1}(e^u)+B.$

7. $\dfrac{\partial}{\partial u}\left(u\dfrac{\partial V}{\partial u}\right)+\dfrac{\partial}{\partial v}\left(v\dfrac{\partial V}{\partial v}\right)+\dfrac{u+v}{4uv}\dfrac{\partial^2V}{\partial\phi^2} = 0.$

Exercises 8(a)

1. $y = e^{-x}.$
2. $y = x+3e^{2x}.$
3. $x = te^{at}.$
4. $x = e^{2t}-\cos 2t-\sin 2t.$
5. $y = 3e^x+4e^x\sin x.$
6. $y = 4e^{-3t}-3e^{-5t}.$
7. $y = (1+6x)e^{-5x}.$
8. $y = e^{-t}\sin 3t.$
9. $x = e^{-kt}\left(\cos mt+\dfrac{k}{m}\sin mt\right).$
10. $y = (2-x)e^{2x}-e^{3x}.$
11. $\theta = \frac{1}{8}\{1-e^{-2t}(\cos 2t-3\sin 2t)\}.$
12. $y = 3\cos t-2e^{-t/2}\cos 3t.$
13. $y = \frac{1}{8}(1+e^{-2x})(\cos 3x+3\sin 3x).$
14. $y = 1-\frac{1}{3}e^{-x}-\frac{2}{3}e^{x/2}\cos\dfrac{\sqrt{3}}{2}x.$
15. $y = 1.$
16. $y = 2x^2+2x+5+e^{2x}.$
17. $2x = e^{-3t}+e^{-t},\ 2y = e^{-3t}-e^{-t}.$
18. $x = e^{2t}(\cos 3t-\sin 3t),\ y = e^{2t}(\cos 3t+2\sin 3t).$
19. $x = 6e^{-t/6}-(t+1)e^{-t}+t-5,\ y = 9e^{-t/6}+(t-2)e^{-t}+t-7.$
20. $36x = 6t+11,\ 36y = -48t-88.$
21. $9y = 4\sin 2x-5\sin x-3x\cos 2x.$

Exercises 8(c)

1. $V = (4/\pi)\sum_{p=1}^{\infty}(2p-1)^{-1}e^{-(2p-1)y}\sin(2p-1)x.$

2. $V = (4C/\pi)\sum_{p=1}^{\infty}(2p-1)^{-1}\operatorname{cosech}(2p-1)\pi\sinh\{(2p-1)\pi y/a\}\sin\{(2p-1)\pi x/a\}.$

4. $V = (2c/\pi)\sum_{p=1}^{\infty}(c^2p^2-1)^{-1}(cp\sin t-\sin cpt)\sin px.$

5. $2\pi V = Kc^2t^2+4K\sum_{p=1}^{\infty}(-1)^p p^{-2}(1-\cos cpt)\cos px.$

Exercises 9(a)

1. $y = x-1+Ce^{-x}$.
4. $z''+z'/x+z/4 = 0$.

Exercises 9(b)

1. 0·4603.
2. 0·197, 0·380, 0·537.
3. ±0·0027, ±0·0213.
4. ±0·0416, ±0·3222.
5. 0·9934, 0·9740.
6. −0·227, 0·184.

Exercises 9(c)

2. 0·333, 0·324, 0·324.
4. 0·0440.
5. 1·238.
7. 0·0200.
8. 1·015.

Exercises 9(d)

1. 0·0187, 0·0703, 0·1488.
2. 0·826, 0·764, 0·774.
3. 0·0027, 0·0214, 0·0724.
4. 0·8392, 0·7489, 0·7115.
5. 0·0053, 0·0440, 0·1606.
6. 1·4188, 1·8710, 2·3520.
7. 1·101, 1·204, 1·314.
8. −0·1005, −0·2041, −0·3144.

Exercises 9(e)

1. 1·1163, 1·1683 1·2290, 1·2980, 1·3752.
2. 0·0716, 0·1674, 0·3184.
3. 3·7504.
4. −0·5751, −2·5759.
5. 0·6481.
6. 1·555, 2·354.
7. 0·100, 0·197, 0·291, 0·380, 0·462, 0·537; 0·462.
8. 1·0407, 1·0769.

Exercises 9(f)

1. $y =$ 1·2199, 1·4789, 1·7748, 2·1040, 2·4622,
   $z =$ 1·1987, 1·3896, 1·5659, 1·7223, 1·8556.
2. $y =$ 1·356, 1·542, 1·768, 2·051, 2·421,
   $z =$ 0·251, 0·321, 0·418, 0·559, 0·772.
3. $y =$ 0·995, 0·982, 0·960, 0·933, 0·901,
   $z =$ 0·005, 0·020, 0·045, 0·080 0·125.
4. $y =$ 1·224, 1·517, 1·916, 2·479, 3·297,
   $z =$ 1·265, 1·690, 2·347, 3·360, 4·946.
5. 0·9604, 0·8463, 0·6711, 0·4554, 0·2239.
6. 0·5052, 1·0853, 1·9571, 3·6111, 7·1989.
7. 0·1974, 0·3799, 0·5370, 0·6640, 0·7616.
8. 1·223, 1·507, 1·883, 2·404, 3·164.
9. 0·2000, 0·3995, 0·5961, 0·7839, 0·9520.
10. 0·9604, 0·8462, 0·6698, 0·4489, 0·2017.

Exercises 10(a)*

1. 26, 44, 58, 78, 96.
2. 57, 95, 108, 95, 53.
3. 16, 9, −10, −27, −23.
4. 0·869, 0·866.
5. 0·249, 0·493, 0·727, 0·946, 1·146, 1·325, 1·478.
6. 0·013 (w/EI).
7. 106.
8. 0·0939.
9. 0·520, 0·843.
10. 0·149, 0·229, 0·235, 0·161.

* In view of the nature of the method of solution, the answers given here are only approximate.

Exercises 10(b)*

1.
```
      79
 86  114  135
110  155  197
```
2. 20, 44; 21, 45.
3. 113.
4. (*a*) 56, (*b*) 54, (*c*) 53, (*d*) 51, (*e*) 46, (*f*) 45, (*g*) 42, (*h*) 35, (*j*) 34.
5. 59.
6. 88, 114, 95, 39.
7. $0{\cdot}88C$, $0{\cdot}55C$, $0{\cdot}13C$.
8. (i) 280, (ii) 271.
9.

ρ \ z	1	2	3
7	69	29	11
6	80	38	15
5	60	28	11

10. 36, 50, 64.

Exercises 11(a)

1. $= Axe^{1/x}$.
4. $(x^2-y^2+Ax+2a^2)^2 = a^2(4a^2-A^2)$; $(y^2-x^2)^2 = 4a^2y^2$.
6. $y = \log_e (A \log_e x+B)$.
8. $y = a \sec (\omega x+\varepsilon)$.

INDEX